Nitrogen Heterocycles in Neurodegenerative Disorders

Neurodegenerative disorders, including Alzheimer's disease, Parkinson's disease, and Huntington's disease, represent a rapidly escalating global health burden with limited disease-modifying therapies. Advances in medicinal chemistry have highlighted nitrogen-containing heterocycles as powerful scaffolds in the search for effective neuroprotective agents. Their structural diversity, target specificity, and favorable pharmacological profiles position them at the forefront of next-generation drug discovery.

Nitrogen Heterocycles in Neurodegenerative Disorders: Chemistry and Therapeutic Potential delivers a comprehensive and research-driven exploration of the chemistry, biological relevance, and translational promise of nitrogen-based heterocyclic compounds. Distinct in its integrative approach, the book bridges synthetic chemistry, molecular pharmacology, and clinical perspectives. It not only discusses established heterocyclic frameworks but also critically evaluates emerging scaffolds, structure–activity relationships, and challenges in blood–brain barrier penetration—offering readers both foundational knowledge and forward-looking insights.

Key Features:

- Detailed overview of neurodegenerative disease pathophysiology
- Exploration of molecular targets and signaling pathways
- Design, synthesis, and SAR studies of key nitrogen heterocycles
- Pharmacokinetic and pharmacodynamic considerations for CNS drugs
- Future perspectives on heterocycle-based neurotherapeutics

This volume serves as an essential reference for medicinal chemists, pharmaceutical scientists, neuroscientists, clinicians, and postgraduate researchers seeking a deeper understanding of heterocyclic strategies in combating neurodegeneration.

Nitrogen Heterocycles in Neurodegenerative Disorders

Chemistry and Therapeutic Potential

Edited by
Shraddha Manish Gupta, Somdutt Mujwar,
Kamal Shah, and Nagendra Singh Chauhan

CRC Press
Taylor & Francis Group
Boca Raton London New York

CRC Press is an imprint of the
Taylor & Francis Group, an **informa** business

Designed cover image: Dr. Somdutt Mujwar

First edition published 2027
by CRC Press
2385 NW Executive Center Drive, Suite 320, Boca Raton FL 33431

and by CRC Press
4 Park Square, Milton Park, Abingdon, Oxon, OX14 4RN

CRC Press is an imprint of Taylor & Francis Group, LLC

ISBN: 978-1-041-29937-0 (hbk)
ISBN: 978-1-041-29938-7 (pbk)
ISBN: 978-1-003-77201-9 (ebk)

DOI: 10.1201/9781003772019

Typeset in Times
by KnowledgeWorks Global Ltd.

Contents

Preface

Neurodegenerative disorders—including Alzheimer's disease, Parkinson's disease, Huntington's disease, and amyotrophic lateral sclerosis—represent some of the most debilitating and rapidly increasing health challenges of the 21st century. Despite remarkable progress in understanding their molecular basis, effective disease-modifying therapies remain limited. In recent years, the exploration of nitrogen-containing heterocyclic compounds has gained significant attention for their diverse chemical versatility and promising therapeutic potential in modulating neurobiological pathways. These privileged scaffolds occupy a central role in modern medicinal chemistry, offering unique opportunities to design novel neuroprotective agents capable of targeting the complex mechanisms underlying neurodegeneration.

This edited volume, *Nitrogen Heterocycles in Neurodegenerative Disorders: Chemistry and Therapeutic Potential*, brings together current advances in the chemistry, molecular pharmacology, and therapeutic applications of nitrogen-based heterocyclic systems. Designed for researchers, academicians, medicinal chemists, pharmaceutical scientists, and clinicians, the book provides a comprehensive and multidisciplinary perspective on the emerging role of these scaffolds in neurotherapeutics.

The chapters, authored by experts from leading academic and research institutions worldwide, collectively cover a broad spectrum of topics—from disease pathology and identification of molecular targets to the synthesis, biological evaluation, and mechanistic understanding of nitrogen heterocycles in brain disorders. Beginning with foundational insights into neurodegenerative disease mechanisms, the book transitions into detailed discussions on key heterocyclic classes such as pyridines, pyrimidines, pyrazoles, quinolines, isoquinolines, indoles, triazoles, and thiazoles. Additional chapters address natural nitrogenous products, and pharmacokinetic and pharmacodynamic considerations. Together, these contributions provide a well-rounded and contemporaneous overview of cutting-edge research in this rapidly evolving field.

As editors, we are grateful to all contributing authors for their scholarly dedication and commitment to advancing scientific knowledge. Their collective expertise has enriched this volume and strengthened its value as a reference for ongoing and future research. We also extend our sincere appreciation to CRC Press for their support and collaboration throughout the development of this book.

We hope that this book serves as a valuable resource, inspiring new ideas and fostering innovative approaches toward the discovery and development of heterocyclic compounds for the treatment of neurodegenerative disorders. As global health systems continue to face the growing burden of neurodegeneration, the need for novel therapeutic interventions has never been greater. It is our belief that the insights presented here will contribute meaningfully to the scientific community's efforts to address these challenges.

Editors
Shraddha M. Gupta
Somdutt Mujwar
Kamal Shah
Nagendra Singh Chauhan

Contributors

Arzine Aziz
Engineering Laboratory of Organometallic, Molecular Materials and Environment
Faculty of Sciences Dhar EL Mahraz
Sidi Mohamed Ben Abdellah University
Fez, Morocco

Hamid Aziz
Department of Chemistry
Quaid-I-Azam University
Islamabad, Pakistan

Khalil Azzaoui
Engineering Laboratory of Organometallic, Molecular Materials, and Environment (LIMOME)
Dhar El Mahraz
Sidi Mohamed Ben Abdellah University
Fez, Morocco

Kosana Sai Chaitanya
Department of Chemistry
Birla Institute of Technology and Science Pilani, Hyderabad Campus
Telangana, India

Pooja A. Chawla
University Institute of Pharmaceutical Sciences and Research
Baba Farid University of Health Sciences
Faridkot, India

Viney Chawla
University Institute of Pharmaceutical Sciences and Research
Baba Farid University of Health Sciences
Faridkot, India

Mohamed El Fadili
LIMAS Laboratory
Faculty of Sciences Dhar El Mahraz
Sidi Mohamed Ben Abdellah University
Fez, Morocco

Menana El Hallaoui
LIMAS Laboratory
Faculty of Sciences Dhar El Mahraz
Sidi Mohamed Ben Abdellah University
Fez, Morocco

Mohammed Er-rajy
LIMAS Laboratory
Faculty of Sciences Dhar El Mahraz
Sidi Mohamed Ben Abdellah University
Fez, Morocco

Estela Melcón Fernández
Department of Biomedical Sciences
University of León
León, Spain

Rafael Balaña Fouce
Department of Biomedical Sciences
University of León
León, Spain

Shraddha M. Gupta
Department of Pharmaceutical Sciences
School of Health Sciences and Technology
University of Petroleum and Energy Studies (UPES)
Dehradun, India

Belkheir Hammouti
Euromed University of Fes (UMF)
Fes, Morocco

Satyaveer Singh Jat
School of Pharmacy
ITM University
Gwalior, India

Ismat Ullah Khan
Department of Chemistry
Quaid-I-Azam University
Islamabad, Pakistan
and
Department of Chemistry
Government Post Graduate College
Lakki Marwat, Pakistan

Devendra Kumar
Center of Excellence in Molecular Biology and Genetics
School of Pharmacy & Technology Management
SVKM's NMIMS University, Mukesh Patel Technology Park
Shirpur, India

Kapil Kumar
Department of Pharmaceutical Sciences,
School of Health Sciences and Technology
University of Petroleum and Energy Studies (UPES)
Dehradun, India

Manoj Kumar
Department of Pharmaceutical Sciences
Guru Jambheshwar University of Science & Technology
Hisar, India

Hao Lu
College of Chemistry and Chemical Engineering
Ningxia Normal University
Guyuan, China

Ehsan Ullah Mughal
Department of Chemistry
University of Gujrat
Gujrat, Pakistan

Somdutt Mujwar
Chitkara College of Pharmacy
Chitkara University
Rajpura, India

Akhtar Munir
Department of Chemistry
Quaid-I-Azam University
Islamabad, Pakistan

Nafeesa Naeem
Department of Chemistry
University of Gujrat
Gujrat, Pakistan

Bhupender Nehra
Department of Pharmaceutical Sciences
Guru Jambheshwar University of Science & Technology
Hisar, India

Renhua Qiu
College of Chemistry and Chemical Engineering
Hunan University
Changsha, China

Saif Ur Rehman
Department of Chemistry
Quaid-I-Azam University
Islamabad, Pakistan

Dinesh D. Rishipathak
Department of Pharmaceutical Chemistry
MET's Institute of Pharmacy
Nashik, India

Aamer Saeed
Department of Chemistry
Quaid-I-Azam University
Islamabad, Pakistan

Murugesan Sankaranarayanan
Medicinal Chemistry Research Laboratory
Department of Pharmacy
Birla Institute of Technology and Science Pilani, Pilani Campus
Pilani, India

Kondapalli Venkata Gowri Chandra Sekhar
Department of Chemistry
Birla Institute of Technology and Science Pilani, Hyderabad Campus
Telangana, India

Kamlesh Sharma
Division of Physical Sciences
School of Arts and Sciences
Karunya Institute of Technology and Sciences (Deemed to be University), Coimbatore, India

Moharir Shreyash Rajendra
Department of Chemistry
Birla Institute of Technology and Science Pilani, Hyderabad Campus
Telangana, India

Pratiksha N. Sonwane
Department of Pharmaceutical Chemistry
MET's Institute of Pharmacy
Nashik, India

Pradyumn Tiwari
School of Pharmacy
ITM University
Gwalior, India

Avanish Tripathi
School of Pharmacy
ITM University
Gwalior, India

Naveed Ullah
Institute of Chemical Sciences
Gomal University
Dera Ismail Khan, Pakistan

Mohammed Zerrouk
Engineering Laboratory of Organometallic, Molecular Materials, and Environment (LIMOME)
Dhar El Mahraz
Sidi Mohamed Ben Abdellah University
Fez, Morocco

Achraf Ziani
Laboratory of Biotechnology, Conservation and Valorization of Bioresources (BCVB)
Research Unit: Api-Phytotherapy, Physiology, Environment and Health
Department of Biology
Faculty of Sciences Dhar El Mahraz
Sidi Mohamed Ben Abdellah University
Fez, Morocco

Editors' Biographies

Shraddha Manish Gupta has more than 18 years of research and teaching experience and is currently Assistant Professor-SG at School of Health Sciences and Technology, UPES, Dehradun, India. She has completed B.Pharm and M.Pharm from NDMVP's College of Pharmacy, Nasik (M.H.), in 2003 and 2006, respectively, and Ph.D. from Oriental University, Indore, India. She secured 96th percentile in GATE-2004 conducted by IIT, Delhi. She worked as Research Assistant in School of Pharmacy, Monash University, Malaysia. She was Mentor in CSIR-SRTP (2020) program under Interdisciplinary Approach for Sustainable Use of Resources project-13. She has written more than 12 articles published in Q1 international journals, two patents (one granted, one published), two edited books with CRC publishers, and 12 book chapters. She is a member of various professional and academic bodies like Life Membership of Association of Pharmaceutical Teachers of India (APTI), Indian Pharmaceutical Association (IPA), and Indian Society of Chemists and Biologists (ISCB).

Somdutt Mujwar is a passionate and dedicated researcher in the field of computational and pharmaceutical chemistry, specializing in drug discovery and development. With a Ph.D. in Bioinformatics, he has made significant contributions to understanding the molecular mechanisms underlying cancer, neurodegenerative, and infectious diseases. Dr. Mujwar's research focuses on exploring the therapeutic potential of natural compounds in treating cancer and neurodegenerative disorders. His work has been published in numerous peer-reviewed journals, showcasing his expertise and dedication to advancing scientific knowledge. Beyond his research endeavors, Dr. Mujwar is committed to education and mentorship. He actively engages in teaching and training programs, guiding aspiring scientists and pharmacists toward academic and professional success.

Driven by a desire to translate scientific discoveries into practical solutions, Dr. Mujwar collaborates with multidisciplinary teams to develop innovative approaches to drug design and delivery. His interdisciplinary approach and collaborative spirit have earned him recognition as a leader in the field. Overall, Dr. Somdutt Mujwar's expertise, dedication, and commitment to scientific excellence make him an asset to the field of Pharmaceutical Sciences, with a profound impact on the advancement of drug discovery and development.

Kamal Shah has more than 18 years of research and teaching experience and is currently working as Professor at the Institute of Pharmaceutical Research, GLA University, Mathura, India. He completed B.Pharm from Rajiv Gandhi Proudyogiki Vishwavidyalaya, Bhopal (M.P.) in 2003. He was a gold medalist in B.Pharm and M.Pharm from the Department of Pharmaceutical Sciences, Sagar University, Sagar, India and Ph.D. from APJ Kalam University, Lucknow, India. He has written more than 90 articles published in national and international journals, 20 book chapters, and five edited books. His Scopus h-index is 17, with 905 citations.

Nagendra Singh Chauhan obtained his M.Pharm. and Ph.D. from Department of Pharmaceutical Sciences, Dr. H.S. Gour University, Sagar in 2006 and 2011. He has around 14 years of research experience. He is presently working as Senior Scientific Officer Grade II and Government Analyst at Drugs Testing Laboratory Avam Anusandhan Kendra, Raipur, Chhattisgarh, India. He has professional expertise in Natural Product Isolation and Phytopharmacology. He has written more than 76 articles published in national and international journals, 35 book chapters, and edited seven books in publishers like Academic Press, CRC, and Wiley. Recently edited books entitled "Advances and Avenues in the Development of Novel Carriers for Bioactives and Biological Agents" and "Natural Products in Vector Borne Disease Management", published by Academic Press. He has written more than 75 articles published in national and international journals, 32 book chapters, and edited five books. He has more than 2700 citations with an h-index of 42 and an i10 index of 38 (Google Scholar) and a Scopus h-index of 15 with 669 citations. He is a member of various professional and academic bodies like Society of Pharmacognosy India, International Natural Product Sciences Taskforce (INPST) society, SILAE: SocietàItalo-Latinoamericana di Etnomedicina (The Scientific Network on Ethnomedicine, Italy), Institutional Human Ethical Committee, Association of Pharmaceutical Teachers of India (APTI), and Indian Pharmacy Graduates' Association.

Foreword

Neurodegenerative disorders represent one of the most persistent global health challenges in the modern world, impacting millions of individuals and imposing substantial emotional, social, and economic burdens. Although remarkable progress has been made in resolving the molecular and cellular mechanisms underlying neurodegeneration, the development of safe and effective neurotherapeutic agents remains an urgent scientific and clinical priority.

Nitrogen-containing heterocycles continue to occupy a central position in medicinal chemistry due to their structural diversity, synthetic accessibility, and broad pharmacological potential. Their ability to modulate complex biological pathways makes them highly attractive scaffolds for designing next-generation neuroprotective agents. This edited volume, *Nitrogen Heterocycles in Neurodegenerative Disorders: Chemistry and Therapeutic Potential*, comes at a crucial time, bringing together comprehensive insights into both the fundamental science and applied aspects of heterocyclic chemistry in neuroprotection.

The book offers an integrative perspective, beginning with foundational discussions on neurodegenerative pathology and extending into focused explorations of key heterocyclic classes. Through contributions from global experts, it presents advanced knowledge on chemical synthesis, molecular mechanisms, pharmacokinetics, preclinical evaluation, and translational potential. By bridging chemistry and neuroscience, this work provides a strong platform for innovation and scientific discovery.

This volume is essential for students, researchers, clinicians, and medicinal chemists engaged in neurotherapeutic research. It not only highlights current advances but also inspires future scientific directions aimed at addressing the unmet needs in neurodegenerative disease management.

I commend the editors, Dr. Shraddha, Dr. Somdutt, Dr. Kamal, Dr. Nagendra, and contributors for their exceptional efforts in producing a scholarly, timely, and impactful reference for the scientific community.

Prof. (Dr.) Thakur Gurjeet Singh
Professor and Dean
Chitkara College of Pharmacy
Chitkara University, Punjab, India

Acknowledgments

The editors express their sincere gratitude to all the contributors whose expertise, dedication, and scholarly commitment made this book possible. Each chapter reflects the hard work, scientific rigor, and collaborative spirit of researchers across chemistry, pharmaceutical sciences, biophysics, and neurobiology. We deeply appreciate the time and effort invested by the authors in preparing high-quality, up-to-date chapters that enrich this volume.

We extend our heartfelt thanks to CRC Press for their guidance, editorial support, and professionalism throughout the development of this book. Their continued encouragement and assistance have been instrumental in bringing this project to completion.

We also acknowledge the academic and research institutions that supported the contributing authors and editors. Their encouragement and provision of resources have played a vital role in advancing the scientific investigations presented in this book.

Finally, we thank our families, colleagues, and students for their constant support, patience, and inspiration during the preparation of this manuscript. Their understanding has been invaluable in achieving the successful completion of this collaborative work.

1 Introduction to Neurodegenerative Disorders and Their Pathophysiology

Pradyumn Tiwari, Avanish Tripathi, Devendra Kumar, and Satyaveer Singh Jat

1.1 INTRODUCTION

Neurodegenerative disorders are a general category of chronic, progressive central, and peripheral nervous system diseases, which are characterized by slow dysfunction and depletion of selectively vulnerable assemblies of neurons and supportive glial cells. The scientific concept of neurodegenerative disorders has changed significantly since the late 1800s and early 2000s, when initial cases of such entities as Alzheimer's disease (AD), Parkinson's disease (PD), and amyotrophic lateral sclerosis (ALS) were described in terms of classical clinical pathology. These conditions tend to develop insidiously and have progressive clinical features, including cognitive impairment, motor disorders, behavioral changes, or motor nerve impairment, which eventually cause significant disability and premature death (Jellinger, 2018). Moreover, the neurodegeneration of particular regions of the brain or spinal cord is a result of the gradual structural and functional degradation of neurons with time. They are generally linked with pathologic modifications in abnormal protein accumulation, mitochondrial dysfunction, oxidative stress, impaired proteostasis, and persistent neuroinflammation, joining pathways of synaptic tragedy and neuronal destruction. Clinically, these diseases are categorized based on the predominant manifestation such as dementias, movement disorders, motor neuron diseases, or the major misfolded protein involved (such as amyloid-β (Aβ), tau, or α-synuclein) but with molecular biology, genetics, and neuroimaging, the arena has changed to be more of a descriptive pathology rather than a mechanistic model of protein misfolding, prion like pathology transmission and gene environment interactions (Dugger & Dickson, 2017).

Cascade of neurodegenerative diseases which impacted negatively on the various signaling pathways and damage the neurons permanently. Neuronal damage can occur through various pathological pathways like phosphorylation of neuron, fibril formation, amyloid β accumulation, etc., as explained in Figure 1.1.

Neurodegenerative disorders are one of the most significant and increasing global population health issues, which are mostly caused by the aging of the population and longer life expectancy. Among the main causes of disability, loss of independence, institutionalization, and decreased quality of life among the older adults of the world are dementias and predominantly AD, and movement disorders, especially PDs. The economic cost related to this is significant, including direct healthcare expenses, long-term care expenses, and indirect expenses in terms of lost productivity and the use of informal caregivers, and worldwide expenses alone on dementia are projected to be in the hundreds of billions of dollars annually. The neurodegenerative disorders are of crucial importance to understand since the current treatments are mostly palliative and fail to prevent or reverse the disease mechanisms (Figure 1.2) (Hou et al., 2019). The complexity and condition have been discussed in detail in Figure 1.2. Disease-modifying management and reliable biomarker use could be essential in developing societal-level disease-preventive, clinical, psychological, and economic management interventions and disease-diagnostic techniques to identify these disorders earlier and prior to their onset and progression (Soto & Pritzkow, 2018).

DOI: 10.1201/9781003772019-1

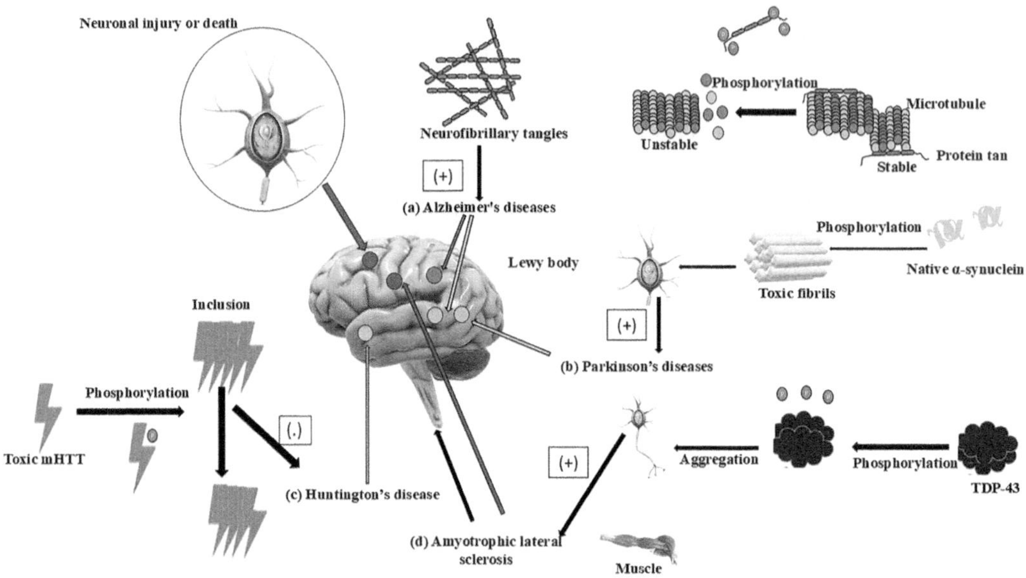

FIGURE 1.1 Pathological changes lead to neurological disorder.

1.2 NEUROANATOMICAL ORGANIZATION

The nervous system is a highly organized network that coordinates sensory input, information processing, and motor output to maintain homeostasis and enable complex behaviors. It is anatomically divided into the central nervous system, consisting of the brain and spinal cord, and the peripheral nervous system, which connects the CNS to the rest of the body. In neurodegenerative diseases, the selective vulnerability of specific neural circuits within this system underlies the characteristic clinical syndromes. In many neurodegenerative disorders, both neuronal dysfunction and maladaptive

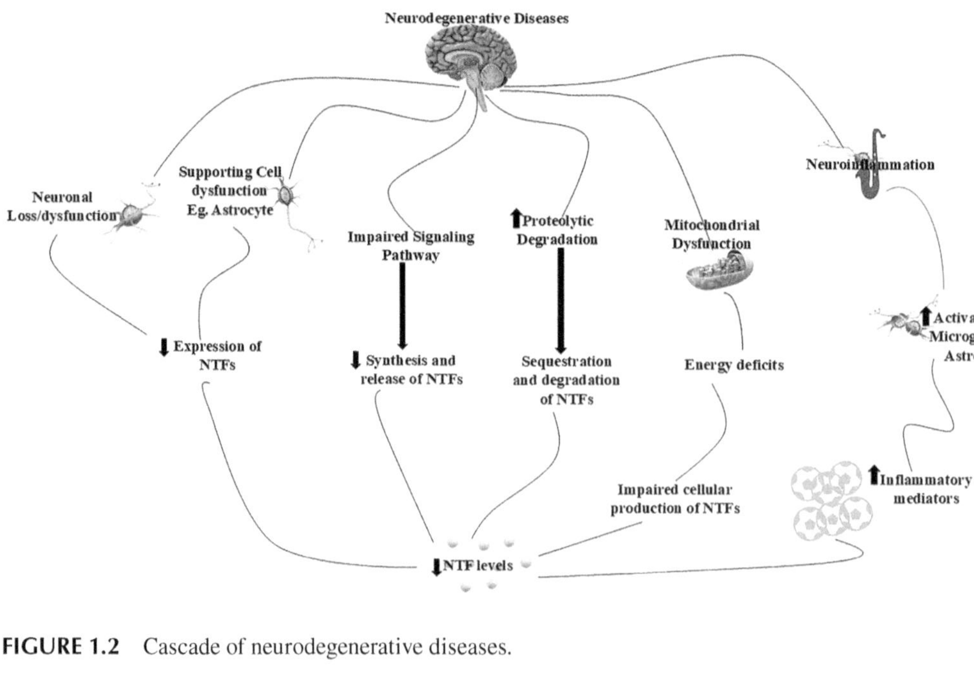

FIGURE 1.2 Cascade of neurodegenerative diseases.

glial responses contribute to disease progression. Neurodegenerative disorders show characteristic patterns of regional vulnerability within the CNS, which correlate with their clinical manifestations (Dorsey & Bloem, 2018). The hippocampus and association cortices are prominently affected in AD, leading to progressive impairment of episodic memory and higher cognitive functions. The substantia nigra pars compacta and related basal ganglia circuits are primarily involved in PD, resulting in bradykinesia, rigidity, and tremor. In contrast, degeneration of upper and lower motor neurons (LMNs) in the motor cortex, brainstem, and spinal cord underlies the weakness and wasting seen in ALS. Other conditions target cerebellar circuits, frontotemporal networks, or specific thalamic and brainstem nuclei, producing distinctive combinations of cognitive, behavioral, and motor symptoms (Jack et al., 2018).

1.3 NEURODEGENERATIVE DISORDERS

Neurodegenerative disorders are heterogeneous disorders that are categorized based on the major clinical expression, distribution of pathology anatomically, or the molecular pathogenesis to enable research and diagnosis. Overlapping or dual categories may be used frequently since numerous diseases have common pathways, such as protein misfolding and neuro-inflammation but have distinct regional susceptibilities (De Strooper & Karran, 2016).

1.3.1 Alzheimer's Disease (AD)

AD, the most common neurodegenerative disorder, features progressive cognitive decline due to amyloid plaques, neurofibrillary tangles (NFTs), and neuronal loss in the hippocampus and entorhinal cortex extending to the association cortices. Pathophysiology involves Aβ oligomer toxicity, tau hyperphosphorylation, neuro-inflammation, and cholinergic deficits (Kovacs et al., 2017).

1.3.2 Parkinson's Disease (PD)

PD manifests with bradykinesia, rigidity, tremor, and postural instability from dopaminergic neuron loss in the substantia nigra pars compacta and Lewy body formation. α-Synuclein aggregation progresses through Braak staging, with mitochondrial dysfunction and oxidative stress playing key roles (Irwin et al., 2021).

1.3.3 Huntington's Disease (HD)

Huntington's disease (HD) is an autosomal dominant triplet-repeat disorder causing striatal medium spiny neuron degeneration, chorea, cognitive decline, and psychiatric symptoms. Mutant huntingtin protein toxicity triggers excitotoxicity, transcriptional dysregulation, and proteasomal impairment (Zetterberg et al., 2023).

1.3.4 Amyotrophic Lateral Sclerosis (ALS)

ALS involves progressive upper and LMN degeneration, leading to muscle weakness, spasticity, and respiratory failure. TDP-43 pathology predominates (95% cases), with SOD1, C9orf72, and FUS mutations in familial forms; excitotoxicity, protein aggregation, and neuro-inflammation drive progression (Bose et al., 2022).

1.3.5 Prion Diseases

Prion diseases (Creutzfeldt–Jakob disease, variant CJD, Gerstmann–Straussler–Scheinker syndrome) feature rapidly progressive dementia, myoclonus, and ataxia from PrP^{Sc} accumulation, causing

TABLE 1.1
Major Types of Neurodegenerative Disorders

S. No.	Name of Disorder	Primary Brain Regions Affected	Key Pathological Features	Major Symptoms
1	**Alzheimer's disease (AD)**	Hippocampus, entorhinal cortex, association cortices	Amyloid-β plaques, neurofibrillary tangles (hyperphosphorylated tau), neuronal loss, cholinergic deficit	Progressive memory loss, cognitive decline, disorientation, behavioral changes
2	**Parkinson's disease (PD)**	Substantia nigra pars compacta, basal ganglia	Dopaminergic neuron loss, Lewy bodies (α-synuclein), mitochondrial dysfunction	Bradykinesia, resting tremor, rigidity, postural instability, gait disturbance
3	**Huntington's disease (HD)**	Striatum (caudate, putamen), cerebral cortex	Mutant huntingtin aggregation, neuronal atrophy, excitotoxicity	Chorea, psychiatric symptoms, cognitive impairment
4	**Amyotrophic lateral sclerosis (ALS)**	Motor cortex, brainstem, spinal motor neurons	Upper and lower motor neuron degeneration, TDP-43 inclusions	Muscle weakness, spasticity, fasciculations, respiratory failure
5	**Prion diseases**	Cerebral cortex, thalamus, cerebellum	PrP^{Sc} accumulation, spongiform changes, neuronal vacuolation	Rapid dementia, myoclonus, ataxia, behavioral changes
6	**Multiple system atrophy (MSA)**	Basal ganglia, cerebellum, brainstem, autonomic centers	α-Synuclein glial cytoplasmic inclusions, neurodegeneration	Parkinsonism, cerebellar ataxia, autonomic dysfunction
7	**Progressive supranuclear palsy (PSP)**	Brainstem, basal ganglia, frontal cortex	Tau aggregation (4R tauopathy), neuronal loss	Postural instability, gaze palsy, rigidity, cognitive decline
8	**Corticobasal degeneration (CBD)**	Cortex and basal ganglia	Asymmetric tau pathology, neuronal loss	Limb apraxia, rigidity, dystonia, cognitive dysfunction

spongiform encephalopathy. Infectious, sporadic, or genetic transmission involves conformational conversion and neuronal vacuolation (Heneka et al., 2015).

Table 1.1 summarises the forms of neurodegenerative disorder and with their detailed and major symptoms. This table also reflects the information about the region of the brain affected and the major causes of diseases.

1.4 ABERRANT PHYSIOLOGICAL PROCESSES

There are converging pathological cascades of neurodegenerative disorders, which engage failure of protein homeostasis, bioenergetics impairments, neuro-inflammation, and circuitry disruption in spite of disease-specific causes. The mechanisms are synergistic and enhance the neuronal susceptibility and disease progression in human diseases such as Alzheimer's, Parkinson's, and ALS (Dong et al., 2020).

A protein that has misfolded and aggregated is called an aggregated protein or precipitated protein. Protein that has folded and aggregated is referred to as aggregated protein or predicated protein.

Protein misfolding produces toxic oligomeric conformers and fibrillar aggregates that disrupt cellular proteostasis, Aβ plaques, tau tangles, α-synuclein, Lewy bodies, and TDP-43 inclusions spread via prototypic misfolding and intercellular seeding. These aggregates interfere with the integrity of the membrane, bury the chaperones, and initiate endoplasmic reticulum stress (Menzies et al., 2015).

1.4.1 Mitochondrial Dysfunction and Reactive Oxidative Stress

Disrupted electron transport in mitochondria of vulnerable neurons, decreased production of ATP, calcium overloading, and release of pro-apoptotic factors; release of reactive oxygen species by dysfunctional complexes I and III damages the DNA, lipids, and proteins. The antioxidant defenses, such as SOD and glutathione, become unproductive, leading to a vicious cycle that is further aggravated due to α-synuclein and mutant huntingtin (Wishart et al., 2021).

1.4.2 Neuro-inflammation and Microglial Activations

Persistent microglial activation and astrogliosis lead to the release of pro-inflammatory cytokines (TNF-α, IL-1β, IL-6), chemokines, and reactive nitrogen species that increase the synaptic loss and blood–brain barrier permeability. Inflammasomes (NLRP3) dysregulation and complement activation are propagating dysregulations, and instead of protective acute responses, maladaptive chronic neurotoxicity ensues amount to maladaptive changes in astrocytes and synapses in the hippocampus (Nicolas et al., 2022).

1.4.3 Excitotoxicity and Glutamate Imbalance

Overactivation of NMDA and AMPA receptors due to excessive glutamate stimulation leads to calcium influx, activation of proteases, and mitochondrial collapse, impaired glutamate uptake by astrocytes, and decreased levels of EAAT2 increase vulnerability. This is one of the pathways that play a major role in ALS and HD, through motor neuron and striatal degeneration· (Goedert et al., 2022).

Impaired autophagy and proteasomal dysfunction are reported to occur in fibrosis-representing anterior cortical lesions of the prefrontal cortex.

1.4.4 Autophagy and Proteasomal Impairment Lysogenic Autophagy and Proteasomal Dysfunction Are Impaired in Lesions of Fibrosis-Representative Anterior cortex during the Prefrontal Cortex

The clearance of aggregate-prone proteins is blocked by autophagosome–lysosome fusion faults and proteasomal blockage; this further causes lysosomal rupture, which triggers the accumulation of damaged organelles in a manner dependent on mTOR hyperactivation and Beclin-1 sequestration, and that again is interchangeable with proteostasis collapse.

1.4.5 Synaptic Pathophysiology and Dysgenesis of Neurons

Depletion of early synaptic vesicles, dendritic loss of spines, and long-term potentiation are all reduced before the apparent death of neurons; oligomeric proteins destabilize PSD-95 scaffolds and AMPA transport. The end outcome of apoptotic, necroptotic, and ferroptotic pathways is somatic shrinkage, axonal dystrophy, and Wallerian degeneration (Selkoe & Hardy, 2023).

1.4.6 Genetic and Epigenetic Reasons for Neurodegeneration

The cascades are initiated by mutations in the APP/PSEN1 (AD), SNCA/LRRK2 (PD), HTT (HD), SOD1/C9orf72 (ALS), and PRNP (prions) and adjusted by epigenetic alterations (DNA methylation and histone acetylation and microRNA). APOEε4 risk forms affect Aβ clearance and the phosphorylation of tau sporadically (Spillantini & Goedert, 2021).

1.5 THE MOLECULAR AND CELLULAROMES

Neurodegenerative molecular processes are centered on abnormal protein processing, aggregation, and propagation, which trigger subsequent cellular stress responses, leading to synapse failure and death. These pathways can have disease-specific pathways but are cross-talking, as in the case of tau seeding in AD/PD or TDP-43 extension in ALS/FTD (Neumann et al., 2018).

1.5.1 The Tau Protein and the Neurofibrillary Tangles Play an Essential Role

One protein, tau, a microtubule-bound protein, hyperphosphorylates and dissociates from microtubules, forming a coiled pair of helical filaments that aggregate into NFTs, which interfere with axonal transport and initiate neurodegeneration. Tangles are closely linked to loss of cognitive function in AD and are also seen in other related tauopathies, such as PSP and CBD, through prion-like templating (Hetz & Saxena, 2022).

1.5.2 A Beta Peptide and Plaque Formation

Based on the cleavage of APP into β- and γ-secretases, the subsequent peptides of APP, Aβ, soluble toxic forms, and extracellular plaques are formed and cause impairment of synaptic plasticity and microglia, as well as tau pathology. Mutations of APP/PSEN1/2 in the family are known to cause AD disease that accelerates the production of Aβ, whereas APOEε4 disrupts the clearance process (Emre et al., 2013).

1.5.3 α-Synuclein and the Lewy Body Formation

The α-synuclein accumulates in oligomeric protofibrils and Lewy bodies, which inhibit dopamine release, mitochondrial activity, and proteostasis in PD, DLB, and MSA. Staging Braak association ascending brainstem-to-cortical dissemination through endocytosis and exosomal release is found with SNCA gene duplication/triplication leading to familial forms (Ahlskog et al., 2011).

1.5.4 Function of TDP-43 and Other Pathological Proteins

In 95% of ALS and 50% of FTLD cases, an RNA-binding protein TDP-43 is mislocalized in the cytoplasm and phosphorylated, which impairs splicing, stress granules, and mitochondrial localization. FUS and other RAN proteins are in C9 or f72 expansions, whereas polyglutamine expansions in HD and SCA cause transcriptional dysregulation (Rascovsky et al., 2011).

1.5.5 Cellular Stress Signaling and Apoptosis

Unresolved endoplasmic reticulum stress triggers UPR (PERK/IRE1/ATF6), which results in apoptosis mediated by CHOP; unresolved mitochondrial outer membrane permeabilization releases cytochrome-c, which activates caspases. Inhibition of autophagy, lysosomal rupture, and necroptosis/MLKL alterations enhances the death in amassed-filled neurons (Brooks et al., 2000).

1.6 CLINICAL MANIFESTATIONS

Neurodegenerative disorders cause clinical presentations due to a selective disruption of circuits in the brain, which establish domain-specific impairments that develop in a predictable progression between stages, which inform diagnosis and treatment. Symptoms are categorized into cognitive, motor, sensory/autonomic, and behavioral domains and overlap with each other, demonstrating common pathologies such as loss of synapses and network failure (McKhann et al., 2011).

1.6.1 There Are a Number of Cognitive and Behavioral Symptoms

Dementias such as AD and FTD exhibit cognitive impairment in terms of amnesia, executive dysfunction, visuospatial impairment, and language impairment, and hippocampal/entorhinal atrophy promotes episodic memory impairment. Changes in behavior involve apathy, agitation, psychosis, and disinhibition, which are correlated with frontal/subcortical and Lewy body dementia with visual hallucinations with attention swings up and down (Blennow & Zetterberg, 2019).

1.6.2 Motor Dysfunction

Parkinsonism (bradykinesia, rigidity, tremor, postural instability), chorea (HD), dystonia, myoclonus (CJD), and spasticity/weakness (ALS/upper motor neuron (UMN)) are predominant motor problems. The loss of nigrostriatal dopamine is the cause of PD akinesia, corticospinal tract degeneration causes UMN signs in ALS, but cerebellar involvement causes ataxia in MSA/SCA (Livingston et al., 2020).

1.6.3 Unawareness and Alteration of Sensory and Autonomic Symptoms According to Insulin Administration

With less severity, there are sensory deficits involving visual hallucinations (PD/DLB), pain (small fiber neuropathy in PD), and olfactory loss (early AD/PD biomarker. Autonomic dysfunction, characterized by orthostatic (postural) hypotension, urinary retention, constipation, erectile dysfunction: the brainstem/noradrenergic degeneration of galaxies becomes autonomic failure, disabling synucleinopathies and advanced AD to a large extent quality of life (Golriz Khatami & Robinson, 2021).

1.6.4 Stage of Development and Disease

The neurodegenerative diseases possess typical patterns of staging and fluctuating prognoses as they are associated with different cascades of pathology and clinical presentation. Development of the AIDS disease consists of the formation of Aβ plaques and NFTs of hyperphosphorylated tau protein. The hierarchical spread of tau NFTs specifically monitored by the Braak staging system of AD begins in the transentorhinal cortex (Stages I–II), extends to limbic structures (Stages III–IV), and ultimately spreads throughout the neocortex (Stages V–VI), and has a strong correlation with cognitive decline. The Consortium to Establish a Registry of AD (CERAD) score measures neuritic plaque density as a complement to the Braak staging and Aβ pathology measurement methods, which chart the neocortical to deeper brain translocation of Aβ. The National Institute on Ageing-Alzheimer's Association research framework incorporates both these pathological hallmarks with clinical symptoms to define AD on a continuum, between a preclinical asymptomatic stage, mild cognitive impairment (MCI), and dementia. This paradigm uses biomarkers, such as cerebrospinal fluid (CSF) Aβ42/tau ratios and amyloid/tau positron emission tomography (PET), to make a biological diagnosis in living persons (Kalia & Lang, 2024).

PD is a disease that is characterized by the accumulation of α-synuclein within cells, and it takes the form of Lewy bodies. Braak staging of PD refers to a progressive series of α-synuclein pathology, which in most instances starts in the lower brainstem (dorsal motor nucleus of the vagus and olfactory bulb), then moves to the substantia nigra, and finally to limbic and neocortical regions as well. This spreading of motor neurons bottom-up explains the initial non-motor symptoms like dysfunction of the olfactory sense and gastrointestinal disorders before the motor symptoms manifest themselves in tremors and bradykinesia. The assessment of PD progression and severity is also often conducted clinically by the Unified Parkinson Disease Rating Scale (UPDRS) or its updated variant (MDS-UPDRS), which assesses motor or non-motor symptoms. Progression may be further classified in terms of when milestone events (walking and balance impairment) occur.

ALS is a disease that is characterized by the progressive degeneration of UMNs as well as LMNs. Staging systems of ALS, like the King staging and Milano Torino staging (MiToS), are used to determine the severity of the involvement of UMN and LMN in other parts of the body, which is very important in the prognostication and design of clinical trials (Hardiman et al., 2017).

1.7 DIAGNOSTIC APPROACHES

The early, accurate identification of neurodegenerative disorders through diagnostic methods incorporates a combination of clinical examination, neuroimaging, fluid biomarkers, and genetic tests to differentiate neurodegenerative diseases and mimics as well as to provide prognostication. Multimodal approaches enhance sensitivity/specificity, and the ATN framework (AD: amyloid/tau/neurodegeneration) is an example of integrated criteria (Hampel et al., 2021).

1.7.1 The Patient Presents with General Health Without Problems Breathing, Shivering, or Muscle Spasms. No Additional Diagnostic Features or Complications Mentioned

Clinical diagnosis is based on an elaborated history, cognitive examination (MMSE/MoCA), and neurological examination of cognition, Parkinsonism (UPDRS), motor (ALSFRS-R), and behavioral (changes). The characteristics to be supported are early memory loss (AD), REM sleep behavior disorder (PD), family history (HD), and progression pattern; probable/definite criteria imply that alternatives must be eliminated by labs/imaging (Beach & Adler, 2020).

1.7.2 Neuroimaging (MRI, PET, CT)

Structural MRI shows atrophy of hippocampal (AD), midbrain (PSP), corticospinal tract (ALS), and FDG-PET demonstrates that there are problems in temporoparietal hypometabolism (AD) and nigrostriatal impairments (PD). Diagnostics Amyloid-PET (florbetapir), tau-PET (flortaucipir), and DaTSCAN affirm proteinopathy and dopamine depletion, whereby a reduction of DAT binding greater than 30% diagnostics PD.

1.7.3 Biomarkers in Blood and Cerebrospinal Fluid (CSF)

AD can be diagnosed using CSF Ab42/tau/p-tau ($A\beta42<500$ pg/mL, p-tau/$A\beta42>0.023$); neurofilament light chains (NfL) detects axonal damage in all disorders. P-tau181/217, GFAP and NfL present the non-invasive screening choices and the plasma Ab42/40 predicts the brain amyloid. (Morris, 2016).

1.7.4 Genetic Testing and Molecular Diagnostics

The mutations that cause the disease are identified by genetic means: APP/PSEN1/2 (AD), SNCA/LRRK2/GBA (PD), HTT CAG>36 (HD), C9orf72/SOD1 (ALS), PRNP (prions). The sequencing panels of the next generation direct the families of cases; the polygenic risk scores (APOEε4) classify the risk which is sporadic.

1.8 STATED IN A JOURNAL ARTICLE, THERE ARE MANY NEWER METHODS OF THERAPY COMPARED TO THE MORE TRADITIONAL ONES. RECENT AND ROLLING THERAPEUTIC TECHNIQUES

Neurodegenerative disorders and their therapeutic approaches can be divided into symptomatic management approaches and disease-modifying therapeutic approaches to tackle underlying pathophysiologic processes, such as protein aggregation and neuro-inflammation, and to which new biologics and gene therapies are beginning to show clinical feasibility. The drugs used in the treatment of AD, like cholinesterase inhibitors (donepezil, rivastigmine), memantine AD motor fluctuations (levodopa/carbidopa). ALS survival tetrabenazine provides only symptomatic relief and are not capable to regenerate the neurons. The non-pharmacological and supportive treatments postpone the process in the prodromal phases. The function of the brain in resistance to medication motor complications is enhanced by cognitive rehabilitation, speech therapy, and multidisciplinary care. Other modern methods, such as stem cell and gene therapy, are limited. The AAV vectors are also under development in gene therapy to lower the negative levels of gene expression in disease-related nervous conditions like ALS and HD. Such methods, such as antisense oligonucleotides of the SOD1, C9orf72, and huntingtin genes, are in Phase 1/2 clinical trials. Moreover, dopaminergic neurons and astrocytes derived from iPSCs are under consideration in the context of the treatment of PD and ALS with the use of stem cell therapies. These cell replacement approaches are meant to provide a replacement of lost neuronal functions and overcome the issues of cell survival, cell integration, and immune rejection. ASOs gene silencing (SOD1/C9orf72, ALS) replacement therapies make progress in stage 1/2 clinical trials; iPSC-created dopaminergic neurons and astrocytes provide cell-based PD/ALS, overcoming engraftment and immune rejection problems (Li & Dong, 2023).

Through nanotechnology, several methods have been developed as efficient vehicles for drugs and gene therapies to the target site with minimal side effects and reduced human side effects.

1.8.1 Nanotechnology/Drug Delivery Systems

Nanotechnology has developed a myriad of strategies as convenient carriers of drug and gene therapies to the target site with minimal side effects and low human side effects.

Nanoparticles improve the use of Ab/tau immunotherapies and neuroprotective in the blood–brain barrier; exosome delivery of miRNAs/siRNAs targeting α-synuclein/TDP-43 enhances selectivity and decreases systemic toxicity.

1.8.2 Disease-Modifying and Neuroprotective Strategies

Monoclonal antibodies to anti-amyloid (lecanemab, donanemab) and anti-tau decrease the plaque/tangle load in early AD; α-synuclein vaccines/antisense oligonucleotides move to PD. GLP-1 agonists, NLRP3 inhibitors, and mitochondrial enhancers are neuroprotective across disorders sharing similar pathways of action.

1.9 CONCLUSION

The relationship between mitochondrial dysfunction and oxidative stress is an extremely close-knit pathological axis of neurodegenerative diseases, including Alzheimer's, Parkinson's, Huntington diseases, and ALS. Neuronal susceptibility is due to intense bioenergetic requirements and complicated morphology. The impairment of mitochondria interferes with ATP production and calcium homeostasis, leading to the overproduction of reactive oxygen species. This redox imbalance floods antioxidant defenses and leads to massive oxidative damage of cellular constituents and is associated with protein misfolding and aggregation, as in the case of Ab and hyperphosphorylated tau in

AD. Synapses, which demand such a lot of energy, are also especially vulnerable, and impaired synaptic plasticity and delivery occur. Oxidative stress also destroys the axonal microtubules and axonal transport, thus interfering with the transport of vital molecules and healthy mitochondria to the remote terminals, and this leads to synaptic failures. Dysfunctional mitochondria discharge mitochondrial DNA (mtDNA) and other damage-related molecular (DAMPs), which cause neuro-inflammation through signaling pathways such as NLRP3 inflammasome. Stressful oxidative and inflammatory responses cause a sustained outer membrane permeabilization of the mitochondrion, where apoptotic processes and the execution of neuronal death by the programmed pathway commence. Moreover, oxidative stress impairs the blood–brain barrier to allow the entry of neurotoxic elements and immune cells, which in turn enhance the generation of ROS, mitochondrial dysfunction, and inflammation. This forms a vicious cycle: impaired mitochondrial activity produces ROS, which overwhelms cellular defenses, worsening mitochondrial dysfunction, aggregation of proteins, impairment of synapses, neuro-inflammation, and cell death. The most important thing is to decipher these complex processes so as to come up with effective, specific treatment measures in relation to these debilitating conditions.

REFERENCES

Ahlskog, J. E., Geda, Y. E., Graff-Radford, N. R., & Petersen, R. C. (2011). Physical exercise as a preventive or disease-modifying treatment of dementia and brain aging. *Mayo Clinic Proceedings*, *86*(9), 876–884.

Beach, T. G., & Adler, C. H. (2020). The challenges of biomarker development in neurodegenerative diseases. *Alzheimer's & Dementia*, *16*(5), 640–654.

Blennow, K., & Zetterberg, H. (2019). Cerebrospinal fluid and plasma biomarkers in Alzheimer disease. *Nature Reviews Neurology*, *15*(3), 137–152.

Bose, K., Beal, M. F., & colleagues. (2022). Mitochondrial dysfunction in neurodegeneration. *Frontiers in Neuroscience*, *16*, 845168.

Brooks, B. R., Miller, R. G., Swash, M., & Munsat, T. L. (2000). El Escorial revisited: Revised criteria for the diagnosis of amyotrophic lateral sclerosis. *Amyotrophic Lateral Sclerosis and Other Motor Neuron Disorders*, *1*(5), 293–299.

De Strooper, B., & Karran, E. (2016). The cellular phase of Alzheimer's disease. *Cell*, *164*(4), 603–615.

Dong, X.-X., Wang, Y., & Qin, Z.-H. (2020). Excitotoxicity in ALS/FTD pathogenesis. *Frontiers in Cellular Neuroscience*, *14*, 577972.

Dorsey, E. R., & Bloem, B. R. (2018). The Parkinson pandemic—A call to action. *JAMA Neurology*, *75*(1), 9–10.

Dugger, B. N., & Dickson, D. W. (2017). Pathology of neurodegenerative diseases. *Cold Spring Harbor Perspectives in Biology*, *9*(7), a028035.

Emre, M., Aarsland, D., Brown, R., Burn, D. J., Duyckaerts, C., Mizuno, Y., Broe, G. A., Cummings, J., Dickson, D. W., Gauthier, S., Goldman, J., Goetz, C., Korczyn, A., Lees, A., Levy, R., Litvan, I., McKeith, I., Olanow, W., Poewe, W., & Quinn, N. (2013). Clinical diagnostic criteria for dementia associated with Parkinson's disease. *Movement Disorders*, *28*(12), 1703–1712.

Goedert, M., Eisenberg, D. S., & Crowther, R. A. (2022). Tau filaments in neurodegeneration. *Nature Reviews Neuroscience*, *23*(10), 599–614.

Golriz Khatami, S., & Robinson, J. L (2021). Precision medicine approaches in neurodegenerative diseases. *Frontiers in Neuroscience*, *15*, 662549.

Hampel, H., Vergallo, A., Aguilar, L. F., Benda, N., Broich, K., Cuello, A. C., Cummings, J., Dubois, B., Federoff, H. J., Fiandaca, M., Frisoni, G. B., Hampel, C., Lista, S., Mead, S., O'Bryant, S. E., Ritchie, C., Scheltens, P., & Younesi, E (2021). The Alzheimer precision medicine initiative. *Journal of Alzheimer's Disease*, *83*(1), 161–198.

Hardiman, O., Al-Chalabi, A., Chio, A., Corr, E. M., Logroscino, G., Robberecht, W., Shaw, P. J., Simmons, Z., & van den Berg, L. H. (2017). Amyotrophic lateral sclerosis. *Nature Reviews Disease Primers*, *3*, 17071.

Heneka, M. T., Carson, M. J., El Khoury, J., Landreth, G. E., Brosseron, F., Feinstein, D. L., Jacobs, A. H., Wyss-Coray, T., Vitorica, J., Ransohoff, R. M., Herrup, K., Frautschy, S. A., Finsen, B., Brown, G. C., Verkhratsky, A., Yamanaka, K., Koistinaho, J., Latz, E., Halle, A., & Petzold, A. (2015). Neuroinflammation in Alzheimer's disease. *The Lancet Neurology*, *14*(4), 388–405.

Hetz, C., & Saxena, S (2022). ER stress in neurodegenerative disease. *Nature Reviews Neurology*, *18*(3), 163–175.

Hou, Y., Dan, X., Babbar, M., Wei, Y., Hasselbalch, S. G., Croteau, D. L., & Bohr, V. A. (2019). Ageing as a risk factor for neurodegenerative disease. *Nature Reviews Neurology*, *15*(10), 565–581.

Irwin, D. J., McMillan, C. T., Toledo, J. B., Arnold, S. E., Shaw, L. M., Wang, L.-S., Van Deerlin, V. M., Lee, E. B., Trojanowski, J. Q., & Grossman, M. (2021). Mapping neurodegeneration across diseases. *Neurobiology of Disease*, *152*, 105289.

Jack, C. R., Bennett, D. A., Blennow, K., Carrillo, M. C., Dunn, B., Haeberlein, S. B., Holtzman, D. M., Jagust, W., Jessen, F., Karlawish, J., Liu, E., Molinuevo, J. L., Montine, T., Phelps, C., Rankin, K. P., Rowe, C. C., Scheltens, P., Siemers, E., Snyder, H. M., & Silverberg, N. (2018). NIA–AA research framework: Toward a biological definition of Alzheimer's disease. *Alzheimer's & Dementia*, *14*(4), 535–562.

Jellinger, K. A. (2018). Pathology of neurodegenerative diseases. *Handbook of Clinical Neurology*, *145*, 123–147.

Kalia, L. V., & Lang, A. E (2024). Parkinson's disease: From pathogenesis to pharmacotherapy. *Nature Reviews Drug Discovery*, *23*(5), 365–386.

Kovacs, G. G., Xie, S. X., Robinson, J. L., Lee, E. B., & Trojanowski, J. Q. (2017). Non-Alzheimer neurodegenerative pathologies. *Cold Spring Harbor Perspectives in Biology*, *9*(2), a026682.

Li, X., & Dong, Q (2023). Machine learning in Alzheimer's disease: Progress and prospects. *Neuroscience Bulletin*, *39*(4), 714–736.

Livingston, G., Huntley, J., Sommerlad, A., Ames, D., Ballard, C., Banerjee, S., Brayne, C., Burns, A., Cohen-Mansfield, J., Cooper, C., Costafreda, S. G., Dias, A., Fox, N., Gitlin, L. N., Howard, R., Kales, H. C., Kivimäki, M., Larson, E. B., Ogunniyi, A., Mukadam, N. (2020). Dementia prevention, intervention, and care: 2020 report of the Lancet Commission. *The Lancet*, *396*(10248), 413–446.

McKhann, G. M., Knopman, D. S., Chertkow, H., Hyman, B. T., Jack, C. R., Kawas, C. H., Klunk, W. E., Koroshetz, W. J., Manly, J. J., Mayeux, R., Mohs, R. C., Morris, J. C., Rossor, M. N., Scheltens, P., Carrillo, M. C., Thies, B., Weintraub, S., & Phelps, C. H. (2011). The diagnosis of dementia due to Alzheimer's disease. *Alzheimer's & Dementia*, *7*(3), 263–269.

Menzies, F. M., Fleming, A., & Rubinsztein, D. C. (2015). Autophagy and neurodegeneration. *Journal of Molecular Biology*, *428*(8), 1773–1793.

Morris, M. C. (2016). Nutrition and risk of dementia. *Annals of the New York Academy of Sciences*, *1367*(1), 31–37.

Neumann, M., Sampathu, D. M., Kwong, L. K., Truax, A. C., Micsenyi, M. C., Chou, T. T., Bruce, J., Schuck, T., Grossman, M., Clark, C. M., McCluskey, L. F., Miller, B. L., Masliah, E., Mackenzie, I. R., Feldman, H., Feiden, W., Kretzschmar, H. A., Trojanowski, J. Q., & Lee, V. M.-Y. (2018). TDP-43 in amyotrophic lateral sclerosis and frontotemporal lobar degeneration. *Cold Spring Harbor Perspectives in Medicine*, *8*(5), a024554.

Nicolas, G., Charbonnier, C., Campion, D., & Brice, A. (2022). Genetics of neurodegeneration. *Nature Reviews Neurology*, *18*(6), 347–362.

Rascovsky, K., Hodges, J. R., Knopman, D., Mendez, M. F., Kramer, J. H., Neuhaus, J., van Swieten, J. C., Seelaar, H., Dopper, E. G. P., Onyike, C. U., Hillis, A. E., Josephs, K. A., Boeve, B. F., Kertesz, A., Seeley, W. W., Rankin, K. P., Johnson, J. K., Gorno-Tempini, M. L., Rosen, H., & Miller, B. L. (2011). Sensitivity of revised diagnostic criteria for the behavioural variant of frontotemporal dementia. *Brain*, *134*(9), 2456–2477.

Selkoe, D. J., & Hardy, J. (2023). Amyloid-β and tau: Dr Jekyll and Mr Hyde in Alzheimer's disease? *Nature Reviews Neurology*, *19*(5), 283–294.

Soto, C., & Pritzkow, S. (2018). Protein misfolding, aggregation, and conformational strains in neurodegenerative diseases. *Nature Neuroscience*, *21*(10), 1332–1340.

Spillantini, M. G., & Goedert, M (2021). Alpha-synuclein in filamentous inclusions of Lewy bodies. *Annals of Neurology*, *89*(6), 1055–1067.

Wishart, T. M., Parson, S. H., & Gillingwater, T. H. (2021). Synaptic pathology in neurodegeneration. *Trends in Neurosciences*, *44*(9), 715–728.

Zetterberg, H., Bendlin, B. B., & Blennow, K. (2023). Protein biomarkers for Alzheimer disease. *Nature Reviews Neurology*, *19*(5), 271–282.

2 Molecular Targets for Neuroprotection

Mohamed El Fadili, Mohammed Er-rajy, Achraf Ziani, Somdutt Mujwar, and Menana El Hallaoui

2.1 INTRODUCTION

Neurodegenerative diseases, including Alzheimer's, Parkinson's, amyotrophic lateral sclerosis, and stroke, are a rising global health challenge due to their complex etiology and progressive nature. Excitotoxicity (Armada-Moreira et al., 2020; Binvignat & Olloquequi, 2020), oxidative stress (Niedzielska et al., 2016; Teleanu et al., 2022), and disruption of the neurotransmitter system (Teleanu et al., 2022) are several aspects integral to the physiological mechanisms of these diseases that influence neuronal survival and cognitive function. Understanding the mechanisms will provide a basis for the rational design of therapies that protect neurons and preserve neural networks.

Among them, a few molecular targets intimately involved in neuronal function and vulnerability have been recognized and studied over the last decades (Houldsworth, 2023a). These are glutamate receptors along with N-methyl-D-aspartate (NMDA) receptors that are responsible for excitatory signaling and synaptic plasticity (Zhang et al., 2025); dopaminergic and cholinergic systems that govern motor, cognitive, and memory processes (Tata et al., 2014); as well as the cellular antioxidant defense system that fights against oxidative damage. Dysfunction in any of these pathways may trigger neurodegenerative alterations through a cascade; hence, they have become the key focus of therapeutic strategies.

Currently, the identification and optimization of neuroprotective agents are increasingly reliant on both computational and experimental approaches (Paul et al., 2010). Molecular docking (Abechi et al., 2024; Aloui et al., 2025; El Fadili, Er-rajy, et al., 2022), quantitative structure–activity relationship (QSAR) modeling (El Fadili, Er-rajy, Imtara, et al., 2023; Er-rajy et al., 2022), and molecular dynamics simulation (El Fadili, Er-rajy, et al., 2022; Lischka et al., 2018) allow for rational design of molecules with high target specificity and predicted efficacy (Ajala et al., 2024; Ed-Dahmani et al., 2025). These computational predictions have also been confirmed by biological activity, pharmacokinetics, and therapeutic potential, through in vitro and in vivo studies (Peters et al., 2020). This chapter is mainly dedicated to the key molecular targets for neuroprotection and the strategies to identify and develop neuroprotective agents. It opens with the dissection of glutamate signaling and the excitotoxicity mediated by NMDA receptors moves through dopaminergic and cholinergic pathways and then discusses the function of antioxidant defenses in the preservation of neuronal integrity. Subsequent to that, the chapter discusses new methods in the field of computational and experimental drug design, highlighting how these two-fold strategies could be useful in the development of the next generation of neuroprotective therapies. Last, but not least, it wraps up by giving the reader a glimpse of some up and coming strategies, multitarget approaches, and technological innovations that may lead to better treatment for neurodegenerative diseases (Dąbrowska, 2019). By combining the mechanistic knowledge of disease with the advancements in design methodologies, the authors offer a detailed understanding of protective neuropharmacology discovery as well as a guide for researchers who want to develop safer and more effective treatments for neurological disorders (Li et al., 2021).

 DOI: 10.1201/9781003772019-2

2.2 GLUTAMATE AND NMDA RECEPTORS

2.2.1 Structure and Function of NMDA Receptors

Glutamate, a neurotransmitter that mediates rapid synaptic transmission, is considered to be the most important among the major excitatory neurotransmitters of the mammalian brain. Among the several types of glutamate receptors, the NMDA receptor is an essential ionotropic receptor subtype that is involved in synaptic plasticity, learning, and memory (Paoletti et al., 2013).

The NMDA receptor is a heterotetrameric complex that normally contains two GluN1 subunits that are combined with two GluN2 (A, D) or GluN3 (A, B) subunits. Each subunit has four chief regions: the amino-terminal domain which is responsible for assembly and allosteric modulation, the ligand-binding domain which identifies glutamate on GluN2 or glycine/D-serine on GluN1, the transmembrane domain that constitutes the ion-conducting channel, and the carboxyl-terminal domain that engages with proteins and signaling cascades inside the cell (Traynelis et al., 2010).

The NMDA receptor activation is made possible by three concurrent conditions: binding of both agonists, glutamate and glycine or D-serine; membrane depolarization to unblock the channel from Mg; and sufficient co-agonist availability. The "coincidence detection" feature of the receptor allows it to combine presynaptic and postsynaptic signals and thus makes it indispensable for learning, synaptic integration, and neuronal development.

2.2.2 Mechanisms of Excitotoxicity

Under regular physiological conditions, the activity of NMDA receptor is tightly regulated. If the extracellular glutamate level increases beyond normal, these receptors get overactivated and thus set off a chain of harmful intracellular events collectively known as excitotoxicity. This is the principal mechanism that leads to neuronal damage and death in numerous neurodegenerative diseases. Overstimulation of NMDA receptors causes an unrestricted entry of Ca^{2+} ions, and this triggers proteases, phospholipases, and endonucleases, which are destructive enzymes, to become activated. These enzymes then further break down the proteins, lipids, and nucleic acids which are the essential cellular components. At the same time, a very high level of Ca^{2+} inside the cell inhibits mitochondrial activity and stimulates the formation of reactive oxygen and nitrogen species, which causes oxidative stress to rise even more (Yu et al., 2023a).

Each of these occurrences leads to the accumulation of ionic disturbances, the exhaustion of ATP, and the deterioration of membrane integrity that finally results in neuronal apoptosis or necrosis. Excitotoxicity has been attributed as a key factor in the development of several neurological disorders, including Alzheimer's disease, Parkinson's disease, epilepsy, ischemic stroke, and schizophrenia; hence, targeting NMDA receptors may be a neuroprotective strategy with great potential for clinical use (Wal et al., 2022).

2.2.3 NMDA Receptor Antagonists

Because excessive activation of NMDA receptors is harmful to neuronal survival, considerable attention has been devoted to developing compounds that can limit their overactivation. These NMDA receptor antagonists act through distinct mechanisms, which are usually differentiated according to their binding site and mode of inhibition. First, competitive antagonists act by binding to the glutamate recognition site on the GluN2 subunit, thereby preventing the natural agonist from attaching. Classic examples include D-AP5 (2-amino-5-phosphonopentanoic acid) and CGS-19755. Then, some compounds interfere with the co-agonist site on the GluN1 subunit. These are referred to as glycine-site antagonists they inhibit receptor activation by glycine or D-serine. Besides that, non-competitive antagonists, or channel blockers, work by binding within the ion pore and physically closing it; thus, ion flux is blocked regardless of agonist binding. MK-801 (dizocilpine) and phencyclidine (PCP) are such examples.

And last but not least, uncompetitive antagonists bind only when the ion channel is open, resulting in voltage-dependent inhibition. Memantine, which is in the clinic used to treat moderate to severe Alzheimer's disease, is a member of the latter category. Because of its moderate affinity and fast dissociation kinetics, it can inhibit excessive NMDA activation without interfering with normal glutamatergic neurotransmission (Lipton, 2006).

These antagonists are highly effective in preventing neurons from excitotoxicity. However, their clinical application has been limited because of the unwanted neuropsychiatric effects resulting from the broad inhibition of NMDA signaling, such as cognitive impairment and hallucinations. Currently, the main emphasis of research is on the creation of subunit-selective, state-dependent, or allosteric modulators which can have a subtle effect on NMDA receptor activity without interfering with its normal function (Hanson et al., 2024; Ugale et al., 2024).

2.3 DOPAMINE RECEPTORS AND NEUROPROTECTION

Dopamine ranks as a key neurotransmitter that is mainly concentrated in the central nervous system (CNS). Dopamine is very involved in regulating various processes such as motility, motivation, emotion, and cognition. Dopamine receptors are activated through either of the two main groups of G protein–coupled receptors: D1-like (D1 and D5) and D2-like (D2, D3, and D4), which, respectively, activate and inhibit adenylate cyclase (AC). However, the dopaminergic system must maintain a balance where both the absence of and an excessive level of dopamine can lead to neurodegenerative and/or psychiatric disorder development. The protective function of dopamine receptors relies on regulating and modulating the intracellular signaling cascades that regulate Ca^{2+} levels, mitochondrial function, and the normal cellular oxidative balance. For example, when the D2 receptor is activated, it is believed to provide neuroprotection by decreasing glutamate-induced excitotoxicity while simultaneously inhibiting pro-apoptotic signals. On the other hand, excessive or diminished dopaminergic activity leads to neuronal cell damage. Therefore, selective modulation of dopamine receptors holds promise as a therapeutic strategy for treating Parkinson's disease, schizophrenia, and drug-induced neurotoxicity (Holstein et al., 2025; Zhang et al., 2024).

2.4 ACETYLCHOLINESTERASE AND CHOLINERGIC TRANSMISSION

The cholinergic system is crucial to cognitive functions such as learning, memory, and attention. The main neurotransmitter in this system is acetylcholine (ACh), and once released from the presynaptic neuron, it is broken down by the enzyme acetylcholinesterase (AChE) into acetic acid and choline; this action of an enzyme helps maintain appropriate synaptic transmission and prevent receptor overstimulation. Cholinergic neurons are lost in some degenerative conditions, such as Alzheimer's disease. The activity of AChE also increases in people with these diseases, leading to dramatic reductions in ACh levels and to an increased degree of decline in cognition. AChE has long been considered a target for increasing cholinergic transmission and improving communication between neurons. Currently, the AChE inhibitors used therapeutically include donepezil, rivastigmine, and galantamine because they temporarily restore the balance of neurotransmitters, thereby alleviating deficits in memory and learning skills. Recent research shows that some of the AChE inhibitors also have neuroprotective and antioxidant properties, which means that their potential therapeutic uses extend beyond symptomatic treatment.

2.5 ANTIOXIDANT DEFENSE SYSTEMS AND OXIDATIVE STRESS

In healthy physiological conditions, a homeostatic equilibrium exists in neurons between the rate of ROS production and the antioxidant defense mechanisms of superoxide dismutase (SOD), catalase (CAT), and glutathione peroxidase (GPx). This provides neuronal protection and helps to maintain redox homeostasis (Uttara et al., 2009). Conversely, the overproduction of ROS, which is generally the result of mitochondrial dysfunction, excitotoxicity, or neuroinflammation, creates oxidative stress as indicated by lipid peroxidation, protein oxidation, and DNA damage. These damaging events disrupt neuronal integrity and have been shown to play a significant role in the pathogenesis of a range of neurodegenerative diseases, including Alzheimer's disease, Parkinson's disease, and Amyotrophic Lateral Sclerosis (ALS). The strengthening of endogenous defense mechanisms and/or the introduction of exogenous antioxidant therapies provides a very effective approach to neuroprotection. Several natural polyphenols, vitamins C and E, as well as various synthetic compounds, possess the ability to quench free radicals, stabilize cellular membranes, and support mitochondrial functions (Jalouli et al., 2025) and, through these mechanisms, promote a neuroprotective effect and neuronal survival from oxidative injury.

2.6 IN SILICO AND EXPERIMENTAL APPROACHES IN NEUROPROTECTIVE DRUG DESIGN

The need for effective neuroprotective agents is increasingly complemented by a combination of computational modeling and experimental evaluation. Such strategies provide more insight into molecular interactions, reduce experimental costs, and accelerate the process of identifying promising therapeutic candidates (El Fadili, Er-rajy, Ali Eltayb, Kara, Imtara, et al., 2023; El Fadili et al., 2024; Haider et al., 2021). Computational methods predict and rationalize drug-target behavior while experimental studies are performed to verify the biological activity and its pharmacological relevance. Together, they constitute the complete framework for the rational design and optimization of neuroprotective molecules (El Fadili, Er-rajy, Abdalla, et al., 2023; El Fadili, Er-rajy, Ali Eltayb, Kara, Assouguem, et al., 2023).

2.6.1 Computational Drug Design Approaches (Docking, QSAR, MD)

These modern in silico methods, including molecular docking, quantitative structure–activity relationship modeling, and molecular dynamics simulations, represent some of the cornerstones of early drug discovery. In this context, molecular docking is a method that enables the prediction of the interaction between small molecules and the active site of a protein, leading to the determination of their binding mode and affinity. The knowledge is helpful in locating the major interactions that are essential for receptor recognition. QSAR modeling, either a two-dimensional or three-dimensional basis, such as CoMFA or CoMSIA, establishes quantitative correlations between chemical structure and biological activity that guide structural optimization to higher potency and selectivity (Khedraoui et al., 2024). Molecular dynamics simulations complement the methods described above by considering the flexibility of both proteins and ligands under near-physiological conditions, hence offering insights into the temporal stability and dynamics of the complex. Coupling these computational tools allows the prioritization of the most promising compounds before synthesis, thereby accelerating the development of new neuroprotective agents (Er-rajy et al., 2023; Zarougui et al., 2023).

2.6.2 Experimental Validation Strategies

Theoretical knowledge gained from computational studies needs to be supported with experimental confirmation of biological activity through initial in vitro bioassays such as receptor binding, enzyme inhibition, and antioxidant activity, which will validate predictions regarding molecular interactions and show pharmacological potential (Mahnashi et al., 2022). Further interested parties can determine if a compound provides neuroprotection through the use of cell culture–based systems to measure neuronal survival, resistance to oxidative stress, or inhibition of apoptosis followed by the use of in vivo animal models to investigate pharmacokinetics, penetration into the CNS, and behavioral effects, thus providing an understanding of the therapeutic potential of a compound in vitro and in vivo (Universidad de Sucre & Pajaro-Castro, 2024). Therefore, by combining predictions obtained from computational studies with those confirmed experimentally, researchers can establish a strong basis for designing and developing new compounds with real potential for providing neuroprotection.

2.7 CONCLUSION AND FUTURE PERSPECTIVES

The molecular mechanisms that determine both the function and dysfunction of neurons will allow researchers to develop effective neuroprotective agents. Excitotoxicity, oxidative stress, and neurodegeneration can be prevented through the modulation of target pathways, including glutamate and NMDA receptors, dopaminergic and cholinergic systems, and antioxidant defense systems. The integration of computational methods, including molecular docking, QSAR, and molecular dynamics simulations with experimental validation, provides researchers with an excellent method for rational drug design and discovering new neuroprotective drugs. Despite advances in computational modeling, however, a number of challenges still exist that hinder researchers from employing their findings to create effective therapeutics. Many of the issues raised by NMDA receptor inhibition and altered neurotransmitter function can be traced back to the different classes of NMDA receptor antagonists and the side effects of these drugs on neuronal physiology. Drug development should be mainly targeted at the discovery of state-dependent or allosteric modulators which allow normal physiological functions to be maintained while abnormal activities are reduced. Using artificial intelligence together with high-throughput screening and advanced imaging techniques will allow scientists to come up with better, more accurate neuroprotective substances. Besides that, it is absolutely vital that we continue to examine multitarget drug strategies for neurodegenerative disorders in order to discover agents capable of reducing excitotoxicity, oxidative stress, and impaired neurotransmission all at once. Understanding the molecular mechanisms underlying neurodegeneration coupled with breakthrough drug design will pave the way for novel neuroprotective therapies developed by extensive experimental work in vitro and in vivo.

REFERENCES

Abechi, S. E., Michael, A. T., Abduljelil, A., Stephen, E., Asipita, O. H., & El Fadili, M. (2024). Virtual screening and pharmacokinetics analysis of inhibitors against tuberculosis: Structure and ligand-based approach. *Scientific African*, *23*, e02085. https://doi.org/10.1016/j.sciaf.2024.e02085

Ajala, A., Eltayb, W. A., Abatyough, T. M., Ejeh, S., El Fadili, M., Otaru, H. A., Edache, E. I., Abdulganiyyu, A. I., Areguamen, O. I., Patil, S. M., & Ramu, R. (2024). In-silico screening and ADMET evaluation of therapeutic MAO-B inhibitors against Parkinson disease. *Intelligent Pharmacy*, *2*(4), 554–564. https://doi.org/10.1016/j.ipha.2023.12.008

Aloui, M., El Fadili, M., Mujwar, S., Er-rajy, M., Abuelizz, H. A., Er-rahmani, S., Zarougui, S., & Menana, E. (2025). In silico design of novel pyridazine derivatives as balanced multifunctional agents against Alzheimer's disease. *Scientific Reports*, *15*(1), 15910. https://doi.org/10.1038/s41598-025-98182-x

Armada-Moreira, A., Gomes, J. I., Pina, C. C., Savchak, O. K., Gonçalves-Ribeiro, J., Rei, N., Pinto, S., Morais, T. P., Martins, R. S., Ribeiro, F. F., Sebastião, A. M., Crunelli, V., & Vaz, S. H. (2020). Going the extra (synaptic) mile: Excitotoxicity as the road toward neurodegenerative diseases. *Frontiers in Cellular Neuroscience*, *14*, 90. https://doi.org/10.3389/fncel.2020.00090

Binvignat, O., & Olloquequi, J. (2020). Excitotoxicity as a target against neurodegenerative processes. *Current Pharmaceutical Design*, *26*(12), 1251–1262. https://doi.org/10.2174/1381612826666200113162641

Dąbrowska, K. (2019). Phage therapy: What factors shape phage pharmacokinetics and bioavailability? Systematic and critical review. *Medicinal Research Reviews*, *39*(5), 2000–2025. https://doi.org/10.1002/med.21572

Ed-Dahmani, I., El Fadili, M., Nouioura, G., Kandsi, F., Atki, Y. E., Abuelizz, H. A., Conte, R., Zahra Lafdil, F., Taleb, A., Abdellaoui, A., & Taleb, M. (2025). *Ferula communis* leaf extract: Antioxidant capacity, UHPLC–MS/MS analysis, and in vivo and in silico toxicity investigations. *Frontiers in Chemistry, 12*, 1485463. https://doi.org/10.3389/fchem.2024.1485463

El Fadili, M., Er-rajy, M., Abdalla, M., Abuelizz, H. A., Zarougui, S., Alkhulaifi, F. M., Alahmady, N. F., Shami, A., & Elhalaoui, M. (2023). In-silico investigations of novel tacrine derivatives potency against Alzheimer's disease. *Scientific African, e02048*. https://doi.org/10.1016/j.sciaf.2023.e02048

El Fadili, M., Er-rajy, M., Ali Eltayb, W., Kara, M., Assouguem, A., Saleh, A., Al Kamaly, O., Zerougui, S., & Elhallaoui, M. (2023). In-silico screening based on molecular simulations of 3,4-disubstituted pyrrolidine sulfonamides as selective and competitive GlyT1 inhibitors. *Arabian Journal of Chemistry, 105105*. https://doi.org/10.1016/j.arabjc.2023.105105

El Fadili, M., Er-rajy, M., Ali Eltayb, W., Kara, M., Imtara, H., Zarougui, S., Al-Hoshani, N., Hamadi, A., & Elhallaoui, M. (2023). An in-silico investigation based on molecular simulations of novel and potential brain-penetrant GluN2B NMDA receptor antagonists as anti-stroke therapeutic agents. *Journal of Biomolecular Structure and Dynamics*, 1–15. https://doi.org/10.1080/07391102.2023.2232024

El Fadili, M., Er-rajy, M., Imtara, H., Kara, M., Zarougui, S., Altwaijry, N., Al kamaly, O., Al Sfouk, A., & Elhallaoui, M. (2022). 3D-QSAR, ADME-Tox in silico prediction and molecular docking studies for modeling the analgesic activity against neuropathic pain of novel NR2B-selective NMDA receptor antagonists. *Processes, 10*(8), 1462. https://doi.org/10.3390/pr10081462

El Fadili, M., Er-rajy, M., Imtara, H., Noman, O. M., Mothana, R. A., Abdullah, S., Zerougui, S., & Elhallaoui, M. (2023). QSAR, ADME-Tox, molecular docking and molecular dynamics simulations of novel selective glycine transporter type 1 inhibitors with memory enhancing properties. *Heliyon, e13706*. https://doi.org/10.1016/j.heliyon.2023.e13706

El Fadili, M., Er-rajy, M., Mujwar, S., Ajala, A., Bouzammit, R., Kara, M., Abuelizz, H. A., Er-rahmani, S., & Elhallaoui, M. (2024). In silico insights into the design of novel NR2B-selective NMDA receptor antagonists: QSAR modeling, ADME-toxicity predictions, molecular docking, and molecular dynamics investigations. *BMC Chemistry, 18*(1), 142. https://doi.org/10.1186/s13065-024-01248-6

El Fadili, M., Er-rajy, M., Kara, M., Assouguem, A., Belhassan, A., Alotaibi, A., Mrabti, N. N., Fidan, H., Ullah, R., Ercisli, S., Zarougui, S., & Elhallaoui, M. (2022). QSAR, ADMET in silico pharmacokinetics, molecular docking and molecular dynamics studies of novel bicyclo (aryl methyl) benzamides as potent GlyT1 inhibitors for the treatment of schizophrenia. *Pharmaceuticals, 15*(6), 670. https://doi.org/10.3390/ph15060670

Er-rajy, M., El Fadili, M., Imtara, H., Saeed, A., Ur Rehman, A., Zarougui, S., Abdullah, S. A., Alahdab, A., Parvez, M. K., & Elhallaoui, M. (2023). 3D-QSAR studies, molecular docking, molecular dynamic simulation, and ADMET proprieties of novel pteridinone derivatives as PLK1 inhibitors for the treatment of prostate cancer. *Life, 13*(1), 127. https://doi.org/10.3390/life13010127

Er-rajy, M., El Fadili, M., Mrabti, N. N., Zarougui, S., & Elhallaoui, M. (2022). QSAR, molecular docking, ADMET properties in silico studies for a series of 7-propanamide benzoxaboroles as potent anticancer agents. *Chinese Journal of Analytical Chemistry, 50*(12), 100163. https://doi.org/10.1016/j.cjac.2022.100163

Haider, M., Chauhan, A., Tariq, S., Pathak, D. P., Siddiqui, N., Ali, S., Pottoo, F. H., & Ali, R. (2021). Application of in silico methods in the design of drugs for neurodegenerative diseases. *Current Topics in Medicinal Chemistry, 21*(11), 995–1011. https://doi.org/10.2174/1568026621666210521164545

Hanson, J. E., Yuan, H., Perszyk, R. E., Banke, T. G., Xing, H., Tsai, M.-C., Menniti, F. S., & Traynelis, S. F. (2024). Therapeutic potential of N-methyl-D-aspartate receptor modulators in psychiatry. *Neuropsychopharmacology, 49*(1), 51–66. https://doi.org/10.1038/s41386-023-01614-3

Holstein, H. J., Bouwknegt, D. G., Visschedijk, M. C., Bulthuis, M. L. C., Reinders-Luinge, M., Schoots, M. H., van Goor, H., Gordijn, S. J., Dijkstra, G., & Bourgonje, A. R. (2025). Exploring biomarkers of systemic oxidative stress and placental insufficiency in pregnant women with inflammatory bowel diseases. *Free Radical Biology and Medicine, 232*, 319–329. https://doi.org/10.1016/j.freeradbiomed.2025.03.017

Houldsworth, A. (2023a). Role of oxidative stress in neurodegenerative disorders: A review of reactive oxygen species and prevention by antioxidants. *Brain Communications, 6*(1), fcad356. https://doi.org/10.1093/braincomms/fcad356

Jalouli, M., Rahman, M. A., Biswas, P., Rahman, H., Harrath, A. H., Lee, I.-S., Kang, S., Choi, J., Park, M. N., & Kim, B. (2025). Targeting natural antioxidant polyphenols to protect neuroinflammation and neurodegenerative diseases: A comprehensive review. *Frontiers in Pharmacology, 16*, 1492517. https://doi.org/10.3389/fphar.2025.1492517

Khedraoui, M., Abchir, O., Nour, H., Yamari, I., Errougui, A., Samadi, A., & Chtita, S. (2024). An in silico study based on QSAR and molecular docking and molecular dynamics simulation for the discovery of novel potent inhibitor against AChE. *Pharmaceuticals*, *17*(7), 830. https://doi.org/10.3390/ph17070830

Li, Y., Xiong, C., Aschenbrenner, A. J., Chang, C., Weiner, M. W., Nosheny, R. L., Mungas, D., Bateman, R. J., Hassenstab, J., Moulder, K. L., & Morris, J. C. (2021). Item response theory analysis of the clinical dementia rating. *Alzheimer's & Dementia*, *17*(3), 534–542. https://doi.org/10.1002/alz.12210

Lipton, S. A. (2006). Paradigm shift in neuroprotection by NMDA receptor blockade: Memantine and beyond. *Nature Reviews Drug Discovery*, *5*(2), 160–170. https://doi.org/10.1038/nrd1958

Lischka, H., Nachtigallová, D., Aquino, A. J. A., Szalay, P. G., Plasser, F., Machado, F. B. C., & Barbatti, M. (2018). Multireference approaches for excited states of molecules. *Chemical Reviews*, *118*(15), 7293–7361. https://doi.org/10.1021/acs.chemrev.8b00244

Mahnashi, M. H., Alshahrani, M. A., Nahari, M. H., Hassan, S. S. ul, Jan, M. S., Ayaz, M., Ullah, F., Alshehri, O. M., Alshehri, M. A., Rashid, U., & Sadiq, A. (2022). In-vitro, in-vivo, molecular docking and ADMET studies of 2-substituted 3,7-dihydroxy-4H-chromen-4-one for oxidative stress, inflammation and Alzheimer's disease. *Metabolites*, *12*(11), 1055. https://doi.org/10.3390/metabo12111055

Niedzielska, E., Smaga, I., Gawlik, M., Moniczewski, A., Stankowicz, P., Pera, J., & Filip, M. (2016). Oxidative stress in neurodegenerative diseases. *Molecular Neurobiology*, *53*(6), 4094–4125. https://doi.org/10.1007/s12035-015-9337-5

Paoletti, P., Bellone, C., & Zhou, Q. (2013). NMDA receptor subunit diversity: Impact on receptor properties, synaptic plasticity and disease. *Nature Reviews Neuroscience*, *14*(6), 383–400. https://doi.org/10.1038/nrn3504

Paul, S. M., Mytelka, D. S., Dunwiddie, C. T., Persinger, C. C., Munos, B. H., Lindborg, S. R., & Schacht, A. L. (2010). How to improve R&D productivity: The pharmaceutical industry's grand challenge. *Nature Reviews Drug Discovery*, *9*(3), 203–214. https://doi.org/10.1038/nrd3078

Peters, S. A., Petersson, C., Blaukat, A., Halle, J.-P., & Dolgos, H. (2020). Prediction of active human dose: Learnings from 20 years of Merck KGaA experience, illustrated by case studies. *Drug Discovery Today*, *25*(5), 909–919. https://doi.org/10.1016/j.drudis.2020.01.002

Tata, A., Velluto, L., D'Angelo, C., & Reale, M. (2014). Cholinergic system dysfunction and neurodegenerative diseases: Cause or effect? *CNS & Neurological Disorders – Drug Targets*, *13*(7), 1294–1303. https://doi.org/10.2174/1871527313666140917121132

Teleanu, D. M., Niculescu, A.-G., Lungu, I. I., Radu, C. I., Vladâcenco, O., Roza, E., Costăchescu, B., Grumezescu, A. M., & Teleanu, R. I. (2022). An overview of oxidative stress, neuroinflammation, and neurodegenerative diseases. *International Journal of Molecular Sciences*, *23*(11), 5938. https://doi.org/10.3390/ijms23115938

Teleanu, R. I., Niculescu, A.-G., Roza, E., Vladâcenco, O., Grumezescu, A. M., & Teleanu, D. M. (2022). Neurotransmitters—Key factors in neurological and neurodegenerative disorders of the central nervous system. *International Journal of Molecular Sciences*, *23*(11), 5954. https://doi.org/10.3390/ijms23115954

Traynelis, S. F., Wollmuth, L. P., McBain, C. J., Menniti, F. S., Vance, K. M., Ogden, K. K., Hansen, K. B., Yuan, H., Myers, S. J., & Dingledine, R. (2010). Glutamate receptor ion channels: Structure, regulation, and function. *Pharmacological Reviews*, *62*(3), 405–496. https://doi.org/10.1124/pr.109.002451

Ugale, V., Deshmukh, R., Lokwani, D., Narayana Reddy, P., Khadse, S., Chaudhari, P., & Kulkarni, P. P. (2024). GluN2B subunit selective N-methyl-D-aspartate receptor ligands: Democratizing recent progress to assist the development of novel neurotherapeutics. *Molecular Diversity*, *28*(3), 1765–1792. https://doi.org/10.1007/s11030-023-10656-0

Universidad de Sucre, & Pajaro-Castro, N. (2024). Computational screening, synthesis and neuroprotective evaluation of small molecule for the treatment of Alzheimer's disease. *Vitae*, *31*(3). https://doi.org/10.17533/udea.vitae.v31n3a354271

Uttara, B., Singh, A., Zamboni, P., & Mahajan, R. (2009). Oxidative stress and neurodegenerative diseases: A review of upstream and downstream antioxidant therapeutic options. *Current Neuropharmacology*, *7*(1), 65–74. https://doi.org/10.2174/157015909787602823

Wal, P., Dwivedi, J., Wal, A., Vig, H., & Singh, Y. (2022). Detailed insight into the pathophysiology and the behavioral complications associated with the Parkinson's disease and its medications. *Future Journal of Pharmaceutical Sciences*, *8*(1), 33. https://doi.org/10.1186/s43094-022-00425-5

Yu, S. P., Jiang, M. Q., Shim, S. S., Pourkhodadad, S., & Wei, L. (2023a). Extrasynaptic NMDA receptors in acute and chronic excitotoxicity: Implications for preventive treatments of ischemic stroke and late-onset Alzheimer's disease. *Molecular Neurodegeneration*, *18*(1), 43. https://doi.org/10.1186/s13024-023-00636-1

Zarougui, S., Er-rajy, M., Faris, A., Imtara, H., El Fadili, M., Al kamaly, O., Zuhair Alshawwa, S., Nasr, F. A., Aloui, M., & Elhallaoui, M. (2023). QSAR, DFT studies, docking molecular and simulation dynamic molecular of 2-styrylquinoline derivatives through their anticancer activity. *Journal of Saudi Chemical Society*, *27*(6), 101728. https://doi.org/10.1016/j.jscs.2023.101728

Zhang, K., Wen, M., Nan, X., Zhao, S., Li, H., Ai, Y., & Zhu, H. (2025). NMDA receptors in neurodegenerative diseases: Mechanisms and emerging therapeutic strategies. *Frontiers in Aging Neuroscience*, *17*, 1604378. https://doi.org/10.3389/fnagi.2025.1604378

Zhang, W., Zhang, X., Lei, M., Zhang, D., Qin, G., Zhou, J., Ji, L., & Chen, L. (2024). Dopamine D2 receptor activation blocks GluA2/ROS positive feedback loop to alienate chronic-migraine-associated pain sensitization. *Antioxidants*, *13*(6), 725. https://doi.org/10.3390/antiox13060725

3 Chemistry of Nitrogen-Containing Heterocyclic Compounds

Ehsan Ullah Mughal and Nafeesa Naeem

3.1 INTRODUCTION

Heterocyclic chemistry represents one of the most vibrant and essential fields in organic synthesis and medicinal chemistry (Eicher et al., 2013). A heterocyclic compound is defined as a cyclic molecule containing at least one atom other than carbon within its ring structure (Quin & Tyrell, 2010). Among all heteroatoms, nitrogen is the most frequently encountered due to its moderate electronegativity, lone pair availability, and ability to participate in aromatic stabilization (Balaban et al., 2004). Nitrogen-containing heterocycles form the structural core of countless biologically active molecules, natural products, and functional materials (Jampilek, 2019; Obaid et al., 2022) (Figure 3.1).

The ubiquitous presence of nitrogen heterocycles in natural metabolites, pharmaceuticals, agrochemicals, and dyes underscores their importance (Mezgebe & Mulugeta, 2022; Ullah et al., 2022). Compounds such as pyridine, quinoline, imidazole, and indole serve as pharmacophores in several therapeutic drugs, including nicotinamide (vitamin B_3), quinine (antimalarial), and tryptamine derivatives (neurotransmitters) (Kabir & Uzzaman, 2022). In addition to medicinal applications, nitrogen heterocycles play a critical role in modern materials science, especially in optoelectronic devices, coordination polymers, and catalysis (Frank & Szőllősi, 2021).

The fundamental interest in nitrogen heterocycles stems from their structural diversity, aromaticity, and tunable reactivity (Eicher et al., 2013). Their chemical behavior is influenced by the position and hybridization of the nitrogen atom(s) in the ring, which affects electron density distribution, basicity, and hydrogen bonding characteristics (Ebenezer et al., 2022). Consequently, the study of

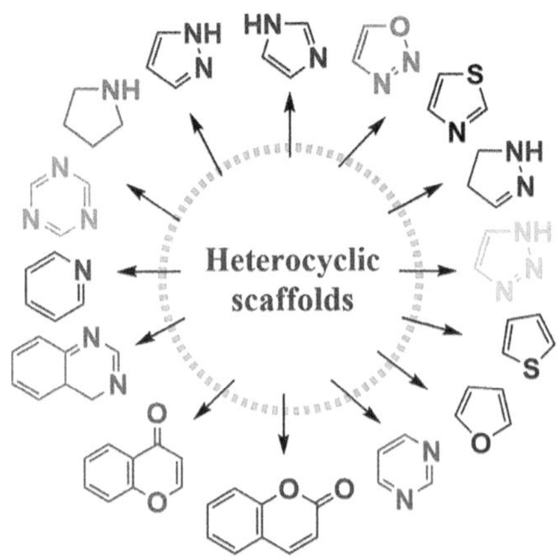

FIGURE 3.1 Representative examples of nitrogen-containing heterocycles. (Reproduced from Ref. Obaid et al. (2022) with permission under a Creative Commons Attribution-Non-Commercial 3.0 Unported Licence.)

DOI: 10.1201/9781003772019-3

nitrogen heterocycles bridges multiple disciplines, integrating aspects of organic synthesis, physical chemistry, and molecular design (Amin et al., 2022).

3.2 STRUCTURAL FEATURES AND NOMENCLATURE

3.2.1 General Characteristics

Nitrogen-containing heterocycles can be distinguished by their ring size, the number of heteroatoms, and their degree of saturation (Table 3.1). The most typical representatives are the five- and six-membered ring systems, although there are also larger macrocyclic structures (Ram et al., 2019). The presence of aromatic heterocyclic compounds such as pyridine and pyrrole allows depositing of the common π-electrons in accordance with Hückel's rule (4n + 2), which gives them an extraordinary thermodynamic stability (Lim et al., 2024).

The nitrogen atom's lone pair in heterocycles is a crucial factor deciding the aromatic character, basicity, and chemical reactivity (Jacobi, 2018). In compounds like pyrrole, the lone pair is involved in the delocalization of the aromatic π-electron, that is, the lone pair is part of a six-membered ring of electrons and thus is not available for protonation or basicity—a property that can be said to be non-basic, essentially. Thus, in contrast, in pyridine, the nitrogen's lone pair is in an sp^2 orbital which is at right angles to the aromatic π system that does not give rise to the nitrogen being a part of the aromaticity, and hence the lone pair remains free to accept a proton, making the basicity measurable, though it is weaker than that of the common amines (Schofield, 2013).

3.2.2 Aromaticity and Electron Delocalization

The aromaticity of nitrogen heterocycles can be evaluated on the basis of resonance and MO consideration. In pyrrole, the nitrogen atom contributes its lone pair to the π-system, creating a six-electron aromatic ring. In contrast, pyridine's nitrogen retains its lone pair in an sp^2 orbital orthogonal to the aromatic plane (Figure 3.2) (Tully, 2025). Fused systems, such as indole, combine both behaviors, leading to rich electronic versatility (Joule & Mills, 2012).

3.2.3 Nomenclature

The naming of nitrogen-containing heterocycles follows the International Union of Pure and Applied Chemistry (IUPAC) conventions. For simple rings, names are derived from the number of ring atoms and the position of nitrogen. Examples include aziridine (three-membered), pyrrole (five-membered), and pyridine (six-membered). Fused-ring systems are named based on parent structures, such as indole (benzopyrrole) or quinoline (benzopyridine) (Nguyen et al., 2025) (Table 3.2).

TABLE 3.1
Comparison of Electronic Properties of Representative Nitrogen Heterocycles

Compound	Hybridization of N	Participation in Aromatic π-System	Basicity (pKa of Conjugate Acid)	Example
Pyrrole	sp^2	Yes	~0	Hemin core
Pyridine	sp^2	No	~5.2	Nicotinamide
Imidazole	Mixed (1 delocalized N, 1 localized N)	Partial	~7	Histidine residue
Indole	sp^2	Yes	Non-basic	Tryptophan
Quinoline	sp^2	No	~4.9	Quinine

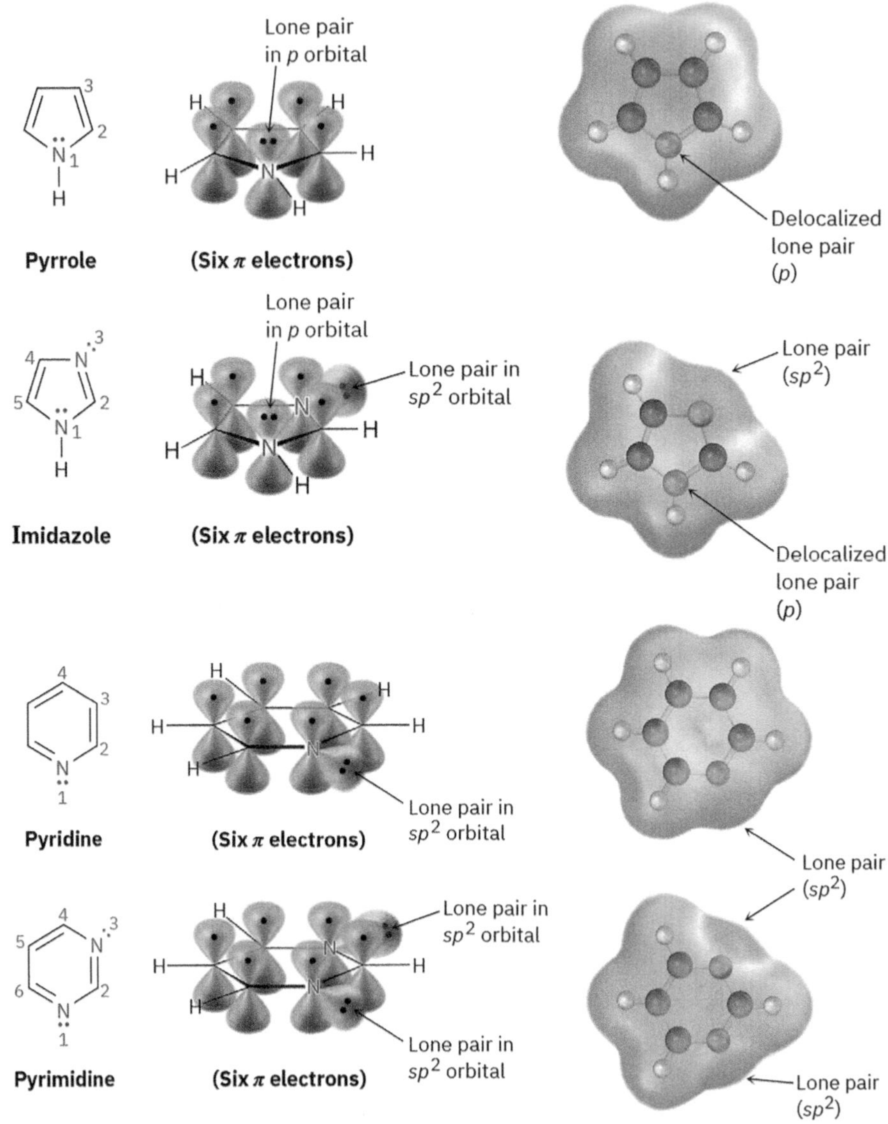

FIGURE 3.2 Molecular orbital representation of pyrrole, imidazole, pyridine, and pyrimidines showing the involvement (or non-involvement) of the nitrogen lone pair in aromatic delocalization. (Reproduced from Ref. Tully (2025) with permission under Creative Commons Attribution-NonCommercial-ShareAlike 4.0 International License.)

TABLE 3.2
Common Nitrogen Heterocycles and Their Basic Structural Characteristics

Ring Size	Generic Name	Example	Formula	Type
3	Aziridine	Ethyleneimine	C_2H_5N	Saturated
5	Pyrrole	NH-pentadienyl	C_4H_5N	Aromatic
6	Pyridine	Hexatrienylamine	C_5H_5N	Aromatic
9	Acridine	Dibenzopyridine	$C_{13}H_9N$	Aromatic
10	Phenanthridine	Tricyclic azine	$C_{13}H_9N$	Aromatic

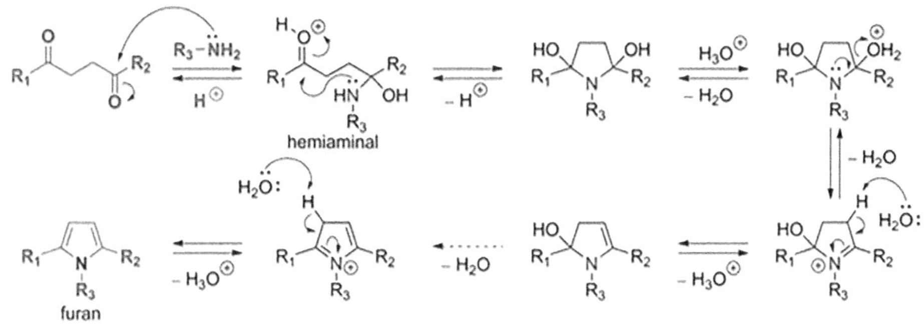

FIGURE 3.3 General Paal–Knorr reaction for pyrrole formation through enamine intermediate. (Reproduced from Ref. Minetto et al. (2005) with permission from Taylor & Francis.)

3.3 SYNTHETIC STRATEGIES

3.3.1 Conventional Synthetic Approaches

Traditional synthetic routes to nitrogen heterocycles rely on cyclization reactions involving nitrogen nucleophiles and unsaturated carbon compounds. Classic methods include the Paal–Knorr synthesis for pyrroles (Figure 3.3) (Balakrishna et al., 2019), the Hantzsch synthesis for pyridines (Faizan et al., 2024) (Figure 3.4), and the Knorr synthesis for pyrazoles (Schrecker et al., 2023). These reactions proceed via condensation and intramolecular cyclization, followed by dehydration to yield the aromatic ring.

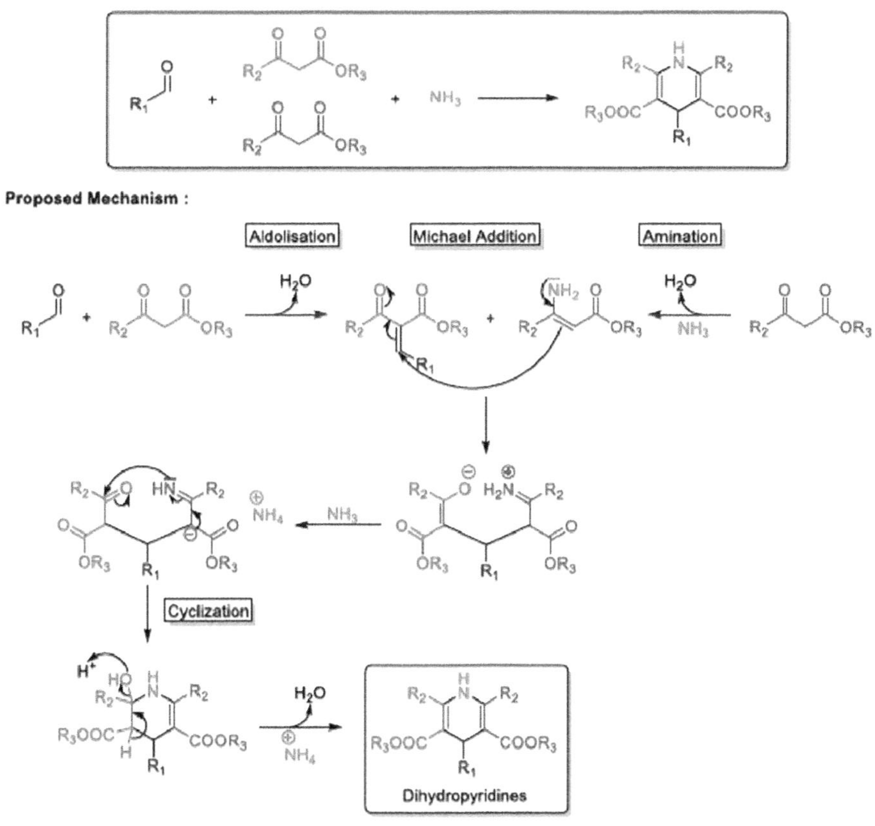

FIGURE 3.4 Hantzsch synthesis for pyridines. (Reproduced from Ref. Vala and Patel (2023) with permission from Elsevier.)

The Bischler–Napieralski reaction (Figure 3.5) and Friedländer synthesis (Figure 3.6) are classical methods for the construction of fused heterocycles such as isoquinolines and quinolines, respectively. These methods typically involve acid-catalyzed cyclization followed by oxidation to restore aromaticity (Chandrasekaran & Sarveswari, 2025; Li et al., 2021; Rajendran et al., 2022).

3.3.2 Metal-Catalyzed and Organocatalytic Methods

Transition metal catalysis has significantly advanced the synthesis of nitrogen heterocycles by offering high regioselectivity and atom economy (Kornet & Müller, 2024; Zhang et al., 2024). Palladium, copper, and nickel complexes are widely employed in cross-coupling and cyclization reactions (Figure 3.7) (Ananthu et al., 2021).

Organocatalysts such as proline, imidazolidinones, and ionic liquids also enable environmentally benign heterocyclizations, especially for enamine or iminium-mediated processes (Li et al., 2018) (Figure 3.8).

3.3.3 Multicomponent and Domino Reactions

Multicomponent reactions (MCRs) provide a route to molecular diversity, generating complex heterocyclic scaffolds in a single synthetic operation. Noteworthy examples include the Ugi, Biginelli (Figure 3.9) (Bosica et al., 2021), and Hantzsch reactions (Faizan et al., 2024) (Table 3.3).

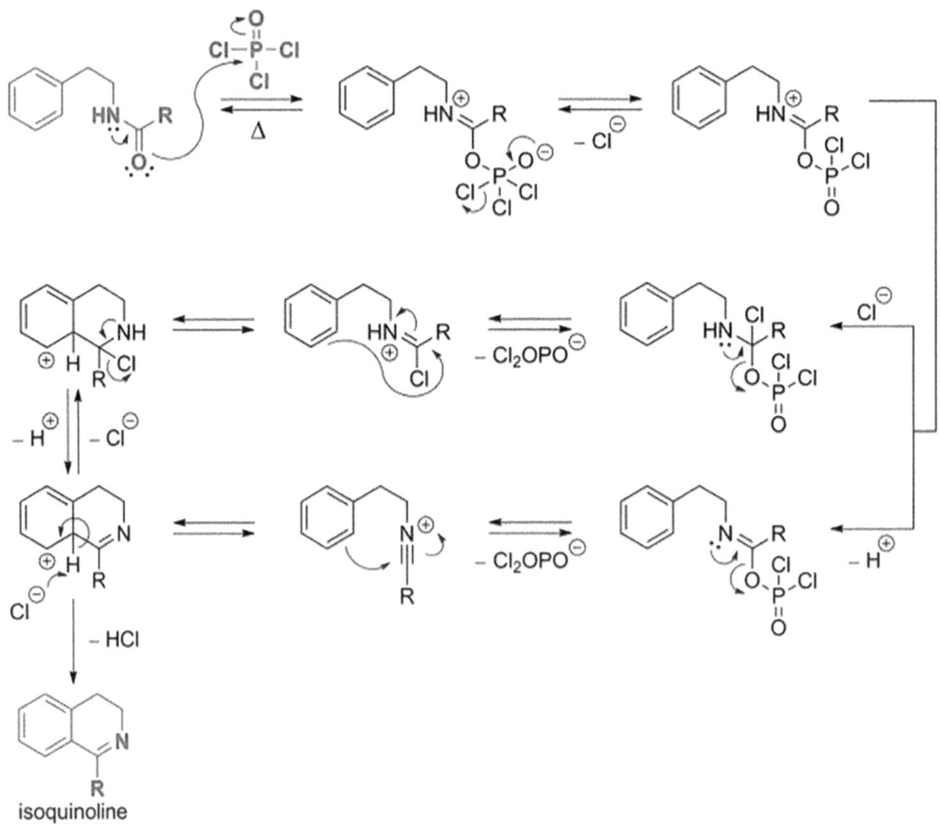

FIGURE 3.5 Bischler–Napieralski reaction for isoquinolines. (Reproduced from Ref. Li et al. (2021) with permission from John Wiley & Sons.)

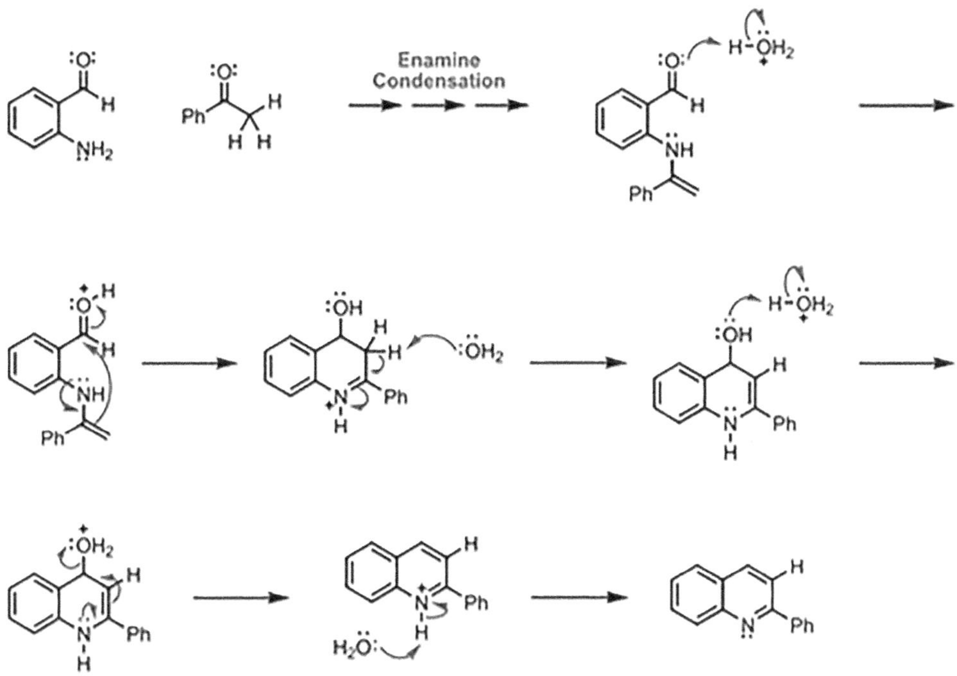

FIGURE 3.6 Friedländer synthesis for quinolines. (Reproduced from Refs. Chandrasekaran and Sarveswari (2025) and Rajendran et al. (2022) with permission from Elsevier.)

3.3.4 Green and Sustainable Approaches

The demand for sustainable chemistry has driven the exploration of solvent-free, microwave-assisted, and bio-catalytic synthesis of heterocycles (Figure 3.10). Solvent-free cyclizations under ultrasound or microwave irradiation significantly reduce reaction time and improve yields. Ionic liquids and deep eutectic solvents serve as recyclable media for heterocycle construction (Adhikari et al., 2022; Maity & Mitra, 2025).

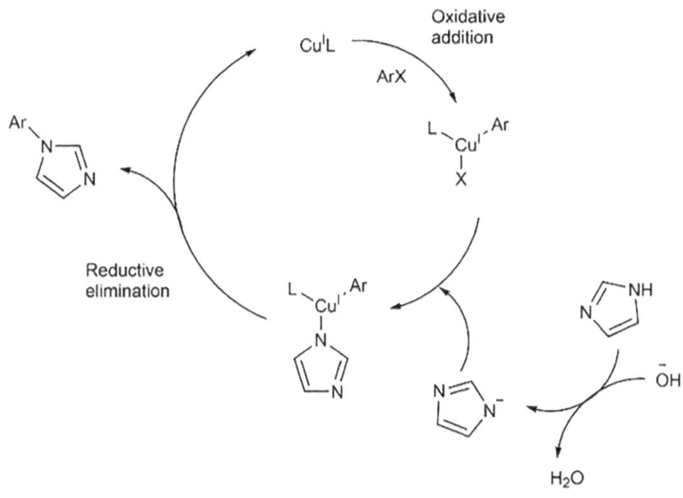

FIGURE 3.7 Pd-Catalyzed C–N coupling for imidazole formation. (Reproduced from Ref. Ananthu et al. (2021) with permission from John Wiley & Sons.)

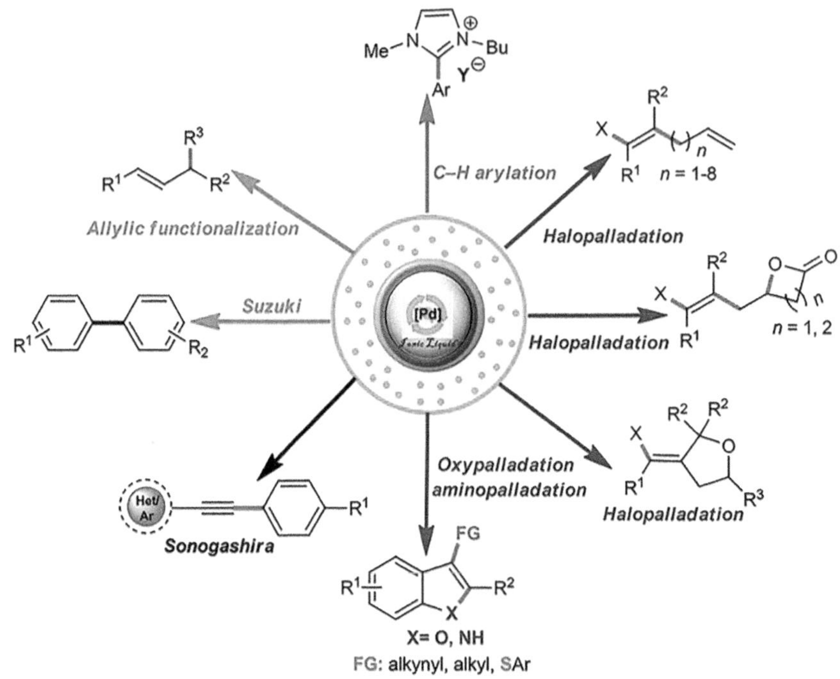

FIGURE 3.8 Representative examples of metal-catalyzed synthesis of nitrogen heterocycles. (Reproduced from Ref. Li et al. (2018) with permission from John Wiley & Sons.)

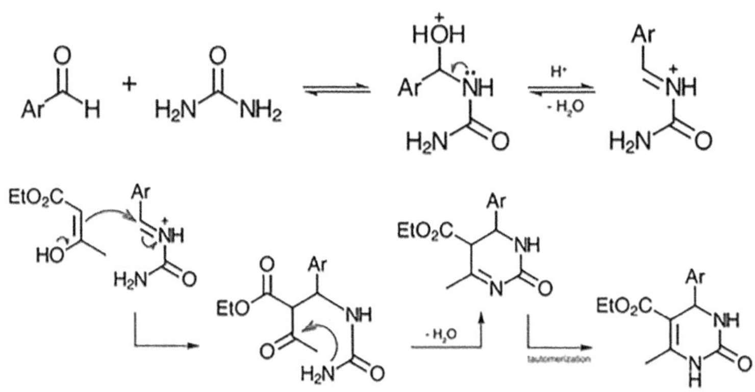

FIGURE 3.9 Biginelli synthesis of 3,4-dihydropyrimidin-2(1*H*)-one. (Reproduced from Ref. Bosica et al. (2021) with permission under the terms and conditions of the Creative Commons Attribution (CC BY) license from *Molecules*.)

TABLE 3.3
Representative Multicomponent Reactions for Nitrogen Heterocycles

Reaction	Starting Materials	Product	Catalyst/Conditions	Reference Scaffold
Biginelli	Aldehyde + β-ketoester + urea	Dihydropyrimidinone	Acidic medium	Pyrimidine
Hantzsch	Aldehyde + β-ketoester + ammonia	1,4-DHP/pyridine	Reflux, ethanol	Pyridine
Ugi	Amine + aldehyde + isocyanide + acid	Peptidomimetic heterocycles	Room temperature	Imidazoles
Gewald	Ketone + malononitrile + sulfur	2-Aminothiophene	Base catalysis	Thiophene analogs

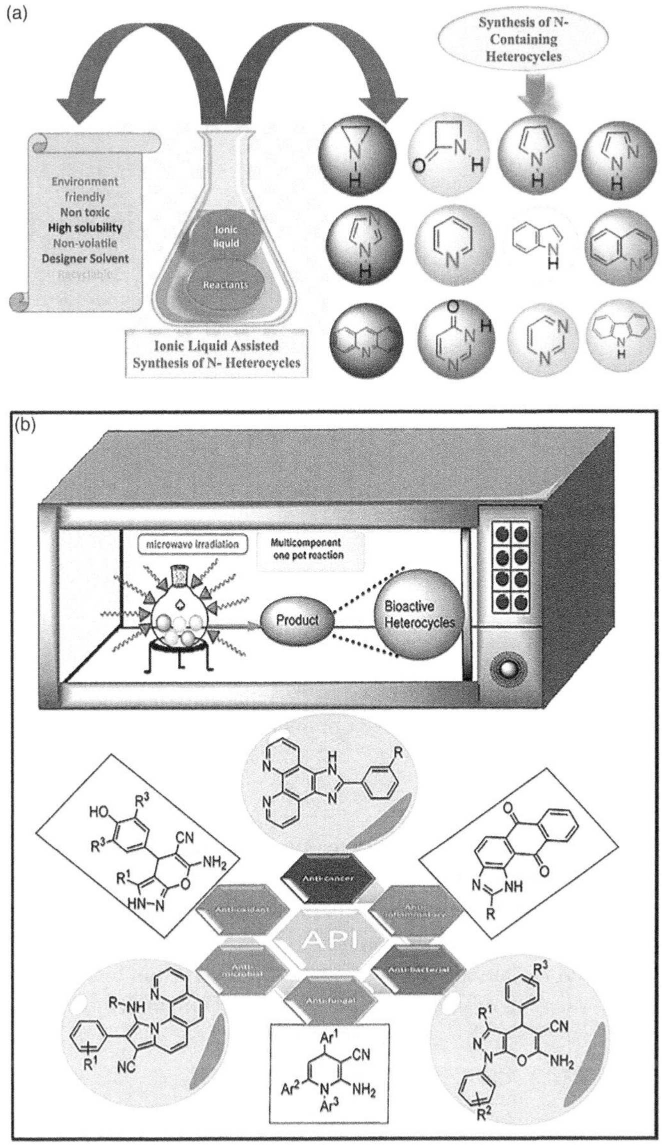

FIGURE 3.10 Schematic illustration of green synthetic approaches for nitrogen heterocycles using (a) ionic liquids and (b) microwave irradiation. (Reproduced from Refs. Adhikari et al. (2022) and Maity and Mitra (2025) with permission from Elsevier.)

3.4 CHEMISTRY OF FIVE-MEMBERED NITROGEN HETEROCYCLES

3.4.1 Pyrrole and Its Derivatives

Pyrrole is an aromatic five-membered heterocycle containing one nitrogen atom whose lone pair participates in the π-system (Fatahala et al., 2022). It is found in heme, chlorophyll, and bile pigments. Typical laboratory synthesis follows the Paal–Knorr condensation, where a 1,4-dicarbonyl compound reacts with ammonia or a primary amine to afford a substituted pyrrole. Pyrroles undergo electrophilic substitution preferentially at the 2-position because the intermediate cation is stabilized by resonance (Ma et al., 2018) (Figure 3.11 and Table 3.4). *N*-alkylation and acylation produce functionalized derivatives used as intermediates for porphyrins, conducting polymers, and pharmaceuticals (Ganesh et al., 2024).

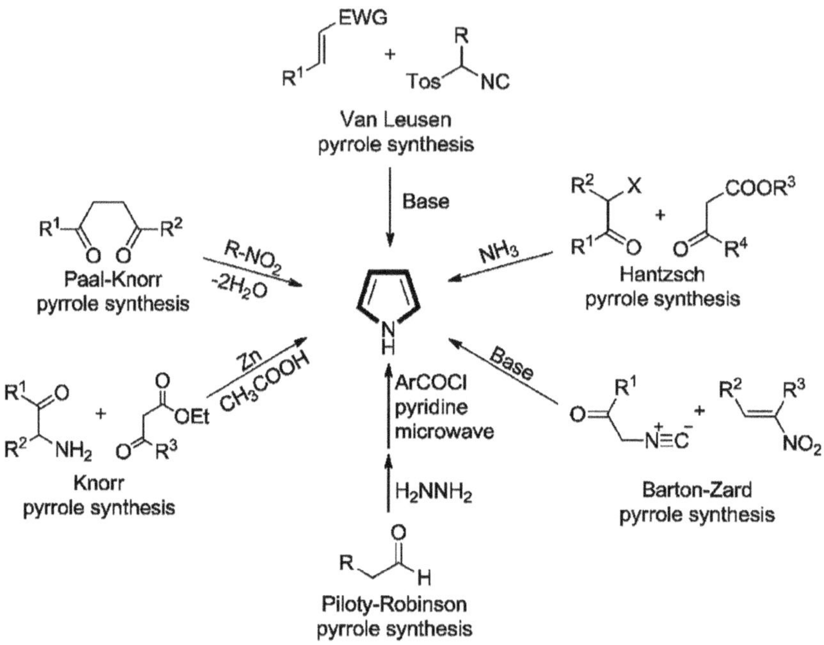

FIGURE 3.11 Different synthetic methods for pyrrole. (Reproduced from Ref. Ma et al. (2018) with permission under the terms and conditions of the Creative Commons Attribution (CC BY) license from *Molecules*.)

3.4.2 Imidazole

Imidazole contains two nitrogens: one contributes to aromaticity and the other behaves as a weak base. This duality provides amphoteric character, allowing proton transfer and metal coordination (Tolomeu & Fraga, 2023). Synthetically, imidazoles are prepared by Debus–Radziszewski condensation of 1,2-dicarbonyls, aldehydes, and ammonia (Figure 3.12 and Table 3.4) (Yu et al., 2023). Benzimidazole is produced by condensation of *o*-phenylenediamine with carboxylic acids or aldehydes. These rings form the core of vitamin B_{12} analogs, antifungal drugs, and corrosion inhibitors.

3.4.3 Pyrazole and Isoxazole

Pyrazole (1,2-diazole) and isoxazole (1,2-oxazole) differ only by the heteroatom sequence (Figures 3.13 and 3.14, Table 3.4) (Moussa et al., 2024). They are synthesized by cyclocondensation of 1,3-diketones with hydrazines or hydroxylamines, respectively (Hu & Szostak, 2015). Substituted pyrazoles display analgesic and anti-inflammatory activities (e.g., celecoxib), while isoxazoles serve as intermediates in β-lactam antibiotics.

TABLE 3.4
Selected Five-Membered Nitrogen Heterocycles and Their Key Features

Heterocycle	Synthesis	Key Reaction Type	Major Application
Pyrrole	Paal–Knorr	Condensation	Porphyrins, polymers
Imidazole	Debus–Radziszewski	Multicomponent	Enzyme active sites
Pyrazole	Hydrazine cyclization	Cyclocondensation	Anti-inflammatory drugs
Triazole	Azide–alkyne cycloaddition	Click reaction	Antifungal agents
Tetrazole	Nitrile + NaN_3	Cycloaddition	Drug bioisosteres

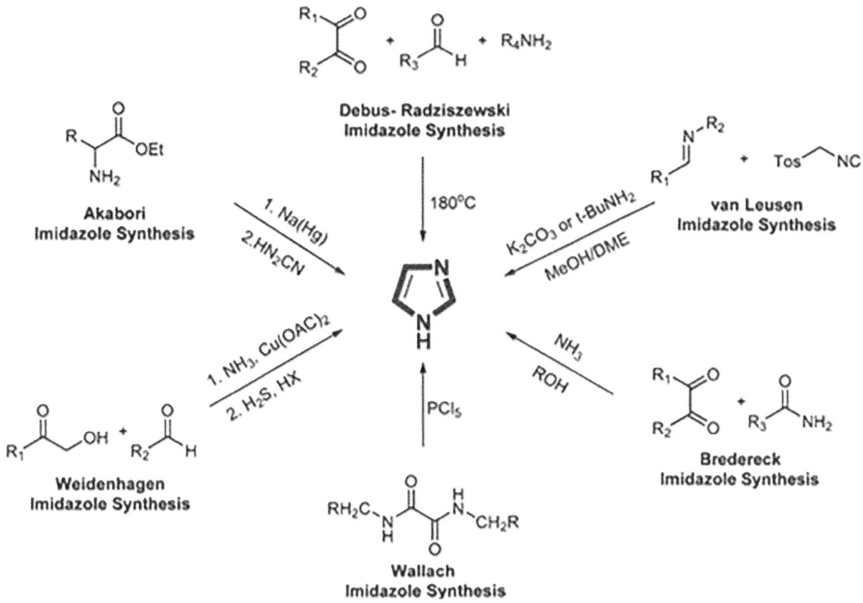

FIGURE 3.12 Different synthetic methods for imidazole. (Reproduced from Ref. Zheng et al. (2020) with permission under the terms and conditions of the Creative Commons Attribution (CC BY) license from *Molecules*.)

3.4.4 Triazoles and Tetrazoles

Triazoles exist in two isomeric forms (1,2,3- and 1,2,4-triazoles). The Huisgen 1,3-dipolar cycloaddition of azides and alkynes ("click chemistry") produces 1,2,3-triazoles under mild Cu(I) catalysis (Gonnet et al., 2021) (Figure 3.15 and Table 3.4).

Tetrazoles are typically obtained by [2 + 3] cycloaddition between nitriles and sodium azide (Figure 3.16 and Table 3.4). They are notable bioisosteres of carboxylic acids and improve drug pharmacokinetics (Kaszás et al., 2022).

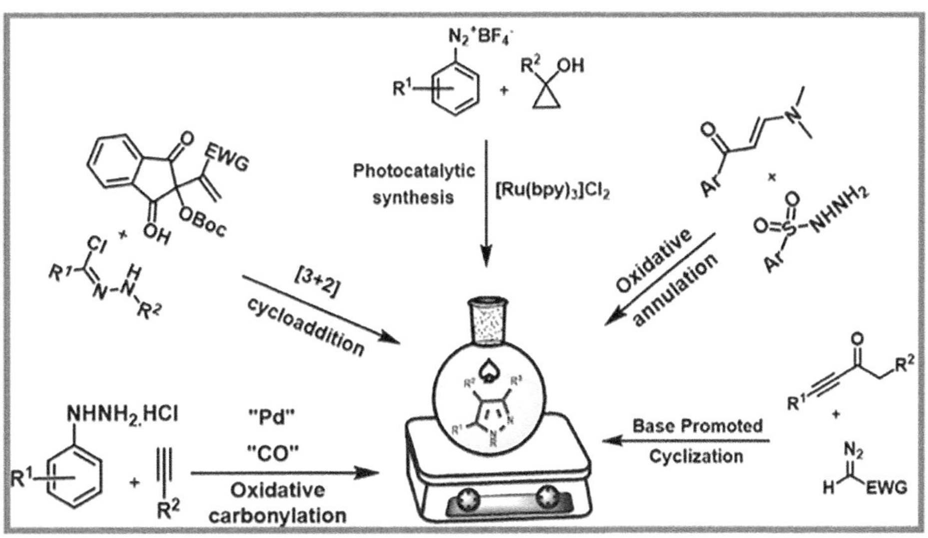

FIGURE 3.13 Different synthetic methods for pyrazole. (Reproduced from Ref. Moussa et al. (2024) under the terms of the Creative Commons (CC BY) license with permission from Elsevier.)

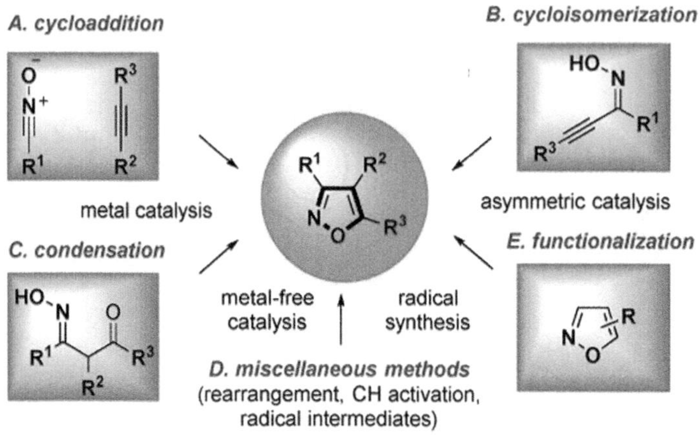

FIGURE 3.14 Different synthetic methods for isoxazole. (Reproduced from Ref. Hu and Szostak (2015) with permission from John Wiley & Sons.)

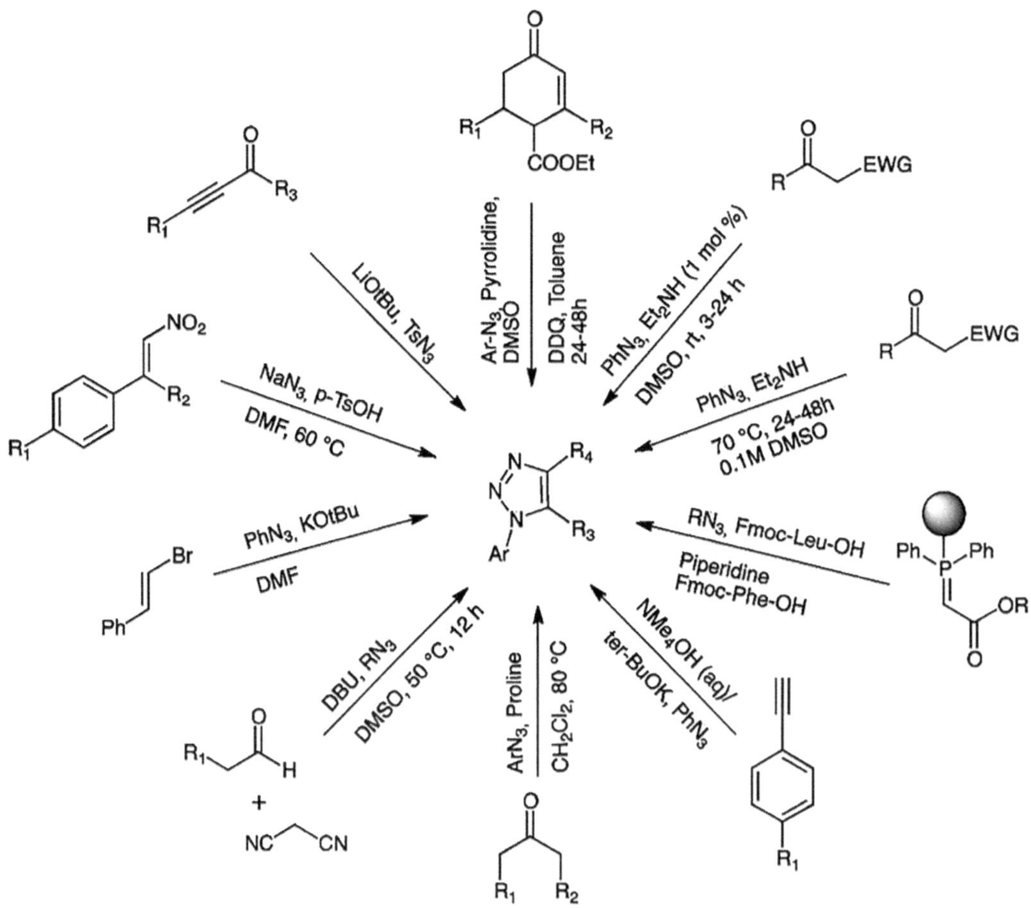

FIGURE 3.15 Different synthesis methods for triazole. (Reproduced from Ref. Gonnet et al. (2021) with permission under the terms and conditions of the Creative Commons Attribution (CC BY) license from *Molecules*.)

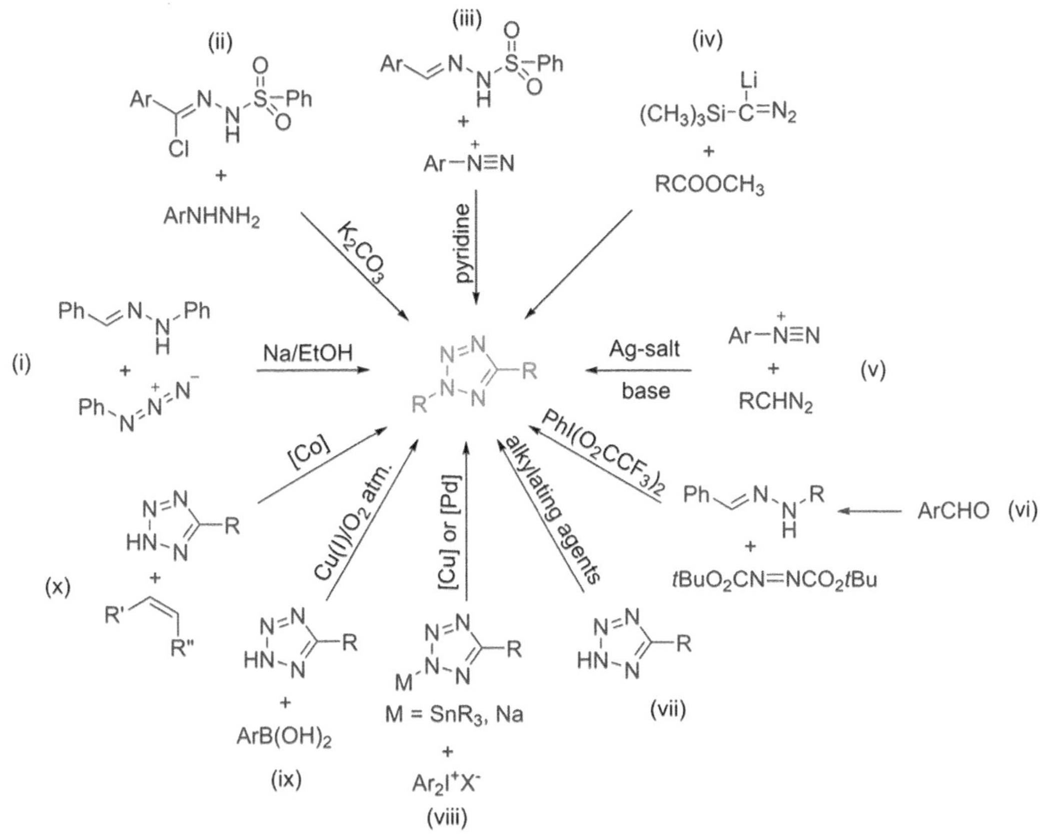

FIGURE 3.16 Different synthesis methods for tetrazole. (Reproduced from Ref. Swami et al. (2021) under a Creative Commons Attribution-Non-Commercial 3.0 Unported Licence with permission from Royal Society of Chemistry.)

3.5 CHEMISTRY OF SIX-MEMBERED NITROGEN HETEROCYCLES

3.5.1 Pyridine and Its Derivatives

Pyridine, the prototype six-membered azine, exhibits basicity due to its non-delocalized lone pair. It is produced industrially by the Chichibabin synthesis from acetaldehyde, formaldehyde, and ammonia (Figure 3.17 and Table 3.5). Electrophilic substitution occurs at the 3-position, whereas nucleophilic substitution prefers the 2- and 4-positions. *N*-oxidation yields pyridine N-oxide, a useful intermediate for functional group transformations.

3.5.2 Pyrimidine

These diazines contain two nitrogen atoms at varying positions, influencing their electron density and reactivity. Pyrimidine is central to nucleic acid bases (cytosine, thymine, uracil). Synthetic routes include condensation of β-dicarbonyls with amidines or urea (Biginelli and Pinner reactions) (Figure 3.18) (Anjirwala et al., 2022). Pyrazine is commonly obtained by oxidative dimerization of α-aminocarbonyl compounds, while pyridazine arises from hydrazine condensation with 1,4-dicarbonyls (Table 3.5).

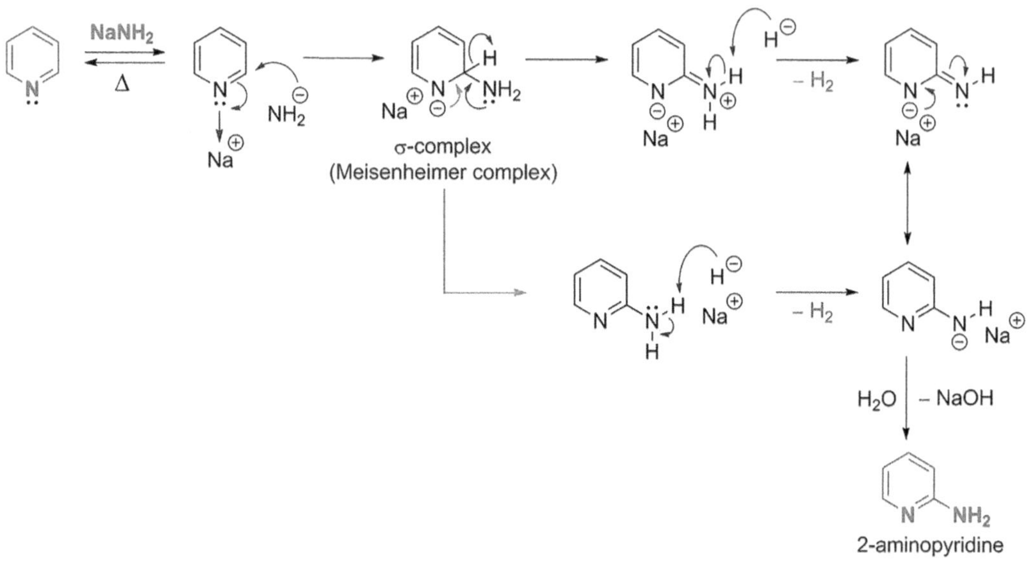

FIGURE 3.17 Chichibabin synthesis of pyridine. (Reproduced from Ref. Rammohan et al. (2023) with permission from Elsevier.)

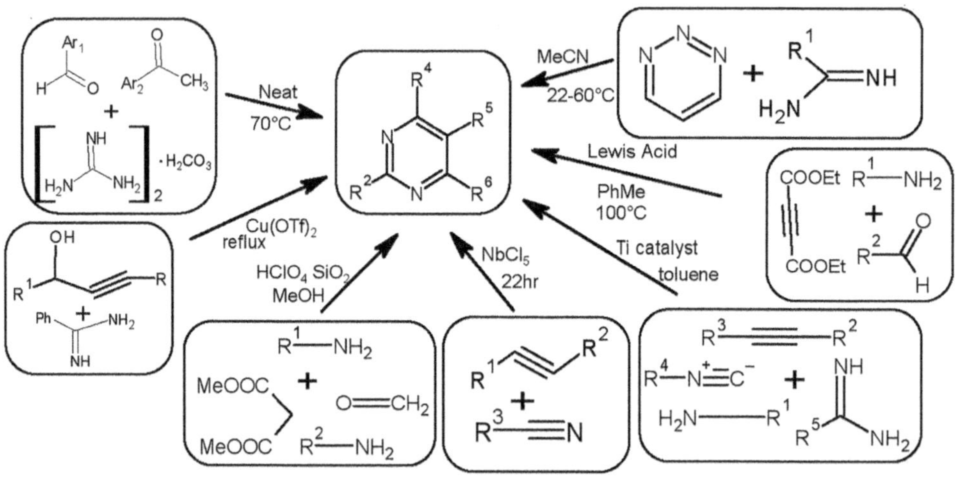

FIGURE 3.18 Different synthesis methods for pyrimidine. (Reproduced from Ref. Islam et al. (2024) with permission from John Wiley & Sons.)

TABLE 3.5
Selected Six-Membered Nitrogen Heterocycles

Ring	Nitrogen Position(s)	Common Synthetic Route	Example
Pyridine	1	Chichibabin condensation	Nicotinamide
Pyrimidine	1,3	Biginelli reaction	Cytosine
Pyrazine	1,4	Dimerization of amino ketones	Phenazine dyes
Pyridazine	1,2	Hydrazine + diketone	Hydralazine (antihypertensive)

3.5.3 Functionalization and Applications

Substituted pyridines act as ligands in coordination chemistry (e.g., terpyridine, bipyridine) and as precursors to NAD^+ analogs. Pyrimidines and their hydrogenated derivatives (1,4-dihydropyrimidinones) show antitumor and antiviral activities (Table 3.5).

3.6 FUSED AND CONDENSED NITROGEN HETEROCYCLES

3.6.1 Indole and Carbazole

Indole is a benzopyrrole system containing both aromatic π-circuits. It is the structural core of tryptophan and serotonin. Common synthetic approaches include the Fischer indole synthesis and the Madelung reaction (Taber & Tirunahari, 2011) (Figure 3.19).

Carbazole, a tricyclic analog, forms through oxidative coupling of diphenylamines or by Cadogan cyclization. These scaffolds exhibit photoluminescence and are essential in OLED emitters (Roy et al., 2012) (Figure 3.20).

3.6.2 Quinoline and Isoquinoline

Quinoline is obtained via Skraup synthesis (aniline + glycerol + oxidizing acid). Isoquinoline forms by the Bischler–Napieralski cyclization of β-phenylethylamides (Figure 3.21) (Al Mehedi, 2021).

3.6.3 Purine and Related Systems

Purine, a fused imidazole–pyrimidine system, constitutes the nucleus of nucleotides (adenine, guanine). Stepwise condensation of formamide derivatives or Traube synthesis can yield purine analogs.

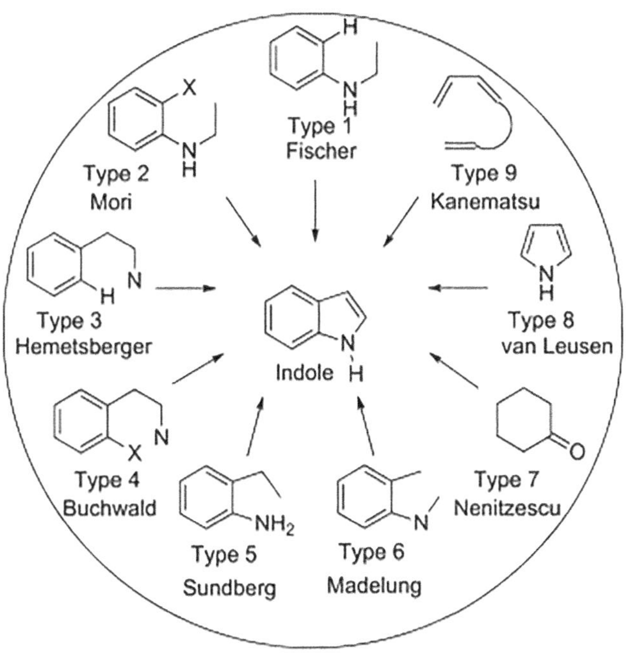

FIGURE 3.19 Different synthesis methods for indole. (Reproduced from Ref. Taber and Tirunahari (2011) with permission from Elsevier.)

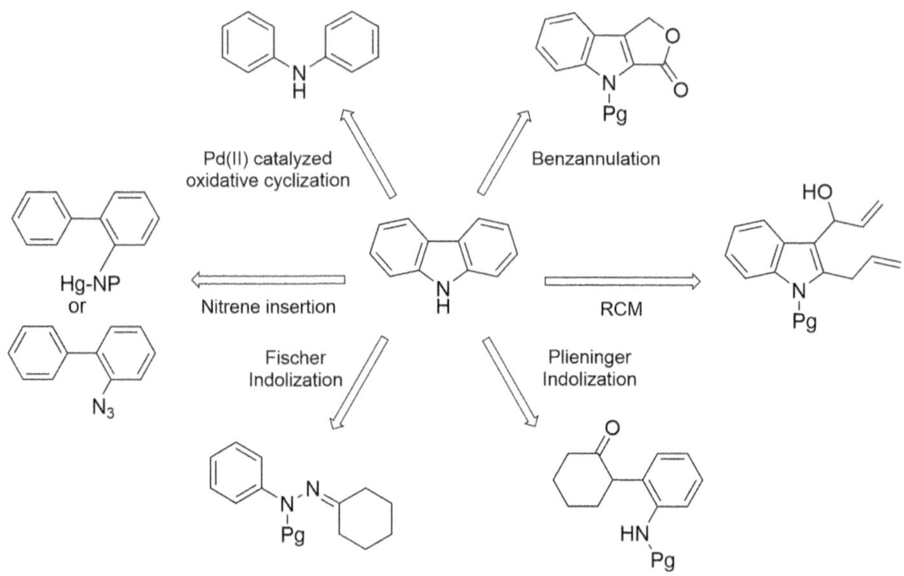

FIGURE 3.20 Different synthesis methods for carbazole. (Reproduced from Ref. Roy et al. (2012) with permission from Elsevier.)

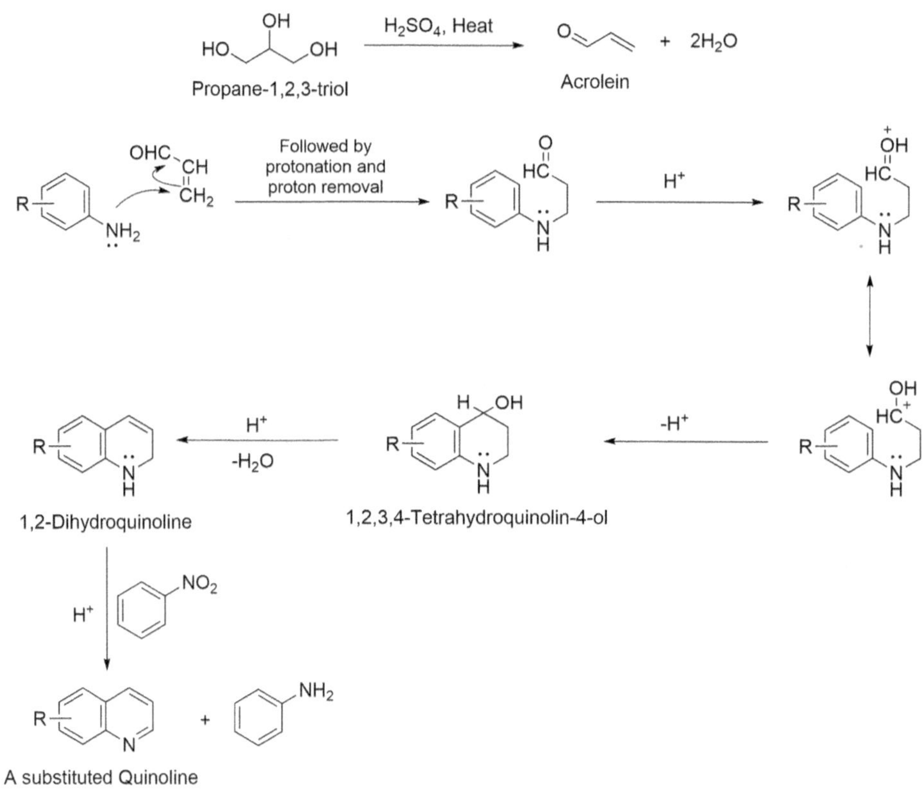

FIGURE 3.21 Mechanistic outline of quinoline formation through Skraup synthesis. (Reproduced from Ref. Ramann and Cowen (2016) with permission under the terms and conditions of the Creative Commons Attribution (CC BY) license from *Molecules*.)

3.6.4 Functional Modifications

Fused heterocycles readily undergo *N*-alkylation, acylation, and electrophilic substitution on the benzene ring, enabling the construction of fluorescent and bioactive molecules.

3.7 REACTIVITY AND MECHANISTIC INSIGHTS

3.7.1 Electrophilic and Nucleophilic Substitution

The electron density in nitrogen heterocycles dictates the site of attack.

- In **pyrrole** and **indole**, electrophiles attack the α-position (Figure 3.22).
- In **pyridine**, nucleophiles prefer the 2- and 4-positions (Figure 3.22) (Blanco et al., 2008; Tully, 2025).
- Mechanistic studies show σ-complex stabilization as the key determining factor.

3.7.2 Metalation and Cross-Coupling Reactions

Directed *ortho*-metalation (DoM) using lithium or magnesium bases allows regioselective functionalization. Modern palladium-catalyzed couplings (Suzuki, Sonogashira, Buchwald–Hartwig) introduce aryl or alkynyl substituents efficiently (Tomar et al., 2022) (Figure 3.23).

3.7.3 Oxidation and Reduction Pathways

Heterocyclic rings can undergo hydrogenation (yielding piperidine or tetrahydroquinoline) or oxidation to *N*-oxides and dioxides. Such transformations alter electronic properties, influencing coordination or biological activity.

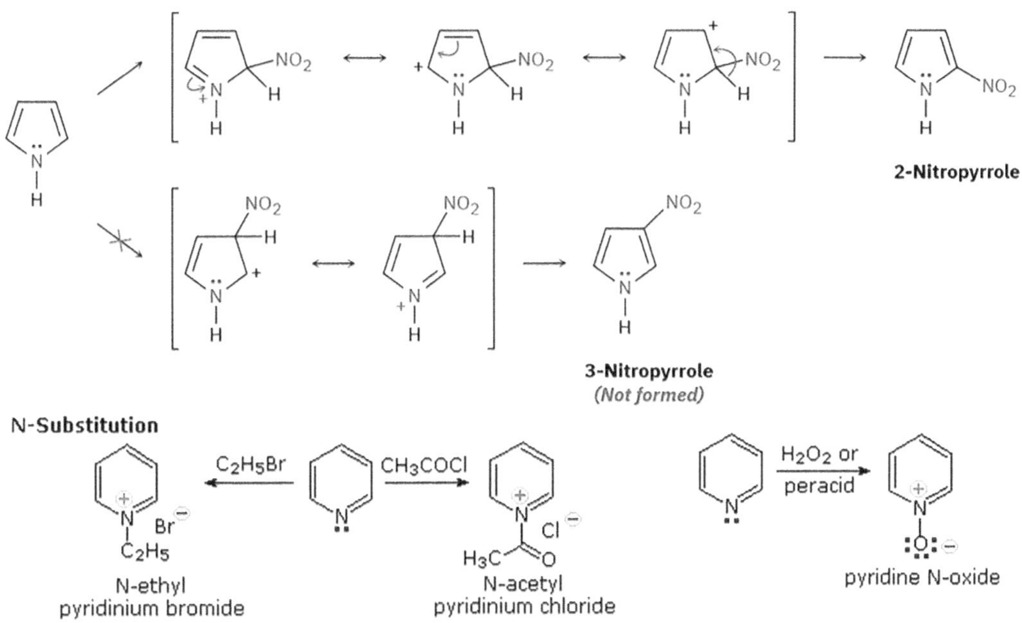

FIGURE 3.22 General patterns of substitution in pyrrole vs. pyridine. (Reproduced from Ref. Al-Issa (2012) with permission under the terms and conditions of the Creative Commons Attribution (CC BY) license from *Molecules*.)

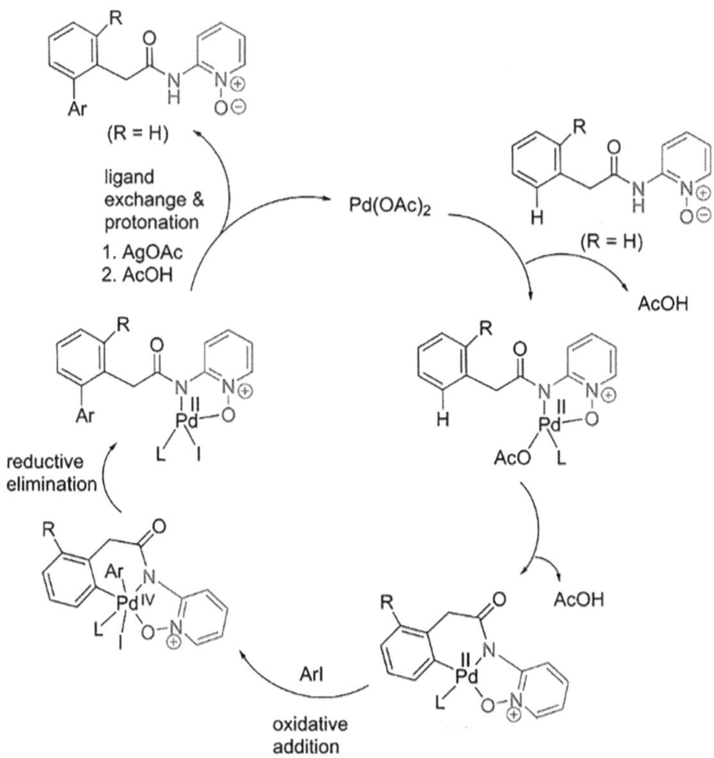

FIGURE 3.23 Pd-catalyzed arylation of pyridine *N*-oxide. (Reproduced from Ref. Tomar et al. (2022) with permission from John Wiley & Sons.)

3.7.4 Tautomerism and Protonation Equilibria

Many azoles (e.g., imidazole, triazole) exhibit prototropic tautomerism. Generally, the ring nitrogen that has the higher basicity (the one that can be compared to pyridine) will be the one attacked in the protonation process. The perception of these equilibriums is very helpful in determining the affinity of the complex established to interact with the metal centers or to be accepted by biological receptors.

3.8 SPECTROSCOPIC AND STRUCTURAL CHARACTERIZATION

Spectroscopic tools confirm structure, substitution patterns, and conjugation (Table 3.6).

- *UV–Vis Spectroscopy*: It reveals π–π^* transitions and aromatic conjugation.
- *IR Spectroscopy*: N–H stretching (3300–3400 cm^{-1}) and C=N stretching (1600 cm^{-1}) bands.

TABLE 3.6
Typical Spectroscopic Features of Selected Nitrogen Heterocycles

Compound	λ_{max} (nm)	IR ν(C=N) cm^{-1}	^{1}H NMR (δ, ppm)	Structural Notes
Pyrrole	250	1580	6.5–7.2	Aromatic, N–H present
Pyridine	260	1595	7.8–8.7	Deshielded protons
Indole	280, 305	1610	6.9–8.0	Dual aromatic ring
Quinoline	322	1605	7.2–8.9	Extended conjugation

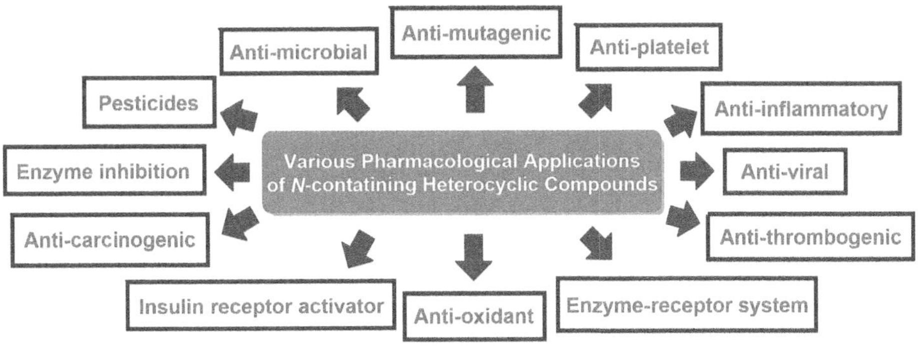

FIGURE 3.24 Representative bioactive nitrogen heterocycles and their pharmacological targets.

- *^1H and ^{13}C NMR*: Chemical shifts depend on nitrogen position and substituent effects.
- *Mass Spectrometry*: Fragmentation often occurs adjacent to nitrogen.
- *X-Ray Crystallography*: It defines planarity and intramolecular hydrogen bonding.

3.9 BIOLOGICAL AND PHARMACOLOGICAL SIGNIFICANCE

Nitrogen heterocycles constitute the backbone of numerous natural and synthetic drugs. Their ability to form hydrogen bonds and ionic interactions facilitates binding with enzymes and receptors (Figure 3.24 and Table 3.7).

- *Antibacterial and Antifungal*: Imidazole, triazole derivatives (e.g., ketoconazole).
- *Anticancer*: Quinoline, acridine, and purine analogs interfere with DNA replication.
- *Antiviral*: Pyrimidine nucleoside analogs (e.g., zidovudine).
- *Neuroactive*: Indole alkaloids (e.g., serotonin, reserpine).

3.10 APPLICATIONS IN ADVANCED MATERIALS

Nitrogen heterocycles extend beyond biology into functional materials (Figure 3.25).

- *Organic Electronics*: Carbazole and pyridine derivatives act as hole-transporting and emissive layers in OLEDs.
- *Photovoltaics*: Indole and quinoxaline donors in dye-sensitized solar cells.
- *Coordination Chemistry*: Bipyridine and terpyridine ligands stabilize transition-metal complexes.
- *Catalysis*: Imidazole and triazole ligands facilitate homogeneous catalysis.
- *Polymers and Sensors*: Poly(pyrrole) and poly(aniline) exhibit electrical conductivity and environmental stability.

TABLE 3.7
Examples of Clinically Relevant Nitrogen Heterocyclic Drugs

Drug	Core Ring	Therapeutic Class	Mechanism
Nicotinamide	Pyridine	Vitamin/coenzyme	Redox metabolism
Caffeine	Purine	CNS stimulant	Adenosine receptor antagonist
Metronidazole	Imidazole	Antibacterial	DNA damage via nitro-reduction
Quinoline analogs	Quinoline	Antimalarial	Heme polymerization inhibition
Albendazole	Benzimidazole	Anthelmintic	β-tubulin binding

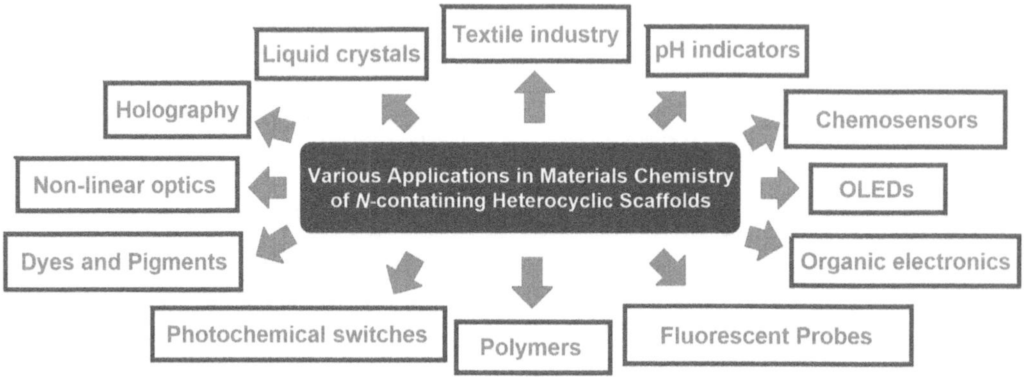

FIGURE 3.25 Examples of nitrogen heterocyclic frameworks used in organic optoelectronic devices.

3.11 FUTURE PERSPECTIVES AND CHALLENGES

The developments in nitrogen heterocyclic chemistry are still powered by sustainability, accurate synthesis, and computer simulation.

- *Use of Green Chemistry*: No solvent and microwave techniques lead to less wastage.
- *Computational Design*: Density functional theory (DFT) and Artificial Intelligence (AI)-assisted screening guide heterocycle discovery.
- *Bio-Orthogonal Chemistry*: Click reactions expand applications in bioconjugation.
- *Functional Materials*: Development of π-extended systems for photonics and catalysis.

Among the challenges that still need to be overcome are the regioselective synthesis of polysubstituted systems, the formulation of environmentally friendly processes that are scalable, and the investigation of non-classical aromaticity.

3.12 CONCLUSION

Nitrogen-containing heterocycles are the main class of molecular structures in organic chemistry, and they have many diverse applications ranging from the most basic theoretical principles to the most complex industrial ones. Their unique electronic properties, tunable chemical reactivity, and biocompatibility have made them essential components in the pharmaceutical industry, catalysis, and the development of advanced functional materials. The synthesis of these molecules has evolved from the use of traditional cyclization methods to the employing of modern catalytic, microwave-assisted, and eco-friendly techniques, thus drastically increasing the variety of structures that can be accessed. In the future, substantial progress is likely to be generated by the synergistic use of computational design tools, resource-saving synthetic protocols, and collaborative interdisciplinary research, which will ultimately lead to the production of new nitrogen heterocycles with better performance and wider application in technology.

REFERENCES

Adhikari, A., Bhakta, S., & Ghosh, T. (2022). Microwave-assisted synthesis of bioactive heterocycles: An overview. *Tetrahedron*, *126*, 133085.

Al-Issa, S. A. (2012). Synthesis of a new series of pyridine and fused pyridine derivatives. *Molecules*, *17*(9), 10902–10915.

Al Mehedi, M. S. (2021). *Development of new methods to synthesize heterocyclic compounds*. Michigan State University.

Amin, A., Qadir, T., Sharma, P. K., Jeelani, I., & Abe, H. (2022). A review on the medicinal and industrial applications of N-containing heterocycles. *The Open Medicinal Chemistry Journal*, *16*(1).
Ananthu, S., Aneeja, T., & Anilkumar, G. (2021). N-arylation of imidazoles: An overview. *ChemistrySelect*, *6*(37), 9794–9805.
Anjirwala, S. N., Parmar, P. S., & Patel, S. K. (2022). Synthetic protocols for non-fused pyrimidines. *Synthetic Communications*, *52*(22), 2079–2121.
Balaban, A. T., Oniciu, D. C., & Katritzky, A. R. (2004). Aromaticity as a cornerstone of heterocyclic chemistry. *Chemical Reviews*, *104*(5), 2777–2812.
Balakrishna, A., Aguiar, A., Sobral, P. J., Wani, M. Y., Almeida e Silva, J., & Sobral, A. J (2019). Paal–Knorr synthesis of pyrroles: From conventional to green synthesis. *Catalysis Reviews*, *61*(1), 84–110.
Blanco, F., O'Donovan, D. H., Alkorta, I., & Elguero, J. (2008). Substitution effects on neutral and protonated pyridine derivatives along the periodic table. *Structural Chemistry*, *19*(2), 339–352.
Bosica, G., Cachia, F., De Nittis, R., & Mariotti, N. (2021). Efficient one-pot synthesis of 3, 4-dihydropyrimidin-2 (1 H)-ones via a three-component Biginelli reaction. *Molecules*, *26*(12), 3753.
Chandrasekaran, I., & Sarveswari, S. (2025). Different catalytic approaches of Friedländer synthesis of quinolines. *Heliyon*, *11*(2).
Ebenezer, O., Jordaan, M. A., Carena, G., Bono, T., Shapi, M., & Tuszynski, J. A. (2022). An overview of the biological evaluation of selected nitrogen-containing heterocycle medicinal chemistry compounds. *International Journal of Molecular Sciences*, *23*(15), 8117.
Eicher, T., Hauptmann, S., & Speicher, A (2013). *The chemistry of heterocycles: Structures, reactions, synthesis, and applications*. John Wiley & Sons.
Faizan, M., Kumar, R., Mazumder, A., Salahuddin, Kukreti, N., Tyagi, P. K., & Kapoor, B. (2024). Hantzsch reaction: The important key for pyridine/dihydropyridine synthesis. *Synthetic Communications*, *54*(15), 1221–1244.
Fatahala, S. S., Mohamed, M. S., Sabry, J. Y., & Mansour, Y. E. E.-D. (2022). Synthesis strategies and medicinal value of pyrrole and its fused heterocyclic compounds. *Medicinal Chemistry*, *18*(10), 1013–1043.
Frank, É., & Szőllősi, G. (2021). Nitrogen-containing heterocycles as significant molecular scaffolds for medicinal and other applications. In (Vol. 26, pp. 4617): MDPI.
Ganesh, B. H., Raj, A. G., Aruchamy, B., Nanjan, P., Drago, C., & Ramani, P. (2024). Pyrrole: A decisive scaffold for the development of therapeutic agents and structure-activity relationship. *ChemMedChem*, *19*(1), e202300447.
Gonnet, L., Baron, M., & Baltas, M. (2021). Synthesis of biologically relevant 1, 2, 3-and 1, 3, 4-triazoles: From classical pathway to green chemistry. *Molecules*, *26*(18), 5667.
Hu, F., & Szostak, M. (2015). Recent developments in the synthesis and reactivity of isoxazoles: Metal catalysis and beyond. *Advanced Synthesis & Catalysis*, *357*(12), 2583–2614.
Islam, M. W., Islam, M. M., Akter, R., Limon, T. R., Vasquez, E. S., Shaikh, M. A. A., & Habib, A. (2024). A review on pyrimidine-based derivatives: Synthesis and their biological application. *Journal of Heterocyclic Chemistry*, *61*(7), 1159–1179.
Jacobi, P. A. (2018). *Introduction to heterocyclic chemistry*. John Wiley & Sons.
Jampilek, J. (2019). Heterocycles in medicinal chemistry. In (Vol. 24, pp. 3839): MDPI.
Joule, J. A., & Mills, K. (2012). *Heterocyclic chemistry at a glance*. John Wiley & Sons.
Kabir, E., & Uzzaman, M. (2022). A review on biological and medicinal impact of heterocyclic compounds. *Results in Chemistry*, *4*, 100606.
Kaszás, T., Szakács, B., Juhász-Tóth, É., Cservenyák, I., Kovács, T., Juhász, L., …, Tóth, M. (2022). Coupling of N-tosylhydrazones with tetrazoles: A regioselective synthesis of 2, 5-disubstituted-2H-tetrazoles. *European Journal of Organic Chemistry*, *2022*(42), e202201103.
Kornet, M. M., & Müller, T. J. (2024). Recent advances in sequentially Pd-catalyzed one-pot syntheses of heterocycles. *Molecules*, *29*(22), 5265.
Li, J., Yang, S., Wu, W., & Jiang, H. (2018). Recent advances in Pd-catalyzed cross-coupling reaction in ionic liquids. *European Journal of Organic Chemistry*, *2018*(11), 1284–1306.
Li, M., Yuan, Y., & Chen, Y. (2021). Bischler-Napieralski cyclization: A versatile reaction towards functional Aza-PAHs and their conjugated polymers. *Chinese Journal of Chemistry*, *39*(11), 3101–3115.
Lim, J. M., Shim, S., Bui, H. T., Kim, J., Kim, H.-J., Hwa, Y., & Cho, S. (2024). Substitution effect of a single nitrogen atom on π-electronic systems of linear polycyclic aromatic hydrocarbons (PAHs): Theoretically visualized coexistence of mono-and polycyclic π-electron delocalization. *Molecules*, *29*(4), 784.
Ma, Z., Ma, Z., & Zhang, D. (2018). Synthesis of multi-substituted pyrrole derivatives through [3+2] cycloaddition with tosylmethyl isocyanides (TosMICs) and electron-deficient compounds. *Molecules*, *23*(10), 2666.

Maity, P., & Mitra, A. K. (2025). Ionic liquid-assisted approaches in the synthesis of nitrogen-containing heterocycles: A focus on 3-to 6-membered rings. *Journal of Ionic Liquids, 05*, 100146–100212.

Mezgebe, K., & Mulugeta, E. (2022). Synthesis and pharmacological activities of azo dye derivatives incorporating heterocyclic scaffolds: A review. *RSC Advances, 12*(40), 25932–25946.

Minetto, G., Raveglia, L. F., Sega, A., & Taddei, M. (2005). *Microwave-assisted Paal–Knorr reaction–three-step regiocontrolled synthesis of polysubstituted furans, pyrroles and thiophenes*. Wiley Online Library.

Moussa, Z., Ramanathan, M., Alharmoozi, S. M., Alkaabi, S. A. S., Al Aryani, S. H. M., Ahmed, S. A., & Al-Masri, H. T. (2024). Recent highlights in the synthesis and biological significance of pyrazole derivatives. *Heliyon, 10*(20).

Nguyen, H. T., Doan, V. T. C., Nguyen, T. H., & Tran, M. H. (2025). Recent advances in metal-free catalysts for the synthesis of N-heterocyclic frameworks focusing on 5-and 6-membered rings: A review. *RSC Advances, 15*(13), 9676–9755.

Obaid, R. J., Naeem, N., Mughal, E. U., Al-Rooqi, M. M., Sadiq, A., Jassas, R. S., …, Ahmed, S. A. (2022). Inhibitory potential of nitrogen, oxygen and sulfur containing heterocyclic scaffolds against acetylcholinesterase and butyrylcholinesterase. *RSC Advances, 12*(31), 19764–19855.

Quin, L. D., & Tyrell, J. A. (2010). *Fundamentals of heterocyclic chemistry: Importance in nature and in the synthesis of pharmaceuticals*. John Wiley & Sons.

Rajendran, S., Sivalingam, K., Karnam Jayarampillai, R. P., Wang, W. L., & Salas, C. O (2022). Friedländer's synthesis of quinolines as a pivotal step in the development of bioactive heterocyclic derivatives in the current era of medicinal chemistry. *Chemical Biology & Drug Design, 100*(6), 1042–1085.

Ram, V. J., Sethi, A., Nath, M., & Pratap, R. (2019). *The chemistry of heterocycles: Nomenclature and chemistry of three to five membered heterocycles*. Elsevier.

Ramann, G. A., & Cowen, B. J. (2016). Recent advances in metal-free quinoline synthesis. *Molecules, 21*(8), 986.

Rammohan, A., Bhaskar, B. V., & Zyryanov, G. V. (2023). Recent developments in the synthesis of pyridine analogues as a potent anti-Alzheimer's therapeutic leads. In *Recent Developments in the Synthesis and Applications of Pyridines* (pp. 411–444). Elsevier.

Roy, J., Jana, A. K., & Mal, D. (2012). Recent trends in the synthesis of carbazoles: An update. *Tetrahedron, 68*(31), 6099–6121.

Schofield, K. (2013). *Hetero-aromatic nitrogen compounds: Pyrroles and pyridines*. Springer.

Schrecker, L., Dickhaut, J., Holtze, C., Staehle, P., Vranceanu, M., Hellgardt, K., & Hii, K. K. M. (2023). Discovery of unexpectedly complex reaction pathways for the Knorr pyrazole synthesis via transient flow. *Reaction Chemistry & Engineering, 8*(1), 41–46.

Swami, S., Sahu, S. N., & Shrivastava, R. (2021). Nanomaterial catalyzed green synthesis of tetrazoles and its derivatives: A review on recent advancements. *RSC Advances, 11*(62), 39058–39086.

Taber, D. F., & Tirunahari, P. K. (2011). Indole synthesis: A review and proposed classification. *Tetrahedron, 67*(38), 7195–7210.

Tolomeu, H. V., & Fraga, C. A. M. (2023). Imidazole: Synthesis, functionalization and physicochemical properties of a privileged structure in medicinal chemistry. *Molecules, 28*(2), 838.

Tomar, R., Kumar, A., Dalal, A., Bhattacharya, D., Singh, P., & Arulananda Babu, S. (2022). Expanding the utility of inexpensive pyridine-N-oxide directing group for the site-selective sp2/sp3 γ-C–H and sp2 δ-C–H functionalization of carboxamides. *Asian Journal of Organic Chemistry, 11*(9), e202200311.

Tully, D. (2025). *Organic chemistry: A tenth edition – OpenStax adaptation*. OpenStax. https://openstax.org/details/books/organic-chemistry

Ullah, F., Khan, T. A., Iltaf, J., Anwar, S., Khan, M. F. A., Khan, M. R., …, Kotwica-Mojzych, K. (2022). Heterocyclic crown ethers with potential biological and pharmacological properties: From synthesis to applications. *Applied Sciences, 12*(3), 1102.

Vala, R. M., & Patel, H. M (2023). Recent developments in the Hantzsch synthesis of dihydropyridines. *Advances in Heterocyclic Chemistry, 141*, 179–208.

Yu, C., Zou, Q., Wang, Q., Zhao, Y., Ran, X., Dong, G., …, Zhou, J. (2023). Silicon solar cell with undoped tin oxide transparent electrode. *Nature Energy, 8*(10), 1119–1125.

Zhang, X., Bi, W., Cao, Z., Shen, J., & Chen, B. (2024). Recent developments in the metal-catalyzed synthesis of nitrogenous heterocyclic compounds. *Molecules, 29*(22), 5458.

Zheng, X., Ma, Z., & Zhang, D. (2020). Synthesis of imidazole-based medicinal molecules utilizing the van Leusen imidazole synthesis. *Pharmaceuticals, 13*(3), 37.

4 Pyridine Derivatives for Neuroprotection

A Comprehensive Review

Hao Lu and Renhua Qiu

4.1 INTRODUCTION

Neurodegenerative diseases (NDs), including Alzheimer's disease (AD), Parkinson's disease (PD), and Huntington's disease (HD), are emerging as a growing global health crisis characterized by progressive neuronal loss and debilitating cognitive and motor impairments (Cai & Jeong, 2020). The multifactorial pathophysiology of these disorders involves complex interactions between oxidative stress, neuroinflammation, protein aggregation, and neurotransmitter dysregulation, posing significant challenges to developing effective disease-modifying therapies. Current treatment options are largely limited to symptomatic relief and do not halt or reverse the underlying neurodegenerative processes. This unmet clinical need has spurred intensive research into novel therapeutic strategies, with a particular focus on multi-target-directed ligands (MTDLs) that can simultaneously modulate several pathological pathways. Among the diverse chemical scaffolds being investigated, pyridine and its derivatives have emerged as a privileged structure in medicinal chemistry, offering a versatile platform for the design and development of potent neuroprotective agents. The pyridine ring's unique electronic properties and its ability to function with a wide array of substituents and fused with other heterocyclic systems have led to the discovery of compounds with a remarkable breadth of biological activities relevant to neuroprotection.

This chapter provides a comprehensive overview of recent research advances on pyridine derivatives as potential neuroprotective agents for treating NDs. The scope encompasses detailed analysis of diverse chemical structures within the pyridine family, including simple pyridines, dihydropyridines, and complex heterocyclic compounds such as pyrimidines and thiophenopyridines. We systematically explore the specific applications of these derivatives in major NDs, emphasizing their therapeutic potential in AD, PD, and HD. The core focus of this chapter is to elucidate the multiple mechanisms through which these compounds exert neuroprotective effects. These mechanisms include potent antioxidant and anti-inflammatory activity, modulation of key neurotransmitter systems (such as cholinergic and serotonergic systems), inhibition of pathological protein aggregation (such as β-amyloid (Aβ) and mutant huntingtin protein), and protection of mitochondrial function. Furthermore, we provide a detailed overview of drug development progress, summarizing key findings from preclinical studies conducted using in vitro and in vivo models. This section highlights the journey of promising compounds from the laboratory to clinical application. A critical analysis of structure–activity relationships (SAR) is also presented, aiming to reveal the connection between the chemical properties of these derivatives and their specific neuroprotective activities, thereby providing important reference points for future research (Leng et al., 2019).

The comprehensive evidence presented in this chapter underscores the tremendous therapeutic potential of pyridine derivatives as multifunctional and potent neuroprotective agents (Stopschinski & Diamond, 2017). Key studies demonstrate that these compounds simultaneously target multiple therapeutic pathways—a decisive advantage given the complex and interrelated pathophysiological mechanisms of NDs. In AD, for instance, pyridine derivatives have been shown to inhibit Aβ

DOI: 10.1201/9781003772019-4

aggregation, mitigate oxidative stress, and protect mitochondrial function, thereby significantly improving cognitive function in animal models. Similarly, novel thiophene[3,2-*b*]pyridine derivatives exhibit a unique ability to reduce mutant huntingtin protein levels by regulating gene splicing, opening promising new avenues for HD therapy. While translating these promising preclinical findings into approved drugs remains a significant challenge, their multidimensional mechanisms of action and structural diversity offer substantial scope for future development.

NDs, such as AD, PD, and HD, are posing an increasingly severe global health challenge (Zhu et al., 2019). These diseases are characterized by the progressive degeneration of specific neuronal populations, leading to a range of disabling symptoms, including cognitive decline, motor dysfunction, and psychiatric disorders. The prevalence of these diseases is rising sharply, driven primarily by the global increase in life expectancy. It is estimated that NDs affect over 1 billion people worldwide, claiming at least 7 million lives annually. AD, the most common form of dementia, is particularly prevalent among the elderly and constitutes a significant socioeconomic burden. Its pathological hallmarks include extracellular deposits of Aβ plaques and intracellular formation of neurofibrillary tangles composed of hyperphosphorylated tau protein. These pathological alterations collectively lead to synaptic dysfunction and neuronal death (Xun et al., 2025). Similarly, PD is characterized by the loss of dopaminergic neurons in the substantia nigra and the presence of Lewy bodies, primarily composed of aggregated α-synuclein. HD is a monogenic disorder caused by an expansion of the CAG trinucleotide repeat sequence in the huntingtin (HTT) gene. This expansion leads to the production of toxic mutant huntingtin (*m*HTT), triggering widespread neuronal degeneration, particularly concentrated in cortical and basal ganglia regions (Agarwal et al., 2019). The profound impact of these diseases on patients, caregivers, and healthcare systems underscores the urgent need for developing more effective therapeutic interventions.

Despite decades of extensive research, therapeutic prospects for NDs remain generally inadequate. Current treatments primarily focus on symptomatic management, aiming to control clinical manifestations without addressing the underlying pathological mechanisms. For example, the core therapeutic approach for AD involves acetylcholinesterase inhibitors (AChEIs), which temporarily improve cognitive symptoms by elevating levels of the neurotransmitter acetylcholine in the brain (Akıncıoğlu & Gülçin, 2020). Although these medications can provide some degree of improvement, they cannot halt or reverse the ongoing loss of neurons. Similarly, the primary treatment for PD involves supplementing dopamine levels with levodopa. While this approach effectively controls motor symptoms in the early stages of the disease, it carries significant long-term side effects and does not prevent disease progression (Mazumder et al., 2018). The development of disease-modifying therapies has been fraught with challenges, primarily due to the multifactorial complexity of these diseases. The "single-target, single-disease" treatment model has proven largely ineffective, as targeting a single pathological pathway often fails to yield significant clinical benefits. This has driven a paradigm shift in drug development, with increasing emphasis on creating multi-targeted drug-like molecules (MTDLs) that simultaneously modulate multiple key pathological mechanisms—such as oxidative stress, neuroinflammation, and protein aggregation. Given the highly complex and interconnected pathways involved in NDs, comprehensive therapeutic approaches are urgently needed. Consequently, identifying novel multimodal drugs has become a top priority in this field.

In the quest for novel neuroprotective agents, pyridine and its derivatives have emerged as a highly promising and versatile class of compounds (Henry, 2004). As a privileged skeleton in medicinal chemistry, the pyridine ring is widely present in diverse bioactive molecules, spanning both natural products and synthetic drugs (Bull et al., 2012). Its unique electronic structure—where the nitrogen atom both functions as a hydrogen bond acceptor and exhibits weak basic properties—makes it an ideal core for designing molecules that interact with multiple biological targets (Vitaku et al., 2014). Pyridine derivatives exhibit remarkable structural diversity, encompassing simple pyridines, dihydropyridines, and complex heterocyclic fused systems such as pyridineimidazoles, thiophenopyridines, and pyrrolypyridines (Lu et al., 2024). This structural diversity enables precise tuning of a compound's physicochemical properties and biological activity, facilitating the development of

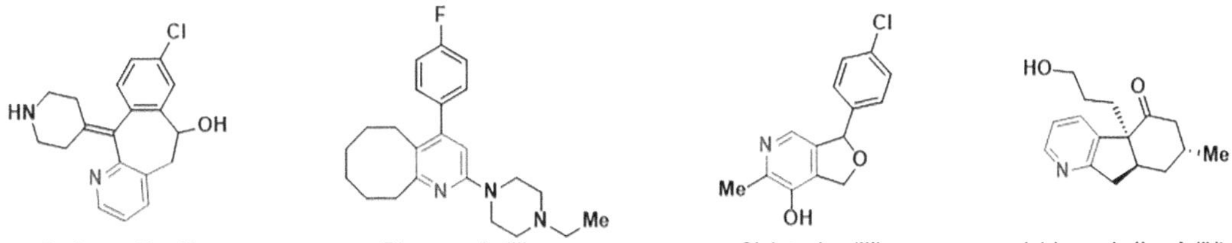

Desloratadine (I) Blonanserin (II) Cicletanine (III) (+)-Lycopladine A (IV)

FIGURE 4.1 Representative biologically active molecules based on fused pyridine scaffolds. (From Lu et al., 2024.)

drugs with optimized therapeutic efficacy, selectivity, and pharmacokinetic characteristics (Allais et al., 2014). Numerous pyridine derivatives have been found to exhibit broad neuroprotective activities, including antioxidant, anti-inflammatory, and anti-apoptotic effects. Other derivatives, such as the pyridineamine derivative PAT, exhibit potent inhibition of Aβ aggregation in AD animal models while significantly improving cognitive function (Zhou et al., 2022). Pyridine derivatives can simultaneously target multiple therapeutic pathways, making them particularly suitable for developing multi-target drugs (MTDLs). This offers a highly promising therapeutic strategy for addressing the complex pathophysiology of NDs (Figure 4.1).

4.2 NEUROPROTECTIVE MECHANISMS, STRUCTURE–ACTIVITY RELATIONSHIPS, AND THERAPEUTIC APPLICATIONS OF PYRIDINE DERIVATIVES IN NEURODEGENERATIVE DISEASES

4.2.1 Neuroprotective Mechanisms of Pyridine Derivatives

4.2.1.1 Scavenging Reactive Oxygen Species (ROS)

Oxidative stress is a hallmark feature of various NDs, arising from an imbalance between reactive oxygen species (ROS) production and cellular antioxidant defense systems. The brain is particularly susceptible to oxidative damage due to its high metabolic rate, high content of polyunsaturated fatty acids, and relatively low levels of antioxidant enzymes. The accumulation of ROS leads to damage of cellular components such as lipids and proteins, ultimately triggering neuronal dysfunction and death. Pyridine derivatives have been demonstrated to possess potent antioxidant properties, capable of both directly scavenging ROS and enhancing cellular antioxidant capacity. For example, a novel class of 4,7-dihydro-2*H*-pyrazolo[3,4-*b*]pyridines has been developed as MTDLs with potent antioxidant properties, capable of scavenging oxygen and nitrogen free radicals. Studies indicate these compounds reduce ROS production in hippocampal slice models of tau hyperphosphorylation, highlighting their potential to mitigate oxidative brain damage (Michalska et al., 2020) (Figure 4.2).

4.2.1.2 Modulation of Inflammatory Pathways

Chronic neuroinflammation, characterized by the activation of microglia and astrocytes along with the release of pro-inflammatory cytokines, represents another key pathological hallmark of

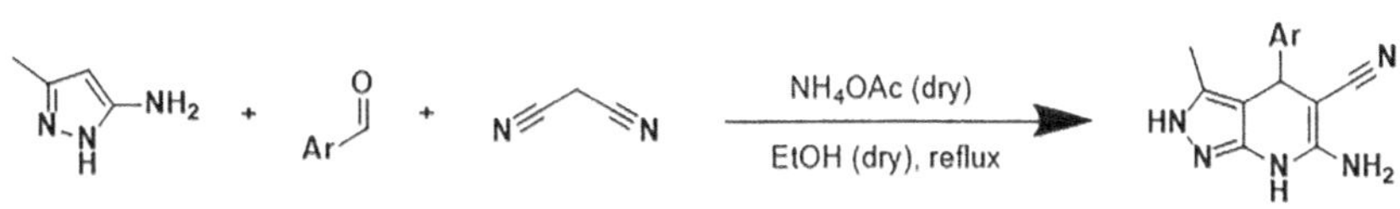

FIGURE 4.2 Multicomponent synthetic approach to the 4,7-dihydro-2*H*-pyrazolo[3,4-*b*]pyridines. (From Michalska et al., 2020.)

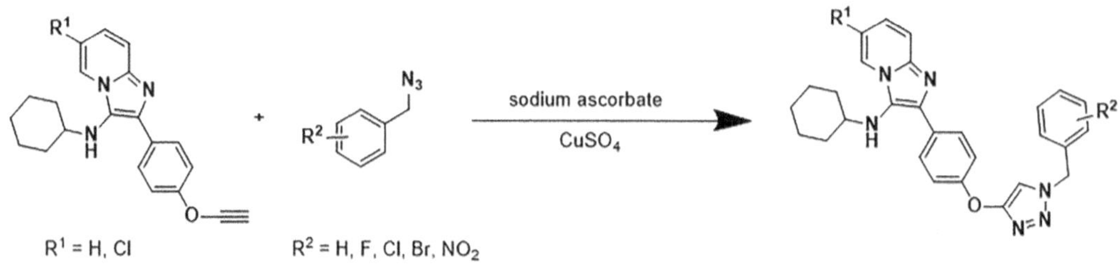

FIGURE 4.3 The synthesis of *N*-cyclohexylimidazo[1,2-*a*]pyridines. (From Haghighijoo et al., 2020.)

NDs. This persistent inflammatory response exacerbates neuronal damage and drives disease progression. Pyridine derivatives exhibit significant anti-inflammatory properties, offering a pathway to modulate this harmful process. The anti-inflammatory potential of *N*-cyclohexylimidazo[1,2-*a*] pyridine derivatives has been investigated (Haghighijoo et al., 2020). Although existing literature does not elaborate on their specific molecular targets, the design concept of these compounds as multi-target drugs for AD indicates that anti-inflammatory activity constitutes a key component of their therapeutic properties. The anti-inflammatory effects of pyridine derivatives may be mediated through multiple mechanisms, including inhibition of *pro*-inflammatory signaling pathways and regulation of inflammatory enzyme activity. By suppressing neuroinflammatory responses, these compounds help create a more favorable environment for neuronal survival and function (Figure 4.3).

4.2.1.3 Inhibition of Cholinesterases (AChE and BuChE)

The cholinergic system plays a crucial role in cognitive functions such as learning and memory, and its dysfunction is a core feature of AD. Degeneration of cholinergic neurons in the basal forebrain leads to a deficiency of the neurotransmitter acetylcholine (ACh), which is the root cause of cognitive decline in AD patients. Acetylcholinesterase (AChE) is responsible for breaking down acetylcholine in the synaptic cleft, and inhibiting this enzyme has become a recognized therapeutic strategy for AD (Kaur & Bansal, 2022). Numerous pyridine derivatives have been identified as potent acetylcholinesterase inhibitors capable of enhancing cholinergic neurotransmission by elevating acetylcholine levels. For instance, the pyridineamine derivative PAT has been demonstrated to inhibit acetylcholinesterase activity in the *Caenorhabditis elegans* model, thereby exerting cognitive-enhancing effects. Beyond acetylcholinesterase, butyrylcholinesterase (BuChE) also plays a role in acetylcholine hydrolysis, particularly in the later stages of AD (Liu et al., 2022). Simultaneous inhibition of acetylcholinesterase and butyrylcholinesterase may enhance cholinergic function more comprehensively and persistently, potentially improving therapeutic outcomes. The development of pyridine-based cholinesterase inhibitors has opened promising avenues for symptomatic treatment of AD, offering potential to improve patients' cognitive function and quality of life.

4.2.1.4 Interaction with Cannabinoid Receptors (CB2)

The endocannabinoid system comprises cannabinoid receptors CB1 and CB2, which play a pivotal role in regulating multiple physiological processes within the central nervous system, including neuroinflammation, neurogenesis, and synaptic plasticity. CB1 receptors are primarily expressed on neurons, while CB2 receptors are predominantly found on immune cells such as microglia. Their activation has been demonstrated to exert neuroprotective and anti-inflammatory effects. Modulating CB2 receptors has emerged as a potential therapeutic strategy for NDs. A series of novel 5,6-dihydropyridine[2′,1′:2,3]imidazo[4,5-*c*]quinoline derivatives have been developed as multi-target drugs for PD, with one key mechanism of action being the modulation of CB2 receptors (Jena et al. 2024). Activation of CB2 receptors on microglia surfaces inhibits pro-inflammatory

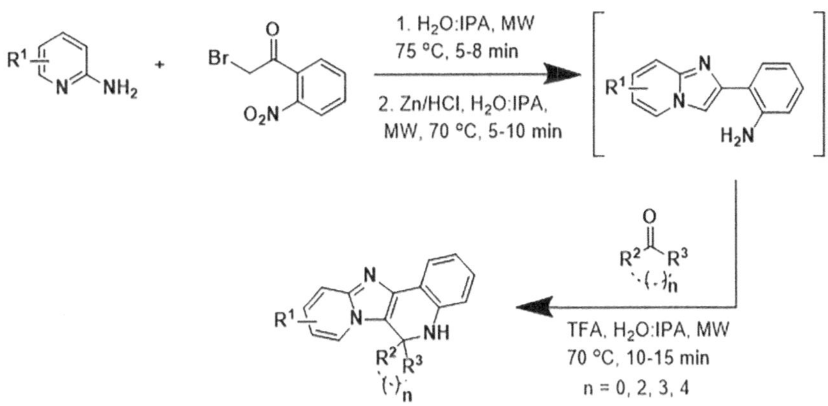

FIGURE 4.4 Microwave-assisted one-pot telescopic approach to quinoline derivatives. (From Jena et al., 2020.)

cytokine release while promoting anti-inflammatory mediators, thereby suppressing neuroinflammation. The development of pyridine derivatives that selectively target CB2 receptors offers a novel neuroprotective approach, holding promise for simultaneously alleviating both inflammatory and neurotoxic aspects of NDs. These compounds' ability to act on multiple targets, including cholinesterase and cannabinoid receptors, further highlights their potential as multimodal therapeutic agents (Figure 4.4).

4.2.2 Applications in Specific Neurodegenerative Diseases

4.2.2.1 Pyridine Amine Derivatives (PAT)

AD is a progressive neurodegenerative disorder characterized by cognitive decline and memory loss. The pyridine derivative 3-bis(pyridin-2-ylmethyl)aminomethyl-5-hydroxybenzyltriphenylphosphonium bromide (PAT) has emerged as a potential therapeutic candidate for AD (Zhu et al., 2019). Research indicates that PAT exerts its neuroprotective effects through multiple mechanisms. It effectively inhibits both spontaneous and metal-induced aggregation of Aβ peptides, a key pathological hallmark of AD. PAT exhibits potent antioxidant activity. In a transgenic *Caenorhabditis elegans* model of Aβ toxicity, PAT was shown to alleviate Aβ-induced paralysis, reduce ROS production, and protect mitochondrial function (Lu et al., 2019). Furthermore, PAT was found to inhibit acetylcholinesterase (AChE) activity. Genetic analysis further confirmed PAT's multimodal mechanism of action, indicating that heat shock proteins participate in its Aβ toxicity-alleviating process (Haider et al., 2015). In vivo studies using APP/PS1 AD model mice revealed that PAT significantly improved memory and cognitive function as assessed by the Morris water maze test. The combined effects of its anti-aggregation, antioxidant, and cholinesterase inhibitory activities position PAT as a highly promising lead compound for developing novel anti-AD drugs (Turdi et al., 2009) (Figure 4.5).

4.2.2.2 Pyrazolo[3,4-b]Pyridine Derivatives

The development of MTDLs represents an effective strategy for treating complex diseases such as AD. Based on the premise that this disease exhibits multifactorial characteristics and necessitates a multi-pronged therapeutic approach, a novel family of 4,7-dihydro-2*H*-pyrazolo[3,4-*b*]pyridine derivatives was designed and synthesized as MTDLs for AD. These derivatives were designed to combine the antioxidant properties and L-type voltage-dependent calcium channel (VDCC) blocking activity of 1,4-dihydropyridine with the glycogen synthase kinase-3β (GSK-3β) inhibitory activity of the pyrazole ring. The compounds obtained exhibit potent multi-target properties. They

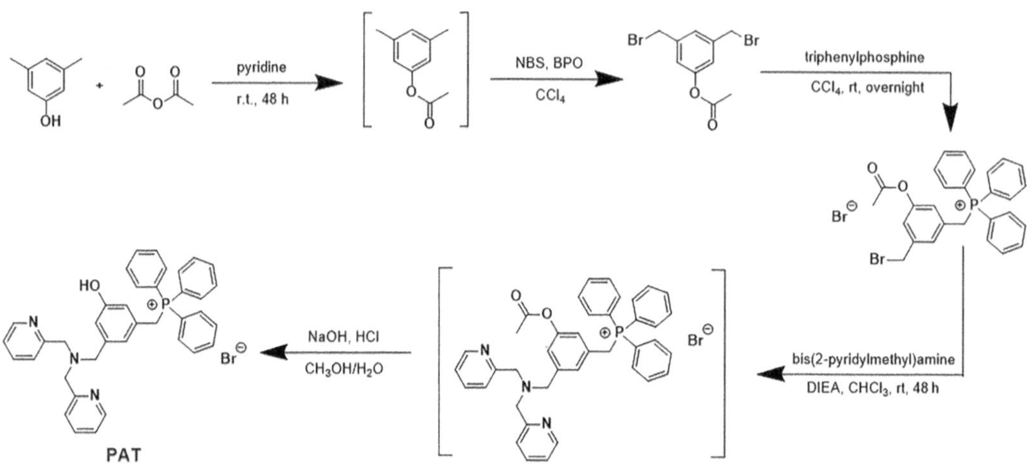

FIGURE 4.5 Synthetic route to PAT. (From Zhu et al., 2019.)

possess strong radical scavenging capabilities, effectively eliminating oxygen and nitrogen free radicals. These compounds also demonstrate significant anti-inflammatory properties and exhibit potent GSK-3β inhibitory activity within the low micromolar concentration range. Furthermore, studies reveal that these compounds exhibit moderate antagonism against L-type voltage-dependent calcium channels (Han et al., 2017). In vitro ND models validated the neuroprotective potential of these derivatives: they demonstrated potent neuroprotective activity against toxicity induced by oxidative stress, tau hyperphosphorylation, and calcium overload.

4.2.2.3 Imidazo[1,2-a]Pyridine Derivatives

Imidazo[1,2-*a*]pyridine derivatives represent another class of pyridine compounds exhibiting significant therapeutic potential for AD (Neves et al., 2010). These derivatives were designed as MTDLs to simultaneously address several key pathological hallmarks of AD. By synthesizing a series of *N*-cyclohexylimidazo[1,2-*a*]pyridine derivatives containing a 1,2,3-triazole moiety, their inhibitory activity against β-site amyloid precursor protein (APP) cleaving enzyme 1 (BACE1) and butyrylcholinesterase (BuChE) at the β-site, along with their antioxidant and metal-chelating properties (Endoori et al., 2021). BACE1 is a key enzyme in the generation of Aβ peptides, and its inhibition represents a major therapeutic target for AD (Youn et al., 2016). Another series of imidazo[1,5-*a*]pyridine derivatives has been developed as 5-HT4 receptor partial agonists for treating cognitive impairment associated with AD (Nirogi et al., 2015). The 5-HT4 receptor is a G protein–coupled receptor widely expressed in the brain and involved in regulating cognitive function (Castro-Alvarez et al., 2021). Research indicates that activation of the 5-HT4 receptor enhances the release of acetylcholine and other neurotransmitters, while promoting non-amyloid processing of the APP. This process generates the sAPPα fragment, which exerts neuroprotective effects. The diverse therapeutic targets of imidazo[1,2-*a*]pyridine derivatives, ranging from enzymes to receptors, highlight their potential as multifunctional therapeutic agents for AD (Figure 4.6).

4.2.2.4 Dihydropyridine Derivatives

PD is a neurodegenerative disorder characterized by progressive loss of dopaminergic neurons in the substantia nigra, leading to motor symptoms such as tremor, rigidity, and bradykinesia (Simuni & Krainc, 2018). Oxidative stress and neuroinflammation represent key pathological hallmarks of PD, making the development of drugs that alleviate these processes an important therapeutic target. This study developed a series of novel 5,6-dihydropyridine[2′,1′:2,3]imidazo[4,5-*c*]quinoline

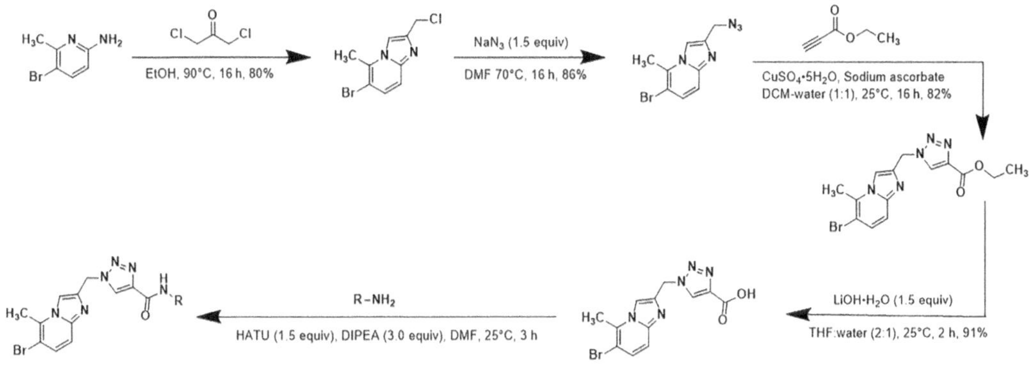

FIGURE 4.6 Synthetic route to novel imidazo[1,2-*a*]pyridines. (From Endoori et al., 2021.)

derivatives as multi-target drugs for treating PD. These derivatives are designed to inhibit cholinesterases (ChEs) and modulate cannabinoid receptor type 2 (CB2), both of which are implicated in the pathophysiology of PD. The neuroprotective effects of these compounds were evaluated in in vitro PD models, specifically against MPP^+-induced toxicity. MPP^+ is a neurotoxin that is taken up by dopaminergic neurons and inhibits mitochondrial complex **I**, leading to ROS production and cell death. Studies revealed that dihydropyridine–imidazole–quinoline derivatives significantly reduced MPP^+-induced ROS production and cell death, with the mechanism attributed to the compound's ability to modulate CB2 receptors. Activation of CB2 receptors on microglia helps suppress neuroinflammatory responses, while inhibition of cholinesterase enhances cholinergic neurotransmission—a function also impaired in PD. These derivatives represent a highly promising class of compounds for developing neuroprotective therapies against PD through their multimodal mechanism targeting both oxidative stress and neuroinflammation. Their ability to act on multiple targets simultaneously is particularly significant, offering a more comprehensive therapeutic strategy for this complex disorder.

4.2.2.5 Thieno[3,2-b]Pyridine Derivatives

HD is a devastating hereditary neurodegenerative disorder caused by an expansion of the CAG trinucleotide repeat sequence in the huntingtin (HTT) gene. This mutation results in the production of mutant huntingtin (*m*HTT) containing an expanded polyglutamine chain, which is toxic to neurons and triggers progressive motor, cognitive, and psychiatric symptoms. Reducing *m*HTT levels is a primary therapeutic target for HD. The development of novel thiophene[3,2-*b*] pyridine derivatives offers a unique and highly promising new approach to achieving this goal (Shen et al., 2023). These compounds effectively reduce the expression levels of both mutant and wild-type HTT by regulating the splicing process of HTT precursor mRNA. By altering splicing, these derivatives promote the production of truncated, non-functional HTT proteins, which are rapidly degraded by cellular mechanisms. This mechanism offers significant advantages as it directly targets the genetic cause of the disease. Thieno[3,2-*b*]pyridine derivatives have been demonstrated to effectively reduce HTT expression levels in both brain and peripheral tissues—a critical requirement for successful HD therapy. Furthermore, these compounds demonstrate improved selectivity against known off-targets of splicing modifiers, which suggests a potentially wider therapeutic window and a more favorable safety profile compared to non-selective splicing modulators. The development of these thieno[3,2-*b*]pyridine derivatives represents a significant breakthrough in the search for a disease-modifying therapy for HD, offering a new and powerful tool to combat this devastating disorder. The ability to selectively lower the levels of the toxic mHTT protein holds the promise of not only alleviating symptoms but also of slowing or even halting the progression of the disease.

4.2.3 Progress in Drug Development

4.2.3.1 In Vitro Models

The early stages of drug development for NDs rely heavily on preclinical research using in vitro models to evaluate the efficacy and safety of potential therapeutic compounds. These models provide a controlled environment for investigating molecular mechanisms of action while enabling cost-effective screening of large compound libraries. Multiple in vitro models have been employed to study the neuroprotective effects of pyridine derivatives. For instance, the neuroprotective properties of 4,7-dihydro-2*H*-pyrazolo[3,4-*b*]pyridine derivatives were evaluated in cell culture models of neurodegeneration induced by oxidative stress, tau hyperphosphorylation, and calcium overload. Studies confirmed these compounds effectively counteracted multiple toxic insults, providing strong evidence for their neuroprotective potential. Another widely used in vitro model is the human neuroblastoma cell line SH-SY5Y, which differentiates into a neuron-like phenotype and is frequently employed to study the effects of neurotoxins and neuroprotective agents (Li & Li, 2025). Beyond cell lines, simpler organisms such as the nematode *Caenorhabditis elegans* are also employed as in vivo models for NDs. *C. elegans*' short lifespan, well-defined nervous system, and ease of genetic manipulation make it an ideal tool for high-throughput screening and investigating the genetic basis of neuroprotection. The pyridine derivative PAT was evaluated in transgenic *C. elegans* models expressing human Aβ peptides. Results demonstrated that this compound alleviates Aβ-induced paralysis symptoms, reduces ROS production, and protects mitochondrial function. These in vitro and in vivo preclinical models are crucial for deepening our understanding of the neuroprotective mechanisms of pyridine derivatives and for screening lead compounds with development potential.

Although in vitro models hold irreplaceable value for initial screening and mechanism studies, evaluating potential therapeutic compounds in more complex in vivo models remains a critical step in the drug development process. In vivo models provide conditions closer to physiological environments, enabling assessment of pharmacokinetic properties such as absorption, distribution, metabolism, and excretion (ADME), while also evaluating therapeutic effects at the whole-organism level. For AD, transgenic mouse models overexpressing human APP and presenilin 1 (PS1) mutations—such as the APP/PS1 mouse model—are widely employed (Zheng et al., 2019). These mice develop Aβ plaques and exhibit cognitive deficits, making them a crucial tool for testing the efficacy of anti-Aβ therapies. The pyridine derivative PAT was evaluated in APP/PS1 mouse models, demonstrating significant improvements in memory and cognitive function as measured by the Morris water maze test. For PD, neurotoxin models are widely employed. The neurotoxin MPTP (1-methyl-4-phenyl-1,2,3,6-tetrahydropyridine) converts to MPP^+ (1-methyl-4-phenylpyridinium) in the brain, selectively destroying dopaminergic neurons to mimic the pathological mechanism of PD. In MPP^+-induced PD models, 5,6-dihydropyridine[2′,1′:2,3]imidazo[4,5-*c*]quinoline derivatives demonstrated significant neuroprotective effects by reducing ROS production and cell death. These in vivo studies provide crucial evidence for the therapeutic potential of pyridine derivatives and represent essential steps toward advancing to human clinical trials.

4.2.3.2 Clinical-Stage Compounds

Although most pyridine derivatives for neuroprotection remain in preclinical development, certain compounds have successfully advanced to clinical trials, marking significant progress in translating laboratory research into potential patient therapies. The most prominent example is nicotinamide riboside (NR), a pyridine compound that serves as a precursor to the essential coenzyme nicotinamide adenine dinucleotide (NAD^+). Given its ability to elevate cellular NAD^+ levels—which decline with age and across various diseases—NR has been proposed as a potential neuroprotective therapeutic strategy for neurodegenerative disorders (Damgaard & Treebak, 2023). A recent randomized double-blind clinical trial named NADPARK evaluated the effects of oral NR supplementation (1000 mg daily for 30 days) in PD patients. The study demonstrated that NR significantly elevated brain NAD levels, accompanied by alterations in brain metabolic patterns and a trend toward mild improvement in clinical symptoms. Additionally, NR treatment correlates with transcriptional upregulation of processes related to mitochondrial, lysosomal, and proteasomal function in peripheral tissues. Another clinical trial is investigating NR's neuroprotective effects in patients with primary open-angle glaucoma, a condition involving optic nerve degeneration. This randomized, double-blind, placebo-controlled study aims to determine whether NR can delay the progressive thinning of the retinal nerve fiber layer—a key indicator of disease progression. These ongoing clinical trials provide valuable insights into the safety, tolerability, and potential efficacy of pyridine-based neuroprotective strategies in humans.

4.2.3.3 Repurposing of Approved Drugs

Drug repurposing (also known as drug repositioning) refers to the process of identifying new therapeutic uses for existing drugs already approved for other indications. Compared to traditional drug development, this approach offers multiple advantages, including shorter development cycles, lower costs, and the benefit of a well-established safety profile. Several pyridine-based drugs already approved for other diseases are undergoing intensive investigation for their potential neuroprotective effects. The most prominent example is nicotinamide—a derivative of vitamin B3 widely used as a dietary supplement (Aboomar et al., 2024). Nicotinamide is a precursor to the coenzyme nicotinamide adenine dinucleotide, which is essential for numerous cellular processes, including energy metabolism, DNA repair, and stress resistance. Extensive research demonstrates that nicotinamide effectively counters multiple forms of neurotoxicity, including oxidative stress, excitotoxicity, and Aβ toxicity. This compound has shown significant therapeutic effects in preclinical models of AD, PD, and HD, with clinical trials currently underway for these conditions. Another example is 4-aminopyridine (4-AP), a potassium channel blocker approved for symptomatic treatment of gait impairment in multiple sclerosis (MS). Although its primary mechanism involves improving nerve conduction by blocking exposed potassium channels in demyelinated axons, growing evidence suggests 4-AP may also possess direct neuroprotective properties (Leussink et al., 2018). The repurposing of these approved pyridine drugs and other analogues offers a promising and efficient strategy for developing novel therapies for NDs.

4.2.4 Chemical Structure Diversity and Structure–Activity Relationships

4.2.4.1 Nicotinic Acid and Nicotinamide

Niacin and its amide derivative, nicotinamide, are fundamental pyridine compounds that play a crucial role in human health (Nomura et al., 2023). They constitute forms of vitamin B3 and serve as essential precursors for synthesizing the coenzyme nicotinamide adenine dinucleotide (NAD^+). NAD^+ participates as a key molecule in numerous cellular processes, including energy metabolism, DNA repair, and gene expression regulation. In the context of NDs, maintaining adequate levels of nicotinamide adenine dinucleotide (NAD^+) is crucial. Research has revealed that NAD^+ levels decline in aging brains and across various neurodegenerative pathologies. This depletion is believed to contribute to mitochondrial dysfunction, oxidative stress, and increased neuronal vulnerability. Nicotinamide, extensively studied for its neuroprotective properties, has been demonstrated to counteract multiple forms of neurotoxic injury, including oxidative stress, excitotoxicity, and the toxic effects of Aβ peptides. The neuroprotective effects of nicotinamide are achieved through multiple mechanisms, including elevating intracellular NAD^+ levels, activating sirtuins (a class of NAD^+-dependent protein deacetylases), and modulating key signaling pathways involved in cell survival and death, such as the SIRT1, mTOR, and AMPK pathways. Another highly promising derivative is NR, a recently discovered vitamin B3 derivative that also converts to NAD^+ in vivo. Research indicates that nicotinic acid (NR) significantly outperforms nicotinamide as an NAD^+ enhancer and is currently being studied as a dietary supplement for healthy aging and the treatment of various age-related diseases, including neurodegenerative disorders. Despite their simple structure, nicotinic acid and nicotinamide play a crucial role, with their ability to regulate fundamental cellular processes establishing them as cornerstones of neuroprotective strategies.

4.2.4.2 Pyridone Derivatives

Pyridone derivatives, characterized by the presence of a carbonyl group at the 2- or 4-position of the pyridine ring, represent another class of important pyridine compounds with potential neuroprotective activity. The pyridone skeleton provides a versatile platform for introducing various substituents, enabling precise modulation of biological activity. This study synthesized a series of novel *N*-benzylpyridine-2-one derivatives and evaluated their neuroprotective effects using a widely applied AD model—the scopolamine-induced cognitive impairment mouse model (Ruan et al., 2024). These derivatives were found to significantly reduce scopolamine-induced increases in acetylcholinesterase (AChE) activity, as well as levels of oxidative stress and nitrosative stress markers. In behavioral tests, these compounds improved muscle grip strength and motor performance. Compounds 28 and 28d exhibited the most pronounced therapeutic effects, demonstrating superior neuroprotective activity compared to donepezil, a standard drug for treating AD. The neuroprotective effects of these *N*-benzylpyridine-2-one derivatives likely arise through synergistic actions of cholinesterase inhibition and antioxidant activity. The pyridinone ring itself may contribute to the antioxidant properties of these compounds, while the *N*-benzyl substituent likely participates in their interaction with the acetylcholinesterase active site. The encouraging results obtained with these pyridinone derivatives indicate that this skeleton holds significant potential for developing novel anti-AD drugs.

4.2.4.3 Pyridoimidazole and Related Fused Heterocycles

Fused heterocyclic systems where the pyridine ring is connected to other aromatic rings such as imidazole, thiophene, or pyrrole constitute a vast and diverse class of compounds exhibiting broad biological activities. The fusion of ring structures forms rigid planar configurations that enhance binding affinity to biological targets and improve pharmacokinetic properties. Pyridine-imidazole derivatives stand out as particularly promising skeletal structures for developing neuroprotective agents. For instance, the imidazole[1,2-*a*]pyridine core has been incorporated into multiple multi-target ligands targeting AD (Endoori et al., 2022). A class of *N*-cyclohexylimidazo[1,2-*a*]pyridine derivatives was designed to simultaneously inhibit BACE1 and butyrylcholinesterase (BuChE), while exhibiting antioxidant and metal-chelating properties. The most active compounds in this series exhibit potent inhibitory activity against both enzymes, with IC50 values in the low micromolar range. Another series of imidazo[1,5-*a*] pyridine derivatives was developed as 5-HT4 receptor partial agonists for treating cognitive impairment associated with AD.

4.2.4.4 Impact of Substituents on Neuroprotective Efficacy

The neuroprotective efficacy of pyridine derivatives is highly dependent on the nature and position of substituents on their pyridine ring or co-cyclic heterocyclic core. Systematic investigation of SAR is crucial for rationally designing more potent and selective neuroprotective agents. Within the *N*-cyclohexylimidazo[1,2-*a*]pyridine derivative series designed as multi-target drugs for AD, it was discovered that the nature of substituents on the benzyl side chain significantly influences biological activity. For example, compounds bearing dichlorine substituents (2,3-Cl_2 and 3,4-Cl_2) on the benzyl ring were found to be the most potent inhibitors of BACE1, with IC_{50} values of approximately 12 and 8.9 μM, respectively. In contrast, compounds bearing a 4-bromo substituent on the benzyl ring exhibited the highest inhibitory activity against butyrylcholinesterase (BuChE), with an IC_{50} value of 2.5 μM. These findings indicate that the electronic and steric properties of substituents play a crucial role in determining a compound's selectivity and potency toward its target. Systematic modification of substituents and evaluation of their impact on biological activity represent key strategies in medicinal chemistry for optimizing the therapeutic potential of lead compounds.

4.2.4.5 Correlation between Structure and Specific Mechanisms

The diverse neuroprotective mechanisms of pyridine derivatives are often associated with their specific structural features. Understanding these correlations is crucial for rationally designing compounds with anticipated mechanisms of action. For example, the antioxidant activity of pyridine derivatives is typically attributed to their electron-donating groups (such as hydroxyl or amino groups), which can scavenge free radicals. Pyridine amine derivatives bearing a benzyl-substituted cyclic hydroxyl group, such as PAT, have demonstrated potent antioxidant capabilities. The metal-chelating properties of certain pyridine derivatives originate from chelating groups (e.g., hydroxyl or carboxyl) adjacent to the pyridine nitrogen atom, which can form stable complexes with metal ions. The ability of pyridine derivatives to modulate kinase activity is often associated with their hinge-binding motifs, such as pyridine or pyrazole ring structures, which form hydrogen bonds with the backbone of the kinase active site. For example, the imidazo[4,5-*b*]pyridine skeleton has been successfully employed in developing potent MLK3 inhibitors. By deciphering the relationship between structure and mechanism of action, medicinal chemists can design and synthesize novel pyridine derivatives with optimized therapeutic properties for treating NDs.

4.3 CONCLUSION AND FUTURE PERSPECTIVES

In summary, pyridine derivatives represent a highly promising and versatile class of compounds demonstrating significant potential in developing novel neuroprotective therapies. Their diverse chemical structures and multidimensional biological activities enable them to effectively address the complex, multifactorial pathological features of NDs. As demonstrated in this review, pyridine derivatives exhibit significant therapeutic effects in preclinical models of multiple NDs, with complementary mechanisms of action, including antioxidant and anti-inflammatory effects, modulation of neurotransmitter systems, and inhibition of protein aggregation. The development of multi-target ligands based on the pyridine skeleton opens exciting new avenues for future research, promising more effective and comprehensive therapeutic approaches for diseases such as AD, PD, and HD. Advances in drug discovery—including the identification of promising lead compounds and the innovative repurposing of existing drugs—lay a solid foundation for continued progress in this field.

Despite significant progress, several challenges remain in developing pyridine derivatives as neuroprotective drugs. One major difficulty lies in translating success in preclinical studies into positive outcomes in clinical trials. The complexity of neurodegenerative diseases and their challenging characteristics to accurately simulate in animal models make it difficult to predict drug efficacy in humans. Another challenge involves optimizing the pharmacokinetic properties of these compounds, particularly their ability to cross the blood–brain barrier, ensuring therapeutic concentrations reach central nervous system targets. However, these challenges also present opportunities for innovation. The development of novel drug delivery systems, such as nanoparticles and liposomes, holds promise for enhancing the brain penetration of pyridine derivatives. Advanced imaging techniques and biomarkers can more effectively monitor disease progression and treatment responses during clinical trials. Furthermore, continued exploration of the chemical space around the pyridine core, guided by SAR studies, will undoubtedly drive the discovery of novel, highly effective neuroprotective agents.

Looking ahead, pyridine derivatives present several key directions worthy of in-depth research in the field of neuroprotection. First, there is an urgent need for more thorough investigation into the specific molecular targets and signaling pathways regulated by these compounds. A thorough understanding of their mechanisms of action will facilitate the rational design of more potent and selective therapeutics. Second, efforts should focus on developing novel pyridine scaffold structures—including novel heterocyclic fusion systems—to expand chemical diversity and explore novel approaches to neuroprotection. Third, integrating pyridine derivatives with other therapeutic modalities such as gene therapy or stem cell therapy could provide synergistic solutions for ND treatment. Finally, applying artificial intelligence and machine learning techniques to optimize pyridine derivative design could accelerate drug discovery and enhance success rates. By advancing these research directions, we will continue to unlock the full therapeutic potential of pyridine derivatives, bringing new hope to patients with NDs.

REFERENCES

Aboomar, N. M., Essam, O., Hassan, A., Bassiouny, A. R., & Arafa, R. K (2024). Exploring a repurposed candidate with dual hIDO1/hTDO2 inhibitory potential for anticancer efficacy identified through pharmacophore-based virtual screening and in vitro evaluation. *Scientific Reports*, *14*, 9386. https://doi.org/10.1038/s41598-024-59353-4

Agarwal, S., Koch, G., Hillis, A. E., Huynh, W., Ward, N. S., Vucic, S., & Kiernan. M. C (2019). Interrogating cortical function with transcranial magnetic stimulation: Insights from neurodegenerative disease and stroke. *Journal of Neurology, NeuroSurgery, and Psychiatry*, *90*, 47–57. https://doi.org/10.1136/jnnp-2017-317371

Akıncıoğlu, H., & Gülçin, İ. (2020). Potent acetylcholinesterase inhibitors: Potential drugs for Alzheimer's disease. *Mini-Reviews in Medicinal Chemistry*, *20*, 703–715. https://doi.org/10.2174/1389557520666200103100521

Allais, C., Grassot, J. M., Rodriguez, J., & Constantieux, T (2014). Metal-free multicomponent syntheses of pyridines. *Chemical Reviews*, *114*, 10829–10868. https://doi.org/10.1021/cr500099b

Bull, J. A., Mousseau, J. J., Pelletier, G., & Charette, A. B (2012). Synthesis of pyridine and dihydropyridine derivatives by regio- and stereose-lective addition to *n*-activated pyridines. *Chemical Reviews*, *112*, 2642–2713. https://doi.org/10.1021/cr200251d

Cai, Q., & Jeong, Y. Y (2020). Mitophagy in Alzheimer's disease and other age-related neurodegenerative diseases. *Cells*, *9*, 150. https://doi.org/10.3390/cells9010150

Castro-Alvarez, A., Chávez-Ángel, E., & Nelson, R (2021). Understanding the molecular basis of 5-HT4 receptor partial agonists through 3D-QSAR studies. *International Journal of Molecular Sciences*, *22*, 3602. https://doi.org/10.3390/ijms22073602

Damgaard, M. V., & Treebak, J. T (2023). What is really known about the effects of nicotinamide riboside supplementation in humans. *Science Advances*, *9*, eadi4862. https://doi.org/10.1126/sciadv.adi4862

Endoori, S., Anna, V. R., Gulipalli, K. C., Bodige, S., Pommidi, A. K., Surapureddi, S. R., & Seelam, N (2022). Synthesis, characterization, and anticancer activity of novel imidazo[1,2-*a*]pyridine linked 1,2,3-triazole derivatives. *Russian Journal of General Chemistry*, *92*, 1775–1784. https://doi.org/10.1134/S1070363222110445

Endoori, S., Gulipalli, K. C., Bodige, S., Ravula, P., & Seelam, N (2021). Design, synthesis, anticancer activity, and in silico studies of novel imidazo[1,2-*a*]pyridine based 1*H*-1,2,3-triazole derivatives. *Journal of Heterocyclic Chemistry*, *58*, 1311–1320. https://doi.org/10.1002/jhet.4259

Haghighijoo, Z., Akrami, S., Saeedi, M., Zonouzi, A., Iraji, A., Larijani, B., Fakherzadeh, H., Sharifi, F., Arzaghi, S. M., Mahdavi, M., & Edraki, N (2020). *N*-cyclohexylimidazo[1,2-*a*]pyridine derivatives as multi-target-directed ligands for treatment of Alzheimer's disease. *Bioorganic Chemistry*, *103*, 104146. https://doi.org/10.1016/j.bioorg.2020.104146

Haider, S., Liaquat, L., Shahzad, S., Sadir, S., Madiha, S., Batool, Z., Tabassum, S., Saleem, S., Naqvi, F., & Perveen, T (2015). A high dose of short term exogenous d-galactose administration in young male rats produces symptoms simulating the natural aging process. *Life Sciences*, *124*, 110–119. https://doi.org/10.1016/j.lfs.2015.01.016

Han, Y.-Y., Wang, X.-D., Liu, L., Guo, H.-M., Cong, W., Yan, W.-W., Huang, J.-N., Xiao, P., & Li, C.-H (2017). L-type VDCCs participate in behavioral-LTP and memory retention. *Neurobiology of Learning and Memory*, *145*, 75–83. https://doi.org/10.1016/j.nlm.2017.08.011

Henry, G. D (2004). De novo synthesis of substituted pyridines. *Tetrahedron*, 2004, *60*, 6043–6061. https://doi.org/10.1016/j.tet.2004.04.043

Jena, S., Gonzalez, G., Vítek, D., Kvasnicová, M., Štěpánková, Š., Strnad, M., Voller, J., & Chanda, K. (2024). *European Journal of Medicinal Chemistry*, *276*, 116592. https://doi.org/10.1016/j.ejmech.2024.116592

Kaur, S., & Bansal. Y. (2022). Design, molecular docking, synthesis and evaluation of xanthoxylin hybrids as dual inhibitors of IL-6 and acetylcholinesterase for Alzheimer's disease. *Bioorganic Chemistry*, *121*, 105670. https://doi.org/10.1016/j.bioorg.2022.105670

Leng, Y., Musiek, E. S., Hu, K., Cappuccio, F. P., & Yaffe. K (2019). Association between circadian rhythms and neurodegenerative diseases. *The Lancet Neurology, 18*, 307–318. https://doi.org/10.1016/S1474-4422(18)30461-7

Leussink, V. I., Montalban, X., & Hartung, H.-P. (2018). Restoring axonal function with 4-aminopyridine: Clinical efficacy in multiple sclerosis and beyond. *CNS Drugs 32*, 637–651. https://doi.org/10.1007/s40263-018-0536-2

Li, W., & Li, Q (2025). Targeting ferroptosis by electroacupuncture offers a therapeutic regimen for brain injury after ischemic stroke. *Neurochemistry International, 190*, 106038. https://doi.org/10.1016/j.neuint.2025.106038

Liu, T., Chen, S., Du, J., Xing, S., Li, R., & Li, Z (2022). Design, synthesis, and biological evaluation of novel (4-(1,2,4-oxadiazol-5-yl)phenyl)-2-aminoacetamide derivatives as multifunctional agents for the treatment of Alzheimer's disease. *European Journal of Medicinal Chemistry, 227*, 113973. https://doi.org/10.1016/j.ejmech.2021.113973

Lu, H., Li, X., Ou, Y.-F., Du, P., Yin, S.-F., Kambe, N., & Qiu, R (2024). Cu(II)/base synergistic promoted [2+1+3] cyclization of cyclic ketones with α,β-unsaturated aldehydes/ketones and NH_4I to access fused pyridines. *Organic Letters, 26*, 7191–7195. https://doi.org/10.1021/acs.orglett.4c02628

Lu, S., Liu, S., Cui, J., Liu, X., Zhao, C., Fan, L., Yin, S., & Hu, H (2019). Combination of patulin and chlorpyrifos synergistically induces hepatotoxicity via inhibition of catalase activity and generation of reactive oxygen species. *Journal of Agricultural and Food Chemistry, 67*, 11474–11480. https://doi.org/10.1021/acs.jafc.9b04814

Mazumder, M. K., Paul, R., Phukan, B. C., Dutta, A., Chakrabarty, J., Bhattacharya, P., & Borah, A (2018). Garcinol, an effective monoamine oxidase-b inhibitor for the treatment of Parkinson's disease. *Medical Hypotheses, 117*, 54–58. https://doi.org/10.1016/j.mehy.2018.06.009

Michalska, P., Mayo, P., Fernandez-Mendivil, C., Tenti, G., Duarte, P., Buendia, I., Ramos, M. T., Lopez, M. G., Menendez, J. C., & Leon, R (2020). Antioxidant, anti-inflammatory and neuroprotective profiles of novel 1,4-dihydropyridine derivatives for the treatment of Alzheimer's disease. *Antioxidants, 9*, 650. https://doi.org/10.3390/antiox9080650

Neves, G., Menegatti, R., Antonio, C. B., Grazziottin, L. R., Vieira, R. O., Rates, S. M.K., Noël, F., Barreiro, E. J., & Fraga, C. A. M. (2010). Searching for multi-target antipsychotics: Discovery of orally active heterocyclic *N*-phenylpiperazine ligands of D_2-like and 5-HT_{1A} receptors. *Bioorganic & Medicinal Chemistry, 18*, 1925–1935. https://doi.org/10.1016/j.bmc.2010.01.040

Nirogi, R., Mohammed, A. R., Shinde, A. K., Bogaraju, N., Gagginapalli, S. R., Ravella, S. R., Kota, L., Bhyrapuneni, G., Muddana, N. R., Benade, V., Palacharla, R. C., Jayarajan, P., Subramanian, R., & Goyal, V. K (2015). Synthesis and SAR of imidazo[1,5-*a*]pyridine derivatives as 5-HT4 receptor partial agonists for the treatment of cognitive disorders associated with Alzheimer's disease. *European Journal of Medicinal Chemistry, 103*, 289–301. https://doi.org/10.1016/j.ejmech.2015.08.051

Nomura, M., Ohuchi, M., Sakamoto, Y., Kudo, K., & Yaku, K. et al (2023). Niacin restriction with NAMPT-inhibition is synthetic lethal to neuroendocrine carcinoma. *Nature Communications, 14*, 8095. https://doi.org/10.1038/s41467-023-43630-3

Ruan, S., Li, J., Ruan, H., Xia, Q., Hou, X., Wang, Z., Guo, T., Zhu, C., Feng, N., & Zhang, Y (2024). Microneedle-mediated nose-to-brain drug delivery for improved Alzheimer's disease treatment. *Journal of Controlled Release, 366*, 712–731. https://doi.org/10.1016/j.jconrel.2024.01.013

Shen, D., Li, X., Guo, S., Xie, X., & Nan, F (2023). Design and synthesis studies of α-methylene-γ-butyrolactones antagonists of GPR52. *Chinese Journal of Organic Chemistry, 43*, 3916–3929. https://doi.org/10.6023/cjoc202303005

Simuni, T., & Krainc, D (2018). High priority publications on Parkinson's disease in 2017. *The Lancet Neurology*, 2018, *17*, 8–10. https://doi.org/10.1016/S1474-4422(17)30416-7

Stopschinski, B. E., & Diamond. M. I. (2017). The prion model for progression and diversity of neurodegenerative diseases. *The Lancet Neurology, 16*, 323. https://doi.org/10.1016/S1474-4422(17)30037-6

Turdi, S., Guo, R., Huff, A. F., Wolf, E. M., Culver, B., & Ren, J (2009). Cardiomyocyte contractile dysfunction in the APPswe/PS1dE9 mouse model of Alzheimer's disease. *PLoS One, 4*, e6033. https://doi.org/10.1371/journal.pone.0006033

Vitaku, E., Smith, D. T., & Njardarson, J. T (2014). Analysis of the structural diversity, substitution patterns, and frequency of nitrogen heterocycles among U.S. FDA approved pharmaceuticals. *Journal of Medicinal Chemistry, 57*, 10257–10274. https://doi.org/10.1021/jm501100b

Xun, Q.-Q., Zhang, J., Li, Y.-P., Li, Y., Ma, Y.-Y., Chen, Z.-B., Ding, L.-P., & Shi, X.-L (2025). Synthesis and biological evaluation of novel pyrrolo[2,3-*b*]pyridine derivatives as potent GSK-3β inhibitors for treating Alzheimer's disease. *European Journal of Medicinal Chemistry, 285*, 117236. https://doi.org/10.1016/j.ejmech.2025.117236

Youn, K., Park, J.-H., Lee, J., Jeong, W.-S., Ho, C.-T., & Jun, M. (2016). The identification of biochanin a as a potent and selective β-site app-cleaving enzyme 1 (Bace1) inhibitor. *Nutrients*, *8*, 637. https://doi.org/10.3390/nu8100637

Zheng, H., Cai, A., Shu, Q., Niu, Y., Xu, P., Li, C., Lin, L., & Gao, H. (2019). Tissue-specific metabolomics analysis identifies the liver as a major organ of metabolic disorders in amyloid precursor protein/presenilin 1 mice of Alzheimer's disease. *Journal of Proteome Research*, *18*, 1218–1227. https://doi.org/10.1021/acs.jproteome.8b00847

Zhou, X.-Y., Zhang, M., Liu, Z., He, J.-H., & Wang, X.-C. (2022). C3-selective trifluorome-thylthiolation and difluoromethylthiolation of pyridines and pyridine drugs via dihydropyridine intermediates. *Journal of the American Chemical Society*, *144*, 14463–14470. https://doi.org/10.1021/jacs.2c06776

Zhu, Z., Yang, T., Zhang, L., Liu, L., Yin, E., Zhang, C., Guo, Z., Xu, C., & Wang, X. (2019). Inhibiting Aβ toxicity in Alzheimer's disease by a pyridine amine derivative. *European Journal of Medicinal Chemistry*, *168*, 330–339. https://doi.org/10.1016/j.ejmech.2019.02.052

5 Pyrimidine-Based Compounds in Neurodegenerative Disease Therapy

Kamlesh Sharma

5.1 INTRODUCTION

Neurodegenerative diseases (NDs) comprise a broad class of progressive conditions marked by the ongoing deterioration and death of nerve cells, resulting in the deterioration of cognitive, sensory, and motor functions across various brain regions (Pathak et al., 2022). Globally, NDs constitute one of the most serious health challenges, contributing significantly to a substantial proportion of disability and death. The World Health Organization (WHO) reports that these conditions are among the second most common causes of illness and mortality globally (Zahra et al., 2019). The most prevalent NDs include Parkinson's disease (PD), Alzheimer's disease (AD), diabetic neuropathy (DN), Huntington's disease (HD), amyotrophic lateral sclerosis (ALS), and multiple sclerosis (MS) (Nishit et al., 2022).

A range of pharmacological agents has been approved for symptomatic management of these disorders. For instance, drugs such as levodopa (Poewe et al., 2010), pramipexole (Aiken, 2007), rasagiline (Mandel et al., 2005), and entacapone (Schrag, 2005) are used in PD, while rivastigmine (Spencer & Noble, 1998), donepezil (Seltzer, 2005), memantine (Jarvis & Figgitt, 2003), and galantamine (Prvulovic et al., 2010) are commonly prescribed for AD. Similarly, riluzole (Wokke, 1996) and edaravone (Jaiswal, 2019) are used for ALS, and pregabalin (Mathieson et al., 2020) for DN. The drugs cladribine (Sipe, 2005) and teriflunomide (Cada et al., 2013) are indicated for MS, and xenazine (Yero & Rey, 2008) is indicated for HD.

Despite the availability of these therapeutic agents, their clinical effectiveness remains suboptimal. Many of them merely provide symptomatic relief without halting disease progression, often displaying poor target specificity and undesirable side effects (Sharma et al., 2014). This scenario highlights the urgent need for innovative molecules capable of modulating key neurodegenerative pathways with enhanced potency and selectivity. Among the diverse chemical scaffolds investigated, pyrimidine has emerged as a versatile and pharmacologically active heterocycle with remarkable therapeutic potential. Pyrimidine-based compounds have been reported to improve cognitive performance, prevent neuronal damage, and modulate neuroprotective mechanisms (Das et al., 2022).

In light of these findings, the present chapter focuses on the structural versatility, pharmacological relevance, and therapeutic potential of pyrimidine derivatives in the management of NDs specifically AD, PD, HD, ALS, DN, and MS.

5.2 THERAPEUTIC APPLICATION OF PYRIMIDINE ANALOGUES AGAINST NEURODEGENERATIVE DISEASES

5.2.1 Parkinson's Disease

PD is a long-lasting neurological condition marked by several pathological mechanisms, including abnormal buildup of α-synuclein, disruptions in neuronal networks, abnormal iron accumulation, mitochondrial impairment, and neuroinflammation (Raj et al., 2021). Given the limited therapeutic

DOI: 10.1201/9781003772019-5

efficacy of currently available antiparkinsonian drugs, there remains a critical need to develop novel and more potent agents capable of targeting the multiple underlying factors contributing to deterioration in PD.

Qiu *et al.* developed an innovative deep-red emitting fluorogenic probe based on pyrimidine derivatives, to monitor H_2O_2 levels in PD-related Drosophila brain tissues. Experimental results revealed that compound **1** (Figure 5.1) exhibited remarkable selectivity for H_2O_2 over other reactive oxygen/nitrogen species together with various amino acids and anions. Confocal fluorescence analysis further confirmed that compound **1** displayed strong responsiveness to hydrogen peroxide within both Drosophila cells and brain tissue. Overall, the findings indicated that this pyrimidine-derived deep-red fluorogenic probe holds significant potential as a therapeutic and diagnostic tool for PD (Qiu et al., 2019).

Kumar and co-workers developed and prepared a range of pyrimidine derivatives with diverse substituents at the 2, 4, and 6 positions to evaluate their potential as monoamine oxidase-B (MAO-B) inhibitors for PD therapy. Among the prepared molecules, **2a** (Figure 5.1) showed significant inhibition of MAO-B ($IC_{50} = 0.38 \pm 0.02$ µM). The introduction of methoxy substituents at the C–4 and C–3 sites of ring B in **2b** enhanced its specificity for MAO-B ($IC_{50} = 0.51 \pm 0.04$ µM). Similarly, **2c** exhibited strong and selective inhibition, achieving a half-maximal inhibitory concentration of 0.48 ± 0.06 µM. Docking simulations further supported the observed bioactivity, revealing favorable binding of the active compounds within the MAO-B binding site. Collectively, these findings suggest that the synthesized pyrimidine derivatives represent promising scaffolds for the development of novel therapeutics for PD (Kumar, Kumar, et al., 2018). Adenosine A_2A receptor-targeting imaging agents have also emerged as another promising approach in PD management.

Vala and colleagues developed and investigated a new set of iodine-123 and fluorine-18 labeled pyrazolo[4,3-e]-1,2,4-triazolo[1,5-c]pyrimidine compounds as potential SPECT and PET imaging agents for targeting brain adenosine A_2A receptor sites. The *in vitro* binding assays revealed that all synthesized ligands exhibited strong affinity toward recombinant human A_2A receptors within the nanomolar range. Among them, [^{123}I]2h (**3b**) and [^{18}F]1b (Figure 5.2) demonstrated

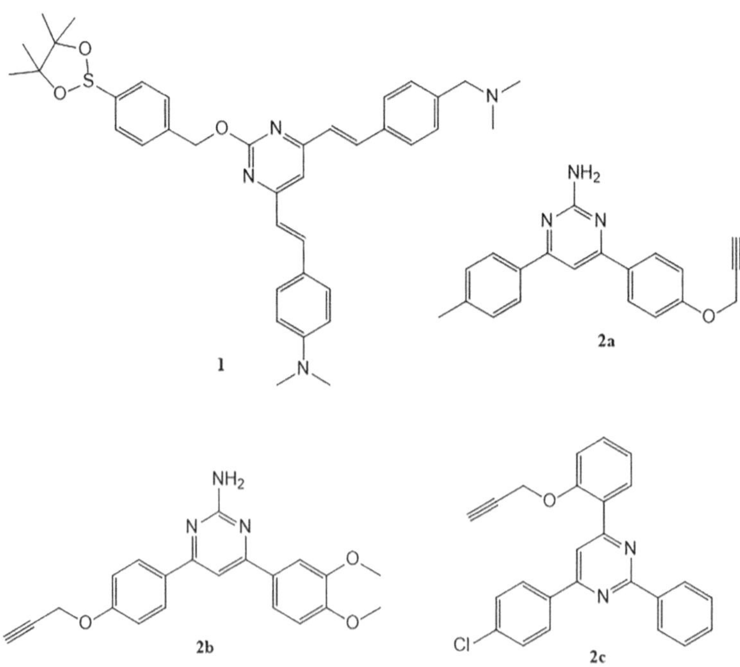

FIGURE 5.1 Structures of compounds **1–2** with potential anti-Parkinson's properties are shown.

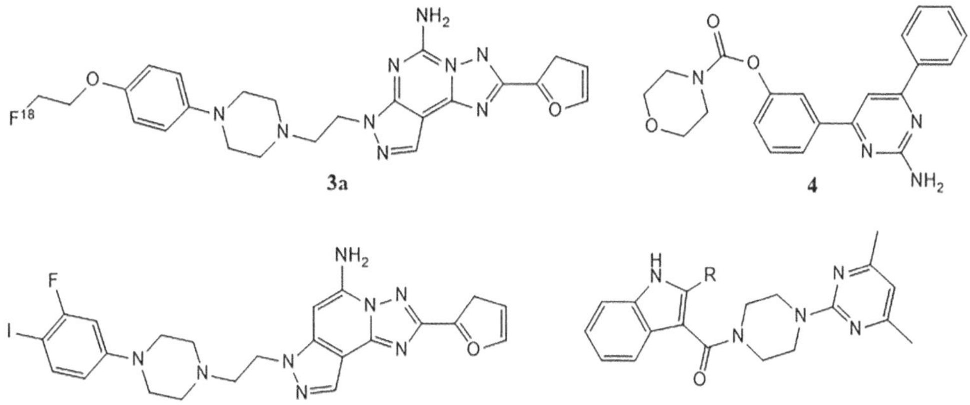

3a 4 3b

5a, R = H
5b, R = Me

FIGURE 5.2 Chemical structure of compounds (**3–5**) as antiparkinsonian agent.

the highest standardized uptake values (SUVs) of 5.5 and 4.1, respectively. Subsequent in vivo evaluations confirmed that both compounds function effectively as PET and SPECT radiotracers, underscoring their potential application in the imaging and diagnosis of PD (Vala et al., 2016).

Robinson *et al.* reported the synthesis and biological evaluation of multiple 2-amino-4,6-diphenylpyrimidine derivatives (Figure 5.2) bearing carbamate substituents as inhibitors targeting adenosine A_2A and A_1 receptor subtypes for PD therapy. The structure–activity relationship (SAR) evaluation revealed that both the size and characteristics of the substituents on the phenyl rings had a greater influence on A_2A receptor affinity than on A_1 receptor affinity. The synthesized compounds exhibited strong binding affinities, displaying A_1K_i values of 0.175–10.7 nM and A_2AK_i values between 1.58 and 451 nM. Among these, compound **4** demonstrated the most promising in vivo activity, highlighting its potential as an effective candidate for the treatment of PD (Robinson et al., 2016).

Shao and colleagues developed a new class of indole–piperazine–pyrimidine derivatives as dual inhibitors targeting human adenosine A_2A and dopamine D_2 receptors for PD treatment. Structural analysis revealed that incorporating a carbonyl linker between the piperazine and indole rings enhanced affinity toward A_2A receptor. Compounds **5a** and **5b** (Figure 5.2) exhibited moderate binding affinities for the human A_2A adenosine receptor, displaying K_i values from 8.7 to 11.2 μM through radioligand-based competition tests. Artificial membrane permeability studies further confirmed that both derivatives displayed significant binding affinity for the dopamine D_2 receptor, with EC_{50} values between 22.50 and 40.2 μM.

5.2.2 Alzheimer's Disease

AD is an age-associated neurodegenerative condition that leads to a severe form of dementia, primarily affecting older population (Akinyemi et al., 2013). It results from progressive deterioration of neurons in the brain, causing impairments in cognition and daily functioning (Gaugler et al., 2022). The hallmark pathological characteristics of AD involve the buildup of amyloid-β (Aβ) plaques and the development of neurofibrillary tangles formed from abnormally phosphorylated tau protein, and a pronounced loss of cholinergic neurotransmission (Majdi et al., 2020). Given the complexity and heterogeneity of its pathophysiology, there is an urgent requirement to develop improved therapeutic strategies and agents to elucidate and target the underlying disease mechanisms. In this context, pyrimidine derivatives have been extensively investigated as potential therapeutic agents for AD.

Han and colleagues synthesized and tested a novel library of 2-(2-oxoethyl)pyrimidine-5-carboxamide compounds as prospective acetylcholinesterase (AChE) inhibitors for AD treatment. Within this series, compound **7** (Figure 5.3) exhibited significant inhibition of AChE, with an IC_{50} of 0.88 ± 0.78 μM. The SAR analysis revealed that the incorporation of p-nitrophenyl and 1,2-dimethoxybenzene in compound **7** resulted in greater AChE inhibitory effectiveness than the control drug huperzine A. Furthermore, docking simulations suggested that compound **7** was able to bind efficiently to both the peripheral anionic site (PAS) and catalytic active site (CAS) of AChE (Han et al., 2022).

Jain and co-workers designed and synthesized a set of 2,4,6-substituted pyrimidine compounds aimed at inhibiting β-secretase (BACE-1) as prospective treatments for AD. These compounds exhibited favorable oral bioavailability and demonstrated efficient penetration across the blood–brain barrier (BBB). FRET-based in vitro assays revealed that compound **8** (Figure 5.3) was the strongest inhibitor, with 73.91% activity and an IC_{50} value of 6.92 μM. Moreover, molecular docking results identified compound **8** as the most active BACE-1 inhibitor within the series. The SAR evaluation indicated that the addition of a bulky benzyloxy substituent at the meta position of one aromatic ring, together with ortho- and para-dichloro groups on the other ring, significantly enhanced BACE-1 inhibition (Jain et al., 2021).

Manzoor et al. reported the design, synthesis, and biological evaluation of a new series of phenyl sulfonyl-pyrimidine carboxylate hybrids as multifunctional candidates for AD therapy. Among them, compounds **9** and **10** (Figure 5.3) demonstrated strong inhibitory effects against AChE with IC_{50} values of 47.33 ± 0.02 and 51.36 ± 0.04 nM, respectively, and against butyrylcholinesterase (BuChE) with IC_{50} values of 159.43 ± 0.72 and 153.3 ± 0.74 nM. Moreover, SAR investigations revealed that the presence of dimethoxy and chloro groups in these molecules enhanced their AChE inhibitory potential in comparison with the reference agent donepezil. Both molecules **10** and **9** exhibited significant neuroprotective effects in neurotoxicity assays using MC-65 cell lines and showed high BBB penetration, with permeability coefficients (Pe) of 23.7×10^{-6} and 24.3×10^{-6} cm/s. Molecular docking studies further confirmed that these derivatives effectively interacted with the active region of AChE, highlighting their promise as lead structures for AD drug development (Manzoor et al., 2021).

Kumar and co-workers prepared and biologically assessed a set of 4,6-diphenylpyrimidine compounds (Figure 5.4) as prospective therapeutics for AD. Among the tested compounds, **11a**

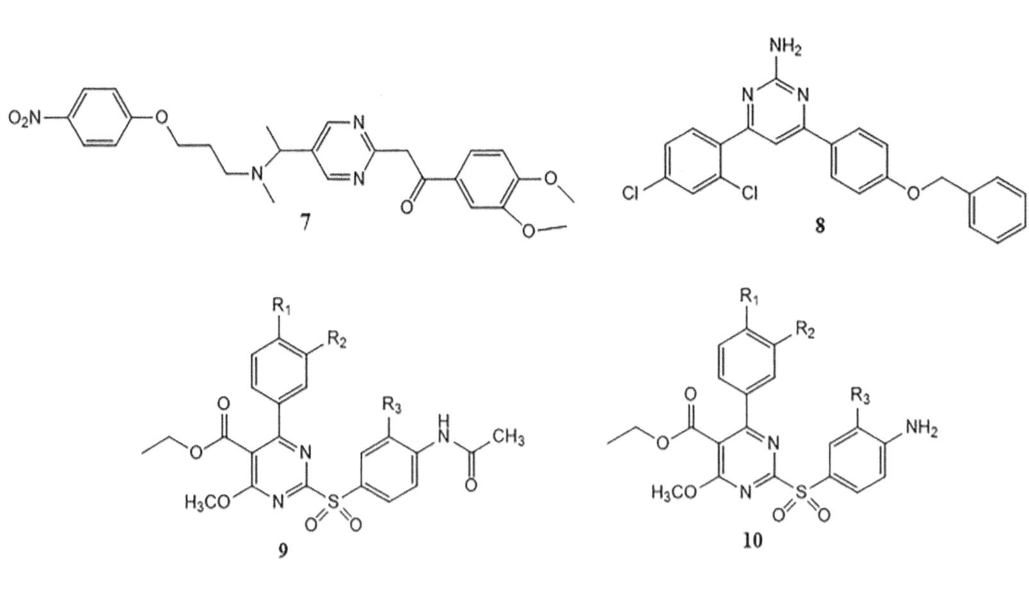

FIGURE 5.3 Structures of molecules **7–10** assessed as potential AD therapeutics.

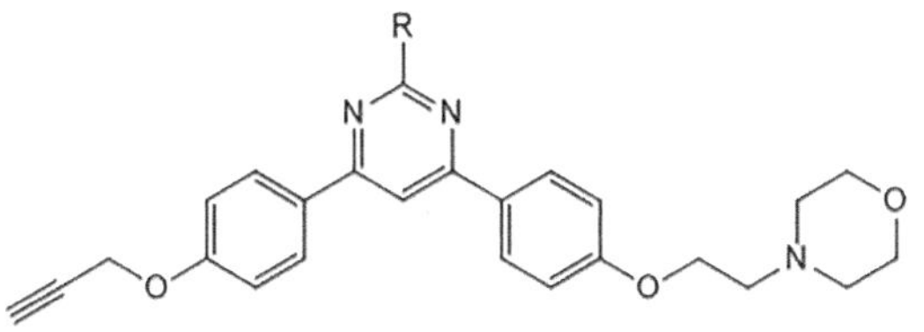

11

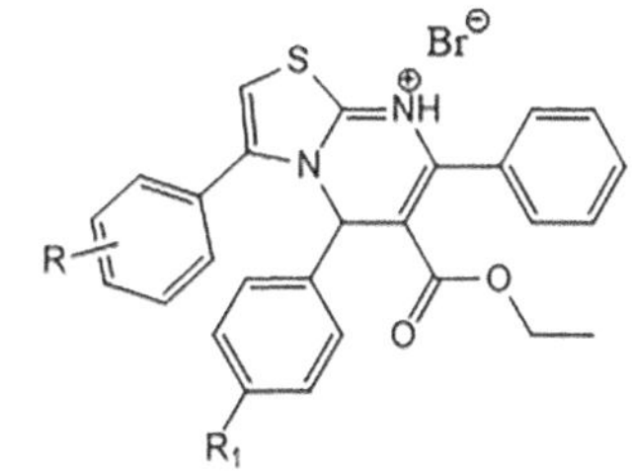

12a, R_1 = H, R= H ; **12b**, R_1 = H, R = $4OCH_3$
12c, R_1 = H, R= F ; **12d**, R_1 = $4OCH_3$, R = H

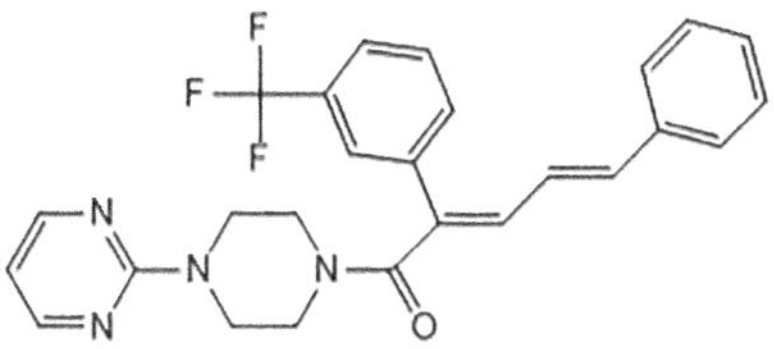

13

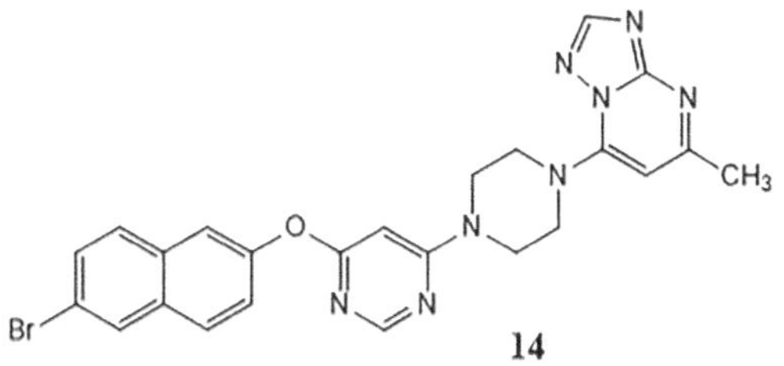

14

FIGURE 5.4 The structures of compounds **11–14** assessed as potential anti-Alzheimer's agents are shown.

exhibited the strongest inhibitory activity toward AChE, BuChE, and MAO-A, with IC_{50} values of 30.46 ± 0.23, 0.66 ± 0.03, and 18.34 ± 0.38 nM, respectively. The derivative **11b** also showed notable inhibition against AChE (IC_{50} = 18.92 ± 0.29 nM) and MAO-A (IC_{50} = 28.33 ± 3.22 nM). The SAR analysis suggested that the presence of an ethyl chain linked to a morpholine ring and a propargyl group at the para positions of both phenyl rings enhanced inhibitory potency toward the MAO-A isoform and AChE (Kumar, Dwivedi, et al., 2018).

Mahgoub and colleagues prepared a group of thiazolo[3,2-a]pyrimidine derivatives (**12a–d,** Figure 5.4) and evaluated their ability to inhibit hAChE using an Ellman-type colorimetric assay. All derivatives demonstrated strong inhibition, with IC_{50} values around 1 μM. Molecular docking analyses further supported these results, showing a stable association of the studied derivatives within the hAChE active site. Collectively, these findings suggest that thiazolo[3,2-a] pyrimidine scaffolds represent promising lead candidates for developing AD therapies (Mahgoub et al., 2019).

Fang and co-workers synthesized and evaluated 28 pyrimidine-based compounds for their anti-neuroinflammatory potential in BV2 microglial cells as prospective AD therapeutics. Among these, compound **13** (Figure 5.4) demonstrated the strongest inhibitory effects on interleukin-1β (IL-1β), nitric oxide (NO), and tumor necrosis factor-α (TNF-α), with IC_{50} values of 0.5, 1.0, and 2.6, respectively. The SAR analysis indicated that the introduction of a CF_3 moiety at the 3-position contributed significantly to the enhanced inhibition of NO production. Additionally, results from a parallel artificial membrane permeability assay revealed that this compound exhibited superior BBB permeability compared to the reference drug resveratrol, suggesting its suitability as a promising agent for AD management (Fang et al., 2018).

Kumar and colleagues developed, synthesized, and biologically evaluated a set of pyrimidine–triazolopyrimidine analogues for their *in vitro* inhibition of AChE in the context of AD. Among the synthesized molecules, compound **14** (Figure 5.4), bearing a 6-bromo-2-naphthol unit and a piperazine-linked triazolopyrimidine fragment, exhibited potent AChE inhibition with IC_{50} of 0.161 ± 0.04 μM. Moreover, the docking analysis confirmed that compound **14** effectively interacted with both the PAS and CAS of AChE, establishing it as one of the most active inhibitors in the series (Kumar, Gill, et al., 2018).

5.2.3 Huntington's Disease

HD is an uncommon, progressively worsening, and ultimately fatal neurological condition marked by impaired movement, behavioral changes, and psychiatric symptoms (Ross & Tabrizi, 2011). The disorder is caused by an unusual expansion of CAG trinucleotide repeats within the *Huntingtin* (*HTT*) gene, which follows an autosomal dominant pattern of inheritance (Arning & Nguyen, 2022). Due to its chronic and debilitating progression, there is a pressing need to discover therapeutic agents that can effectively target the molecular mechanisms underlying HD.

One of the major metabolic pathway linked to HD involves kynurenine processing, which represents the main mechanism for tryptophan breakdown in mammals (Tutakhail et al., 2020). Modulating the enzymes of this pathway presents a promising therapeutic strategy for HD management. Toledo-Sherman and co-workers designed and synthesized a set of pyrimidine-based carboxylic acid compounds that showed strong inhibitory activity against kynurenine monooxygenase (KMO). The SAR analysis revealed that these compounds effectively inhibited KMO activity in both cellular and biochemical assays, with notable selectivity over other enzymes in the kynurenine pathway. Among them, compound **15** (Figure 5.5) emerged as the most potent inhibitor of human (hKMO), mouse (mKMO), and rat KMO, exhibiting IC_{50} values of 0.0005, 0.0051, and 0.004 μM, respectively (Toledo-Sherman et al., 2015).

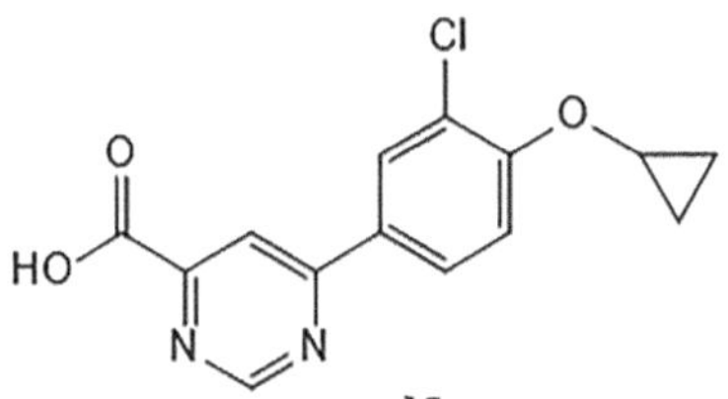

15

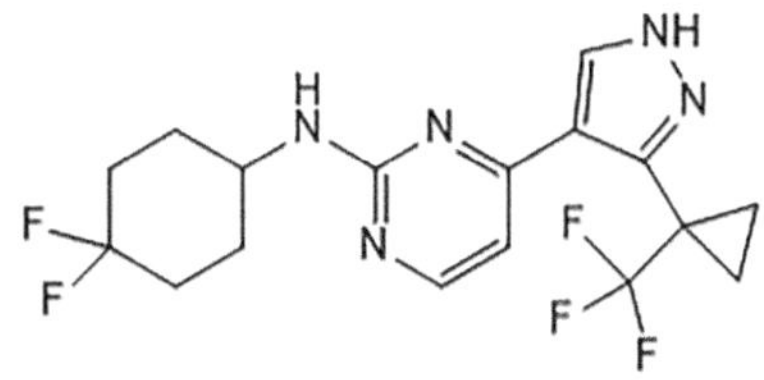

16

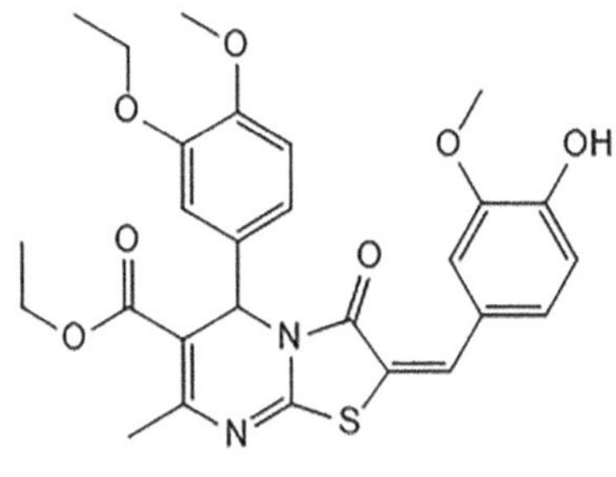

17

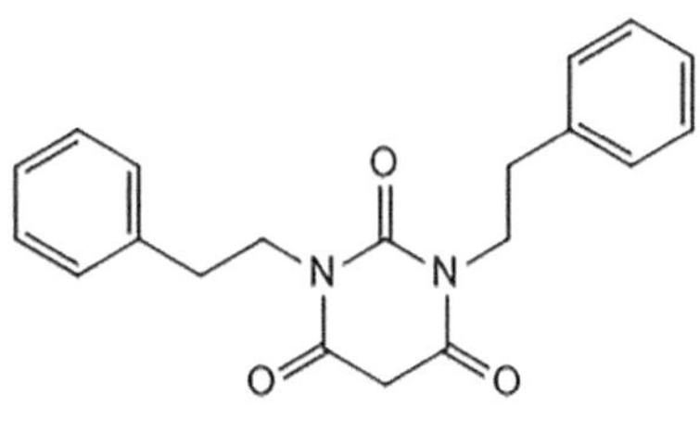

18

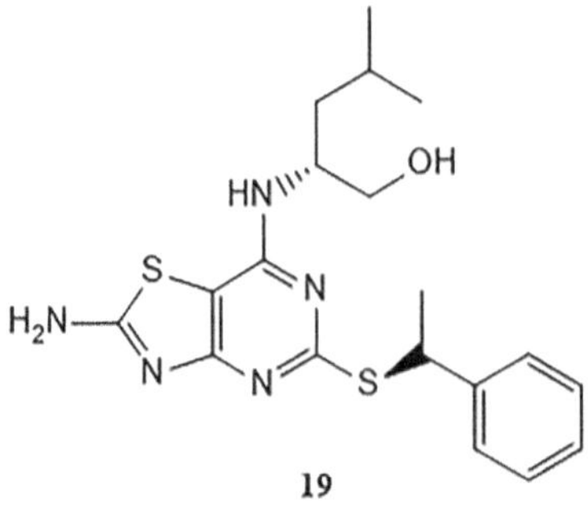

19

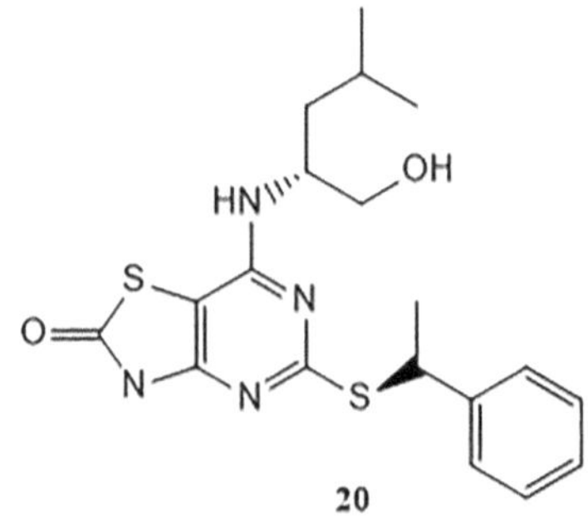

20

FIGURE 5.5 Chemical structure of compounds **15–20** for the management of HD, ALS, MS, and DN.

Another critical signaling cascade involved in HD is the c-Jun N-terminal kinase (JNK) signaling pathway, which becomes activated in response to mutant HTT and promotes neuronal damage and degeneration (Lontay et al., 2020). Wityak and colleagues examined a group of pyrimidine-derived pan-JNK inhibitors through X-ray crystallographic studies and computational modeling approaches to explore their therapeutic potential against HD. Among the synthesized analogues, compound **16** (Figure 5.5) demonstrated the most potent activity in ADME evaluations. When compound **16** was given together with the pan–cytochrome P450 blocker 1-aminobenzotriazole, its concentration in the brain rose to levels sufficient for therapeutic activity, leading to clear effects on pharmacodynamic markers in mice (Wityak et al., 2015).

5.2.4 Diabetic Neuropathy

DN is a frequent and disabling neurological complication of diabetes, characterized by progressive nerve malnutrition and irreversible peripheral nerve damage (Khan et al., 2025). Clinically, DN manifests as sensory loss, pain, tingling, numbness, and motor dysfunction, substantially impairing patients' quality of life. Despite its high prevalence, existing treatments mainly provide symptomatic relief and limited efficacy in halting disease progression. This lack of effective therapies emphasizes the importance of identifying new biological targets and developing mechanism-based treatment strategies for effective management of DN.

The Kv channels are essential regulators of neuronal physiology, playing critical roles in controlling neuronal excitability, action potential repolarization, synaptic integration, and neurotransmitter release. Dysregulation or aberrant activation of specific Kv channel subtypes has been increasingly associated with the progression of various neurodegenerative conditions, such as DN (Tussupbekova, 2025). In particular, Kv2.1 channels are known to influence neuronal survival and mitochondrial homeostasis, making them attractive targets for therapeutic intervention. Accordingly, selective inhibition of Kv channels has emerged as a promising strategy to alleviate neuropathic symptoms and prevent neuronal damage in diabetic conditions.

In this context, Zhu and co-workers evaluated SP6616 (compound **17**, Figure 5.5), a selective Kv2.1 channel inhibitor incorporating a pyrimidine scaffold, for its neuroprotective potential in DN. *In vivo* testing in diabetic mouse models indicated that treatment with SP6616 significantly improved mechanical sensitivity, increased motor nerve conduction speed, and restored thermal responsiveness, indicating substantial functional recovery of peripheral nerves. Mechanistic investigations further revealed that SP6616 improved mitochondrial activity by modulating the Kv2.1/CaMKKβ/AMPK/PGC-1α signaling pathway, which is crucial for maintaining mitochondrial biogenesis and energy metabolism (Zhu et al., 2020). Collectively, these findings highlight SP6616 and pyrimidine-based Kv2.1 inhibitors as promising drug candidates for DN therapy.

5.2.5 Amyotrophic Lateral Sclerosis

ALS, commonly known as Lou Gehrig's disease, is a severe and continuously progressive neurological condition marked by the loss of both upper and lower motor neurons in the brain and spinal cord (Talbott et al., 2016). The loss of these motor neurons leads to a gradual decline in voluntary muscle control, manifesting clinically as muscle weakness, stiffness, cramps, and fasciculations, which progressively worsen over time. As the disease progresses, patients develop severe paralysis, including weakening of the respiratory muscles, and respiratory failure becomes the primary cause of death. The clinical course of ALS is typically rapid, and the majority of affected individuals die within two to four years following the onset of symptoms, although a small subset of patients may survive longer (Hardiman et al., 2011). Globally, ALS has an estimated annual incidence of approximately 1.5–2.0 cases per 100,000 people, indicating a considerable clinical and social burden despite being a rare condition (Wolfson et al., 2023).

The pathogenesis of ALS is highly complex and multifactorial, involving a convergence of several interrelated molecular and cellular mechanisms. These include glutamate-mediated excitotoxicity, mitochondrial dysfunction, oxidative stress, chronic neuroinflammation, dysregulation of RNA metabolism, abnormal protein aggregation and misfolding, and defects in intracellular as well as nucleocytoplasmic transport (Ramesh & Pandey, 2017). Among these pathogenic pathways, oxidative stress is considered particularly critical, particularly in inherited forms of ALS associated with mutations in the superoxide dismutase 1 (SOD1) gene. Mutant or misfolded SOD1 proteins acquire toxic properties that promote excessive generation of ROS, leading to mitochondrial impairment, oxidative damage, and eventual motor neuron death.

Despite advances in understanding ALS pathophysiology, current therapeutic options remain largely palliative and offer only modest survival benefits. Consequently, there is a pressing need to create new disease-modifying treatments that directly target the underlying molecular drivers of neurodegeneration. In this context, Xia and co-workers conducted a comprehensive study of pyrimidine-2,4,6-trione (PYT) derivatives and structurally related analogues as potential inhibitors of SOD1 for ALS treatment. Among the compounds synthesized and evaluated, compound **18** (Figure 5.5) emerged as a particularly promising lead. This compound exhibited strong pharmacological potency with an EC_{50} value of 1.68 μM, indicating effective inhibition of SOD1 activity. Importantly, it also demonstrated efficient BBB penetration, as reflected by a favorable blood-to-brain concentration ratio of 3:1 (150:50 μM), along with high oral bioavailability (98%) and minimal toxicity (Xia et al., 2011). Collectively, these findings highlight PYT scaffolds as valuable and promising lead structures for developing novel agents intended to slow or modify ALS progression.

5.2.6 Multiple Sclerosis

MS is a long-term, immune-driven disease of the central nervous system marked by ongoing inflammation, loss of myelin, and gradual neuronal degeneration (Mey et al., 2023). These pathological processes disrupt neuronal signal transmission, resulting in diverse neurological manifestations such as muscle weakness, loss of coordination, sensory deficits, and, in advanced stages, permanent paralysis (Machado et al., 2020). Despite disease-modifying treatments, many individuals still show ongoing disease progression, and there are treatment-related limitations, highlighting the critical need for innovative therapeutic strategies that more effectively target the mechanisms underlying neuroinflammation and neuronal damage.

Chemokine signaling has been recognized as a key factor in the development of MS, particularly through its involvement in the activation of microglia and immune cells, trafficking within the CNS. The CX3CR1, a chemokine receptor predominantly expressed by microglia, has a key function in regulating neuroimmune interactions and inflammatory responses. Accumulating evidence indicates that dysregulation of CX3CR1 signaling contributes to chronic neuroinflammation, neuronal degeneration, and motor impairment, whereas pharmacological inhibition of this receptor can reduce neuroinflammatory damage and improve neurological outcomes in experimental models of MS (Ridderstad Wollberg et al., 2014).

Based on this concept, Karlström and colleagues developed and prepared two related libraries (series A and B) of 7-amino-5-thio-thiazolo[4,5-d]pyrimidine compounds as new CX3CR1 antagonists with potential application in MS therapy. The SAR investigations demonstrated that targeted chemical modifications were critical for optimizing biological activity. In particular, the incorporation of a leucinal group at the 7-position together with α-methyl-branched benzyl substituents at the 5-position significantly enhanced receptor affinity, selectivity, and overall physicochemical properties. These structural optimizations led to the identification of compound **19** from series A and compound **20** from series B as promising lead molecules with improved drug-like characteristics (Karlstrom et al., 2013). The structures of compounds **19** and **20** are given in Figure 5.5.

Overall, these findings emphasize the potential of CX3CR1 inhibitors in MS treatment. By targeting microglia-mediated neuroinflammation, such compounds offer a promising approach for mitigating disease progression, reducing neuronal damage, and improving motor function in patients with MS.

5.3 FUTURE ASPECTS

ALS, also called Lou Gehrig's disease, is a fast-progressing neurodegenerative condition marked by the loss of both lower and upper motor neurons in the central nervous system (Tsvetkova et al., 2023). The degeneration of these neurons results in muscle weakness, cramps, twitching, and progressive paralysis, and most patients die within two to four years after symptoms begin, primarily because of respiratory failure (Vita et al., 2020). Although ALS is relatively rare, it has a significant clinical impact, with an estimated yearly incidence of about 1.5–2.0 per 100,000 individuals (Wolfson et al., 2023).

The pathogenesis of ALS is multifactorial and involves a complex interplay of mechanisms, including mitochondrial dysfunction, oxidative stress, glutamate-mediated excitotoxicity, neuroinflammation, impaired RNA metabolism, protein misfolding and buildup, together with impaired intracellular and nucleocytoplasmic trafficking (Ramesh & Pandey, 2017). Among these, oxidative stress is particularly prominent, most prominently in hereditary ALS caused by SOD1 mutations, which contribute to neuronal degeneration and disease advancement.

Given the lack of curative therapies, there is a critical need for novel disease-modifying agents. In this context, Xia et al. investigated PYT derivatives and related analogues as potential SOD1-targeted inhibitors for ALS therapy. Notably, compound **18** (Figure 5.5) demonstrated strong potency (EC_{50} = 1.68 μM), effective BBB penetration with a favorable blood-to-brain concentration ratio of 3:1 (150:50 μM), excellent oral bioavailability (98%) with minimal toxicity (Xia et al., 2011). These findings highlight PYT scaffolds as promising leads for developing new therapeutic agents against ALS.

5.4 CONCLUSION

This review underscores the substantial promise of pyrimidine analogues as a versatile and powerful class of therapeutic candidates for the management of neurodegenerative disorders. An integrated analysis of the available biological and pharmacological data reveals that pyrimidine-based compounds act as potent multi-target modulators, capable of intervening at several critical biochemical and molecular pathways implicated in neurodegeneration.

Importantly, these compounds exhibit inhibitory or modulatory activity against a wide spectrum of disease-relevant targets, including cholinesterases (ChE) involved in cholinergic dysfunction, Aβ aggregation associated with protein misfolding and plaque formation, and MAO-B, a key enzyme linked to oxidative stress and dopaminergic neuronal loss. In addition, their activity against KMO highlights a capacity to regulate neuroinflammatory and excitotoxic pathways within the kynurenine pathway. Modulation of SOD1 further suggests a role in mitigating oxidative damage, while interactions with Kv channels indicate potential to normalize neuronal excitability and synaptic signaling. The ability of certain pyrimidine analogues to target CX3CR1, a chemokine receptor central to neuron–microglia communication, also emphasizes their relevance in controlling neuroinflammation and immune-mediated neurotoxicity.

Collectively, these findings position pyrimidine scaffolds as structurally adaptable and pharmacologically rich platforms capable of addressing the multifactorial nature of NDs. Unlike single-target agents, pyrimidine-based multi-target-directed ligands (MTDLs) offer the strategic advantage of simultaneously modulating interconnected pathological processes such as neurotransmitter imbalance, protein aggregation, mitochondrial dysfunction, oxidative stress, and neuroinflammation. Consequently, pyrimidine analogues provide a robust and rational framework for the design and development of next-generation neurotherapeutics, with the potential to achieve enhanced efficacy, disease-modifying effects, and improved clinical outcomes in complex neurodegenerative conditions.

ACKNOWLEDGMENT

The author is thankful to the School of Basic and Applied Sciences, SGT University, Gurugram, Haryana, INDIA, for providing essential facilities.

REFERENCES

Aiken, C. B. (2007). Pramipexole in psychiatry: A systematic review of the literature. *Journal of Clinical Psychiatry*, *68*(8), 1230–1236.

Arning, L., & Nguyen, H. P. (2022). Huntington disease update: New insights into the role of repeat instability in disease pathogenesis. *Medizinische Genetik*, *33*(4), 293–300.

Cada, D. J., Demaris, K., Levien, T. L., & Baker, D. E. (2013). Teriflunomide. *Hospital Pharmacy*, *48*(3), 231–240.

Das, S., Akbar, S., Ahmed, B., Dewangan, R. P., Iqubal, A., Pottoo, F. H., & Joseph, A. (2022). Structural activity relationship based medicinal perspectives of pyrimidine derivatives as anti-Alzheimer's agents: A comprehensive review. *CNS & Neurological Disorders-Drug Targets-CNS & Neurological Disorders)*, *21*(10), 926–939.

Fang, Y., Xia, W., Cheng, B., Hua, P., Zhou, H., Gu, Q., & Xu, J. (2018). Design, synthesis, and biological evaluation of compounds with a new scaffold as anti-neuroinflammatory agents for the treatment of Alzheimer's disease. *European Journal of Medicinal Chemistry*, *149*, 129–138.

Gaugler, J., James, B., Johnson, T., Reimer, J., Solis, M., Weuve, J., ..., Hohman, T. J. (2022). 2022 Alzheimer's disease facts and figures. *Alzheimers & Dementia*, *18*(4), 700–789.

Han, C., Wei, B.-B., Shang, P.-P., Guo, X.-Y., Bai, L.-G., & Ma, Z.-Y. (2022). Design, synthesis and evaluation of 2-(2-oxoethyl) pyrimidine-5-carboxamide derivatives as acetylcholinesterase inhibitors. *Bioorganic & Medicinal Chemistry Letters*, *72*, 128873.

Hardiman, O., Van Den Berg, L. H., & Kiernan, M. C. (2011). Clinical diagnosis and management of amyotrophic lateral sclerosis. *Nature Reviews Neurology*, *7*(11), 639–649.

Jain, P., Wadhwa, P. K., & Jadhav, H. R. (2021). Design, synthesis and evaluation of 2, 4, 6-substituted pyrimidine derivatives as BACE-1 inhibitor: Plausible lead for Alzheimer's disease. *Medicinal Chemistry*, *17*(10), 1194–1206.

Jaiswal, M. K. (2019). Riluzole and edaravone: A tale of two amyotrophic lateral sclerosis drugs. *Medicinal Research Reviews*, *39*(2), 733–748.

Jarvis, B., & Figgitt, D. P. (2003). Memantine. *Drugs & Aging*, *20*(6), 465–476.

Karlstrom, S., Nordvall, G., Sohn, D., Hettman, A., Turek, D., Åhlin, K., ..., Lo-Alfredsson, Y. (2013). Substituted 7-amino-5-thio-thiazolo [4, 5-d] pyrimidines as potent and selective antagonists of the fractalkine receptor (CX3CR1). *Journal of Medicinal Chemistry*, *56*(8), 3177–3190.

Khan, S., Ahmad, S., Khan, M., Lohani, M., Khan, M. S., & Haneef, M. (2025). Diabetic peripheral neuropathy: Navigating controversies and pioneering advances. *Advancements in Life Sciences*, *12*(1), 01–12.

Kumar, B., Dwivedi, A. R., Sarkar, B., Gupta, S. K., Krishnamurthy, S., Mantha, A. K., ..., Kumar, V. (2018). 4, 6-diphenylpyrimidine derivatives as dual inhibitors of monoamine oxidase and acetylcholinesterase for the treatment of Alzheimer's disease. *ACS Chemical Neuroscience*, *10*(1), 252–265.

Kumar, B., Kumar, M., Dwivedi, A. R., & Kumar, V. (2018). Synthesis, biological evaluation and molecular modeling studies of propargyl-containing 2, 4, 6-trisubstituted pyrimidine derivatives as potential anti-Parkinson agents. *ChemMedChem*, *13*(7), 705–712.

Kumar, J., Gill, A., Shaikh, M., Singh, A., Shandilya, A., Jameel, E., ..., Jayaram, B. (2018). Pyrimidine-triazolopyrimidine and pyrimidine-pyridine hybrids as potential acetylcholinesterase inhibitors for Alzheimer's disease. *ChemistrySelect*, *3*(2), 736–747.

Lontay, B., Kiss, A., Virág, L., & Tar, K. (2020). How do post-translational modifications influence the pathomechanistic landscape of Huntington's disease? A comprehensive review. *International Journal of Molecular Sciences*, *21*(12), 4282.

Machado, S. B., Cabral, R., Murade, N., Ares, N. C., Scorcine, C., & Fragoso, Y. D. (2020). Dietary habits in a group of patients with multiple sclerosis are similar to those of healthy control subjects. *Arquivos de neuro-psiquiatria*, *78*, 638–641.

Mahgoub, M. Y., Elmaghraby, A. M., Harb, A.-E. A., Ferreira da Silva, J. L., Justino, G. C., & Marques, M. M (2019). Synthesis, crystal structure, and biological evaluation of fused thiazolo [3, 2-a] pyrimidines as new acetylcholinesterase inhibitors. *Molecules*, *24*(12), 2306.

Majdi, A., Sadigh-Eteghad, S., Rahigh Aghsan, S., Farajdokht, F., Vatandoust, S. M., Namvaran, A., & Mahmoudi, J. (2020). Amyloid-β, tau, and the cholinergic system in Alzheimer's disease: Seeking direction in a tangle of clues. *Reviews in the Neurosciences*, *31*(4), 391–413.

Mandel, S., Weinreb, O., Amit, T., & Youdim, M. B. (2005). Mechanism of neuroprotective action of the anti-Parkinson drug rasagiline and its derivatives. *Brain Research Reviews*, *48*(2), 379–387.

Manzoor, S., Prajapati, S. K., Majumdar, S., Raza, M. K., Gabr, M. T., Kumar, S., ... Krishnamurthy, S. (2021). Discovery of new phenyl sulfonyl-pyrimidine carboxylate derivatives as the potential multi-target drugs with effective anti-Alzheimer's action: Design, synthesis, crystal structure and in-vitro biological evaluation. *European Journal of Medicinal Chemistry*, *215*, 113224.

Mathieson, S., Lin, C.-W. C., Underwood, M., & Eldabe, S. (2020). Pregabalin and gabapentin for pain. *BMJ*, *369*.

Mey, G. M., Mahajan, K. R., & DeSilva, T. M. (2023). Neurodegeneration in multiple sclerosis. *WIREs Mechanisms of Disease*, *15*(1), e1583.

Nishit, P., Kumar, V. S., Ishi, T., Lokesh, A., Cao, H., & Sanjib, B. (2022). Neurodegenerative disorders of Alzheimer, parkinsonism, amyotrophic lateral sclerosis and multiple sclerosis: An early diagnostic approach for precision treatment. *Metabolic Brain Disease*, *37*(1), 67–104.

Akinyemi, R. O., Mukaetova-Ladinska, E. B., Attems, J., Ihara, M., & N. Kalaria (2013). Vascular risk factors and neurodegeneration in ageing related dementias: Alzheimer's disease and vascular dementia. *Current Alzheimer Research*, *10*(6), 642–653.

Pathak, N., Vimal, S. K., Tandon, I., Agrawal, L., Hongyi, C., & Bhattacharyya, S. (2022). Neurodegenerative disorders of Alzheimer, parkinsonism, amyotrophic lateral sclerosis and multiple sclerosis: An early diagnostic approach for precision treatment. *Metabolic Brain Disease*, *37*(1), 67–104.

Poewe, W., Antonini, A., Zijlmans, J. C., Burkhard, P. R., & Vingerhoets, F. (2010). Levodopa in the treatment of Parkinson's disease: An old drug still going strong. *Clinical Interventions in Aging*, 229–238.

Prvulovic, D., Hampel, H., & Pantel, J. (2010). Galantamine for Alzheimer's disease. *Expert Opinion on Drug Metabolism & Toxicology*, *6*(3), 345–354.

Qiu, X., Xin, C., Qin, W., Li, Z., Zhang, D., Zhang, G., ..., Li, L. (2019). A novel pyrimidine based deep-red fluorogenic probe for detecting hydrogen peroxide in Parkinson's disease models. *Talanta*, *199*, 628–633.

Raj, K., Kaur, P., Gupta, G., & Singh, S. (2021). Metals associated neurodegeneration in Parkinson's disease: Insight to physiological, pathological mechanisms and management. *Neuroscience Letters*, *753*, 135873.

Ramesh, N., & Pandey, U. B (2017). Autophagy dysregulation in ALS: When protein aggregates get out of hand. *Frontiers in Molecular Neuroscience*, *10*, 263.

Ridderstad Wollberg, A., Ericsson-Dahlstrand, A., Juréus, A., Ekerot, P., Simon, S., Nilsson, M., ..., Sunnemark, D. (2014). Pharmacological inhibition of the chemokine receptor CX3CR1 attenuates disease in a chronic-relapsing rat model for multiple sclerosis. *Proceedings of the National Academy of Sciences*, *111*(14), 5409–5414.

Robinson, S. J., Petzer, J. P., Rousseau, A. L., Terre'Blanche, G., Petzer, A., & Lourens, A. C. (2016). Carbamate substituted 2-amino-4, 6-diphenylpyrimidines as adenosine receptor antagonists. *Bioorganic & Medicinal Chemistry Letters*, *26*(3), 734–738.

Ross, C. A., & Tabrizi, S. J. (2011). Huntington's disease: From molecular pathogenesis to clinical treatment. *The Lancet Neurology*, *10*(1), 83–98.

Schrag, A. (2005). Entacapone in the treatment of Parkinson's disease. *The Lancet Neurology*, *4*(6), 366–370.

Seltzer, B. (2005). Donepezil: A review. *Expert Opinion on Drug Metabolism & Toxicology*, *1*(3), 527–536.

Sharma, V., Chitranshi, N., & Agarwal, A. K. (2014). Significance and biological importance of pyrimidine in the microbial world. *International Journal of Medicinal Chemistry*, *2014*(1), 202784.

Sipe, J. C. (2005). Cladribine for multiple sclerosis: Review and current status. *Expert Review of Neurotherapeutics*, *5*(6), 721–727.

Spencer, C. M., & Noble, S. (1998). Rivastigmine: A review of its use in Alzheimer's disease. *Drugs & Aging*, *13*(5), 391–411.

Talbott, E. O., Malek, A. M., & Lacomis, D. (2016). The epidemiology of amyotrophic lateral sclerosis. *Handbook of Clinical Neurology*, *138*, 225–238.

Toledo-Sherman, L. M., Prime, M. E., Mrzljak, L., Beconi, M. G., Beresford, A., Brookfield, F. A., ..., Dijkman, U. (2015). Development of a series of aryl pyrimidine kynurenine monooxygenase inhibitors as potential therapeutic agents for the treatment of Huntington's disease. *Journal of Medicinal Chemistry*, *58*(3), 1159–1183.

Tsvetkova, D., Ivanova, S., & Obreshkova, D. (2023). Neurodegenerative multioethiology Lou Gehrig's disease–genetic mutations, pharmacological mechanisms and applications of Rilusole. *International Journal of Pharmaceutical Research and Allied Sciences*, *12*(3-2023), 61–70.

Tussupbekova, G. (2025). Voltage-gated potassium channels (Kv7): Molecular structure, regulation, and role in pathogenesis. *Engineered Science*, *36*, 1632.

Tutakhail, A., Boulet, L., Khabil, S., Nazari, Q. A., Hamid, H., & Coudoré, F. (2020). Neuropathology of kynurenine pathway of tryptophan metabolism. *Current Pharmacology Reports*, *6*(1), 8–23.

Vala, C., Morley, T. J., Zhang, X., Papin, C., Tavares, A. A. S., Lee, H. S., …, Baldwin, R. M (2016). Synthesis and in vivo evaluation of fluorine-18 and iodine-123 pyrazolo [4, 3-e]-1, 2, 4-triazolo [1, 5-c] pyrimidine derivatives as PET and SPECT radiotracers for mapping A2A receptors. *ChemMedChem*, *11*(17), 1936–1943.

Vita, G., Mongini, T., Racca, F., Ruggeri, P., Versaci, A., Vianello, A., & Vita, G. L. (2020). Respiratory emergencies in neurological diseases. *Decision Algorithms for Emergency Neurology*, 413–434.

Wityak, J., McGee, K. F., Conlon, M. P., Song, R. H., Duffy, B. C., Clayton, B., …, Haber, J. (2015). Lead optimization toward proof-of-concept tools for Huntington's disease within a 4-(1 H-pyrazol-4-yl) pyrimidine class of pan-JNK inhibitors. *Journal of Medicinal Chemistry*, *58*(7), 2967–2987.

Wokke, J. (1996). Riluzole. *The Lancet*, *348*(9030), 795–799.

Wolfson, C., Gauvin, D. E., Ishola, F., & Oskoui, M. (2023). Global prevalence and incidence of amyotrophic lateral sclerosis: A systematic review. *Neurology*, *101*(6), e613–e623.

Xia, G., Benmohamed, R., Kim, J., Arvanites, A. C., Morimoto, R. I., Ferrante, R. J., …, Silverman, R. B. (2011). Pyrimidine-2, 4, 6-trione derivatives and their inhibition of mutant SOD1-dependent protein aggregation. Toward a treatment for amyotrophic lateral sclerosis. *Journal of Medicinal Chemistry*, *54*(7), 2409–2421.

Yero, T., & Rey, J. A. (2008). Tetrabenazine (xenazine), an FDA-approved treatment option for Huntington's disease–related chorea. *Pharmacy and Therapeutics*, *33*(12), 690.

Zahra, W., Rai, S. N., Birla, H., Singh, S. S., Dilnashin, H., Rathore, A. S., & Singh, S. P. (2019). The global economic impact of neurodegenerative diseases: Opportunities and challenges. *Bioeconomy for Sustainable Development*, 333–345.

Zhu, X., Chen, Y., Xu, X., Xu, X., Lu, Y., Huang, X., …, Shen, X. (2020). SP6616 as a Kv2. 1 inhibitor efficiently ameliorates peripheral neuropathy in diabetic mice. *EBioMedicine*, *61*.

6 Pyrazole Compounds
Potential Therapeutic Agents for Neurodegeneration

Bhupender Nehra, Manoj Kumar, Viney Chawla, and Pooja A. Chawla

6.1 INTRODUCTION

Pyrazole-based drugs have potent therapeutic roles in neurodegenerative illnesses, i.e., being more widely acknowledged with time. These conditions are defined by progressive and permanent loss of neurons which exert an increasing burden on global health, especially with increased number of aged people worldwide (Ryan et al., 2020). Further, multifactorial mechanisms causing neuronal degeneration are not addressed by conventional neurotherapeutics which simply offer symptomatic alleviation. So, necessity for novel chemical scaffolds that can simultaneously modulate several disease pathways has become apparent (Khan & Iturria-Medina, 2024). Pyrazole is a five-membered heterocycle with unique electrical and structural characteristics and become well-known in medicinal chemistry due to its wide bioactivity and capacity to participate in a variety of molecular interactions (Gao et al., 2021). Pyrazole-based molecules have drawn attention for their neuroprotective (Eldehna et al., 2025), anti-inflammatory, antioxidant (Feng et al., 2025) and enzyme inhibitory properties in neurodegeneration (Narayanan et al., 2021). Because of their structural adaptability, they can be specifically designed to target key elements of neurodegenerative pathology including protein aggregation, enzyme dysregulation, mitochondrial malfunction and synaptic imbalance (Chen et al., 2022). Hereby, chemical, pharmacological and molecular aspects of neuroprotective potential of pyrazole compounds are discussed in detail in this chapter along with modern design techniques and translational viewpoints to influence their future in neuropharmacology.

6.2 OVERVIEW OF NEURODEGENERATIVE DISORDERS

A group of long-term, progressive illnesses known as neurodegenerative disorders are characterized by degeneration of neurons in certain brain areas which leads to deficits in cognition, movement and behavior. Protein misfolding, oxidative stress, neuroinflammation, mitochondrial failure and synaptic dysfunction are some common mechanisms associated with conditions like Alzheimer's disease, Parkinson's disease, Huntington's disease and amyotrophic lateral sclerosis (Benatar et al., 2022), as depicted in Figure 6.1. Many illnesses develop slowly over a long period prior to any symptoms, i.e., starting with minute molecular changes. Although genetic predispositions, environmental exposures, metabolic impairment and epigenetic changes play an important part, age is still most significant risk factor. Globally, neurodegenerative illnesses are alarmingly on increasing prevalence as per current epidemiological trends which are partially caused by longer lifespans or either by lifestyle-related risk factors like metabolic syndrome, vascular comorbidities and chronic inflammation (Smolen et al., 2023). Moreover, therapeutic intervention is difficult due to complicated etiology and biological heterogeneity of many illnesses which highlights need for substances that can simultaneously modulate several pathogenic pathways. Pyrazole derivatives have such potential *via* interaction with a variety of biological targets to modify neural homeostasis (Hassan & Aboulthana, 2023).

DOI: 10.1201/9781003772019-6

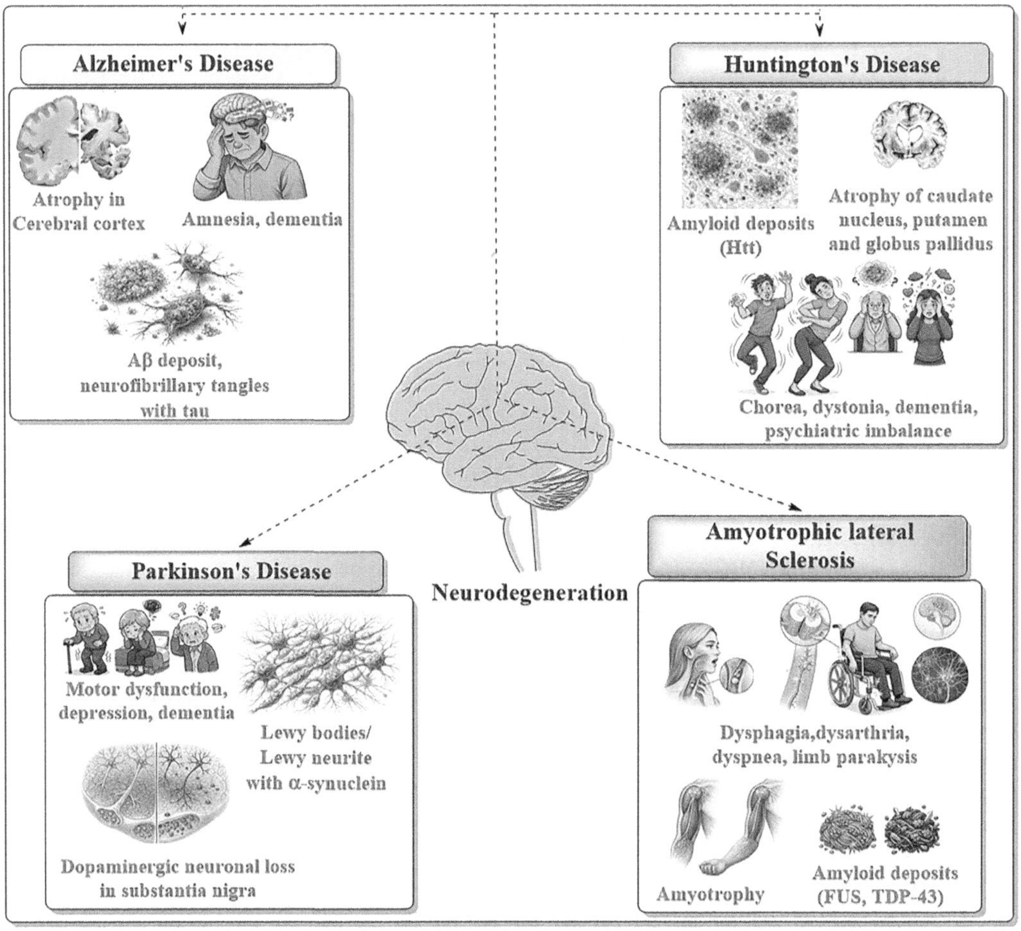

FIGURE 6.1 Various neurodegenerative diseases and associated symptoms.

6.2.1 Pathophysiological Mechanisms Underlying Neurodegeneration

Neuronal damage and functional deterioration in neurodegeneration are controlled by a number of interconnected pathways. One of most common causes is oxidative stress which damages cellular macromolecules and reduces mitochondrial function due to an excess of reactive O- and N-species. Defects in fission, fusion, biogenesis and clearance are among other aspects of mitochondrial dynamics, exacerbate apoptosis and energy failure (Kesari et al., 2020). In next, neuroinflammation is another pivotal mechanism that spreads neuronal injury through uncontrolled generation of cytokines, chemokines and neurotoxic mediators due to long-term activation of microglia and astrocytes (Paciello et al., 2024). Protein misfolding and aggregation are a common pathogenic mechanism in many neurodegenerative disorders (Risen et al., 2024). Moreover, amyloid-β (Montagne et al., 2021), tau (Leuzy et al., 2020), α-synuclein (Smith et al., 2023) and huntingtin are examples of abnormal proteins (as shown in Figure 6.2) that form insoluble aggregates to cause interfere with axonal transport, alter cellular functions and activate stress signaling pathways (Machiela et al., 2020). Impaired autophagy–lysosomal pathways, excitotoxicity from excessive glutamate signaling and dysregulated iron, copper, zinc metal homeostasis are further factors (Ashraf et al., 2020; Kim et al., 2022). By convergence of these mechanisms, neuronal survival pathways break down leads to synaptic failure and eventual loss of brain networks. Multifunctional compounds like pyrazole analogues are very useful since they can simultaneously address several interrelated pathogenic processes (Kanwal et al., 2025).

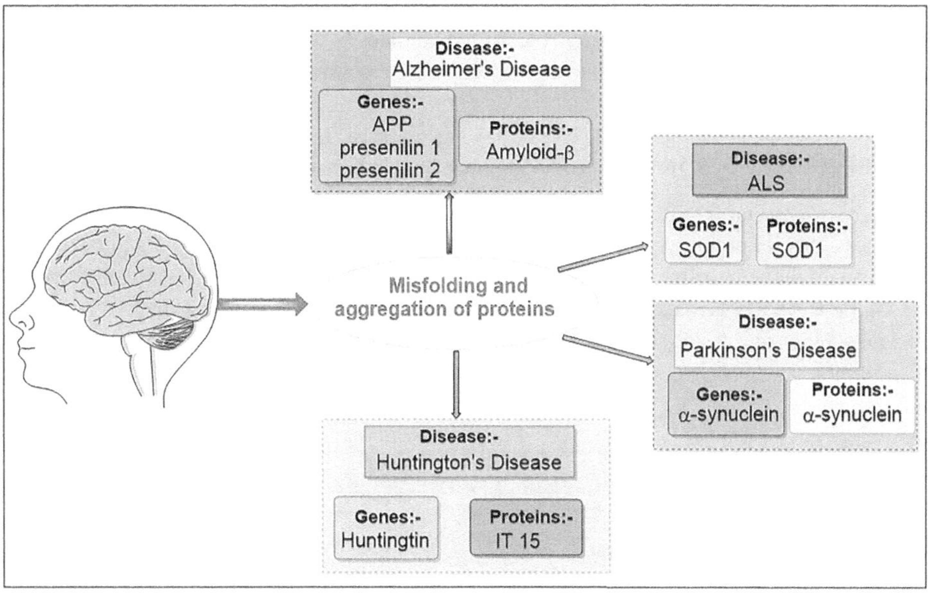

FIGURE 6.2 Involvement of protein misfolding and aggregation in neurodegeneration.

6.2.2 Current Therapeutic Landscape and Limitations

Instead of preventing or reverse neuronal death, current treatments for neurodegenerative illnesses mostly concentrate on symptom relief. For example, Parkinson's disease treatment focusses on dopamine-replacement techniques while Alzheimer's disease treatment uses cholinesterase inhibitors and NMDA receptor modulators. These treatments do not address underlying biological issues including protein aggregation, chronic inflammation and mitochondrial degeneration, although they temporarily enhance quality of life (Parambi et al., 2020). Furthermore, majority of modern medications have limited brain penetration due to restrictive blood–brain barrier (BBB) and several have side effects that make long-term use difficult (Cortes-Flores et al., 2024). Furthermore, lack of disease-modifying medications and failure of several clinical trials focused on a single pathogenic pathway highlight inadequacy of monofunctional treatments for illnesses, i.e., intrinsically complex. These drawbacks emphasize urgent need for novel chemical compounds with greater safety margins, increased brain permeability and multitarget action. Because of their versatile structure and capacity to influence a variety of biological targets, pyrazole derivatives are viable substitutes that could possibly be able to overcome number of drawbacks of existing neurotherapeutic approaches (Roecker et al., 2023).

6.2.3 Emerging Role of Heterocyclic Molecules in Neuroprotection

Heterocyclic scaffolds are major building blocks of many neuroactive medicines because of their property to successfully cross cellular barriers, mimic endogenous metabolites and engage in physiologically relevant interactions (Zheng et al., 2025). Heterocycles like indoles (Sun et al., 2022), pyridines (Haghighijoo et al., 2020), thiazoles (Hussain et al., 2025), triazoles (Liao et al., 2020) and pyrazoles have emerged as crucial frameworks in creation of neuroprotective compounds in recent years (Maqbool et al., 2020). They can alter signaling pathways related to oxidative defense, inflammatory management, mitochondrial stability, synaptic modulation and apoptotic signaling by virtue of their intrinsic structural variety. In this context, the ability of N-containing heterocycles to coordinate metal ions and engage in H-bonds and π–π interactions within receptor pockets makes them very important (Satish & Shinde, 2025). Several studies reveal that adding electron-rich or electron-withdrawing groups to heterocyclic rings can significantly increase their affinity and specificity for neurological targets. Hence, investigation of pyrazole derivatives as effective neuroprotective candidates is strongly justified by growing evidence that heterocyclic chemicals can modulate neurodegenerative pathways (Nikiforov et al., 2023).

6.2.4 Significance of Pyrazole as a Pharmacophore in Drug Discovery

Pyrazole possess special five-membered ring with two adjacent nitrogen atoms that gives it favorable electrical flexibility, H-bonding potential and metabolic durability which renders it a preferred pharmacophore in medicinal chemistry (Norton et al., 2021). Also, exact insertion of substituents that improve pharmacokinetic and pharmacodynamic characteristics is made possible by its adaptable structure (Abdel-Maksoud et al., 2025). Pyrazole compounds have a wide range of biological actions such as kinase inhibition, anti-inflammatory, antioxidant and anticancer effects as depicted in Figure 6.3.

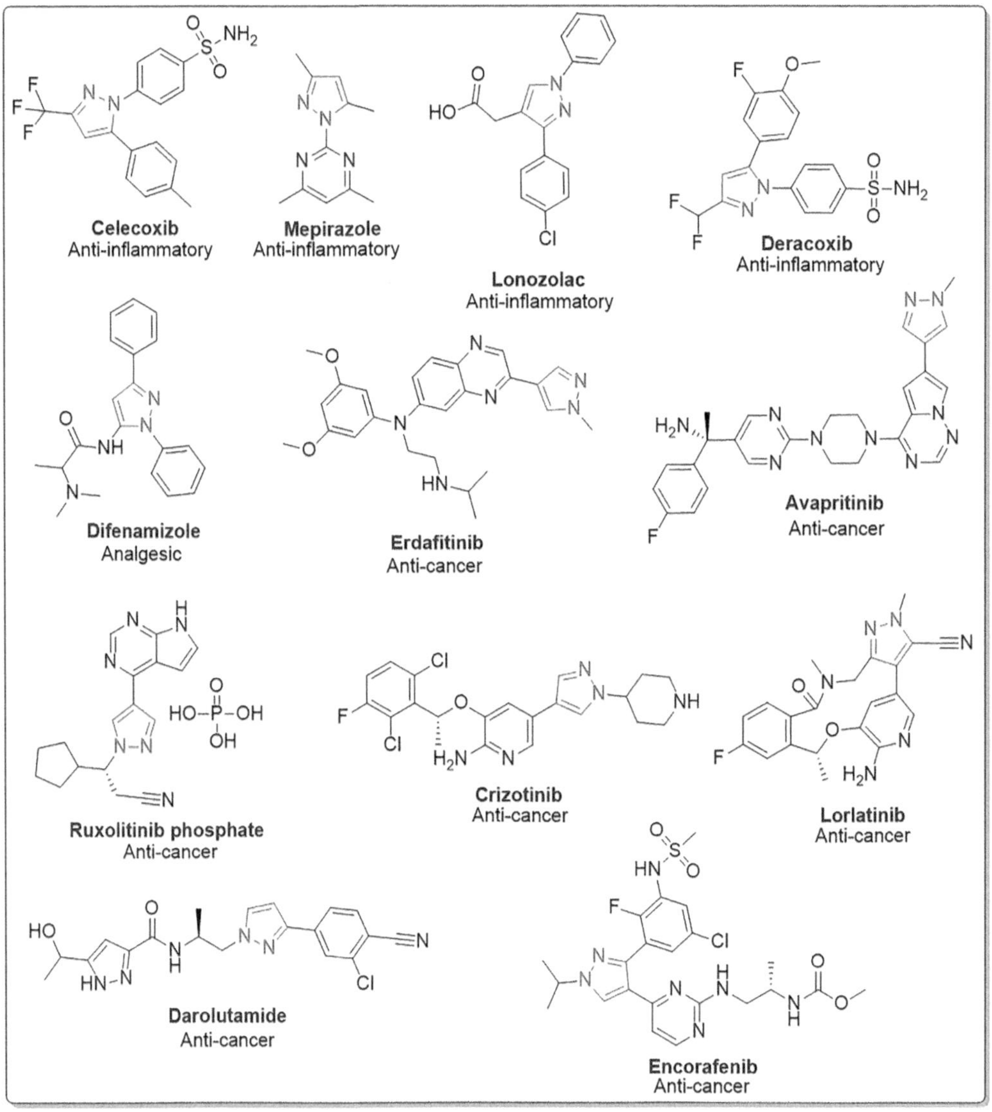

FIGURE 6.3 Pyrazole-containing drugs with anti-inflammatory and anticancer properties.

TABLE 6.1
Various Clinical Trials Linked with Therapeutic Significances of Pyrazole-Based Molecules

Entry	Trial code	Sponsored By	Sexes Eligible	Ages Eligible	Accepts Healthy Volunteers	Last Update Posted
1.	NCT07224607	Cedars-Sinai Medical Center	All	55–89 Years	No	4 November, 2025
2.	NCT07221110	Ege University	All	18–75 Years	No	27 October, 2025
3.	NCT06853743	Oslo University Hospital	All	18–80 Years	No	19 March, 2025
4.	NCT07010705	University of Rochester	All	25–60 Years	Yes	15 September, 2025
5.	NCT07204977	National Institute of Neurological Disorders and Stroke (NINDS)	All	18–99 Years	No	4 November, 2025
6.	NCT07110207	Centre Hospitalier Universitaire Dijon	All	18 Years and older	No	24 August, 2025
7.	NCT07197866	Teva Branded Pharmaceutical Products R&D LLC	All	30 Years and older	No	24 October, 2025
8.	NCT05124392	Massachusetts General Hospital	All	18–85 Years	No	11 September, 2025
9.	NCT07223489	Medstar Health Research Institute	All	18 Years and older	No	31 October, 2025
10.	NCT07136844	Centre Hospitalier Universitaire de Liege	All	18 Years and older	No	22 August, 2025
11.	NCT07122908	Korea University Anam Hospital	All	65 Years and older	No	14 August, 2025
12.	NCT07029581	ACADIA Pharmaceuticals Inc.	All	55–84 Years	No	21 October, 2025

These substances exhibit exceptional adaptability in context of neurodegeneration and target several important molecular pathways for neuronal survival. They are very useful to create multitarget-directed ligands due to their capacity to inhibit cholinesterase, scavenge free radicals, regulate monoamine oxidase (MAO) activity and stop protein aggregation (Krátký et al., 2021). Additionally, pyrazoles frequently exhibit favorable physicochemical properties for penetration of CNS such as adequate molecular weight and mild lipophilicity. Their potential as lead candidates for development of disease-modifying neurotherapeutic drugs (address multifaceted nature of neurodegeneration) is increased by these characteristics (El Fadili et al., 2025). Moreover, certain clinical trials are shown in Table 6.1 based on pyrazole scaffold.

6.3 CHEMISTRY AND STRUCTURAL CHARACTERISTICS OF PYRAZOLE SCAFFOLD

Pyrazole scaffold exhibits significant resonance stabilization which enhances its chemical robustness and ability to engage in diverse intermolecular interactions. Further, presence of two nitrogen atoms positioned at first and second positions permits tautomerism, protonation and metal coordination. These all features influence biological binding properties (AbuSalim & Lash, 2023). Pyrazole derivatives comprise unique substitution pattern that have a significant impact on their medicinal efficacy. Important structural alteration opportunities are provided by N1, C3, C4 and C5 positions which allow addition of aromatic rings, heterocycles, alkyl chains, halogens or electron-modulating groups. Substituent effects strongly influence electronic distribution. Herein, EDGs enhance electron density while EWGs increase acidity and modify pharmacodynamic behavior (El Bakri et al., 2024). By modification of molecular flexibility, polarity or steric bulk, different functional groups can dramatically change biological activity. Additionally, pyrazoles can fuse with other heterocycles to generate bicyclic or tricyclic complexes with better metabolic stability or increased binding to neurological enzymes (Abdelazeem et al., 2024). Depending on intended therapeutic application, these changes can maximize cholinesterase inhibition, antioxidant capacity or anti-aggregation capabilities. This flexibility allows medicinal chemists to design derivatives with customized lipophilicity and stereochemical orientation that ultimately influencing receptor affinity and metabolic stability. The inherent balance between aromaticity and electron density distribution within pyrazole ring makes it a particularly attractive scaffold to construct multitarget neuroprotective agents (Bajad et al., 2025).

6.3.1 Methods of Synthesis of Pyrazole Analogues

Pyrazole derivatives can be generated *via* a number of synthetic pathways that allows precise integration of functional groups required for biological activity. For instance, condensing 1,3-dicarbonyl compounds with hydrazine or its derivatives to create substituted pyrazoles through cyclization is one of the most popular synthetic methods (Hassan & Al-Sabawi, 2023). Asymmetrical or totally substituted derivatives can be formed with the help of variations of this technique such as Knorr synthesis (Schrecker et al., 2023). Further, reaction of α,β-unsaturated ketones or nitriles with hydrazines in mild to moderate circumstances is another well-known synthetic strategy (Betcke et al., 2024). Moreover, modern chemistry has broadened its coverage through catalytic and environmentally friendly techniques including microwave-assisted synthesis (Desai et al., 2021), transition-metal-catalyzed cycloadditions (Jeong & Joo, 2021), solvent-free conditions (Annes et al., 2021) and solid-supported reagents that increase yields and shorten reaction durations (Kharde et al., 2025). Additionally, multicomponent reactions and click chemistry approaches are useful to create intricate pyrazole compounds with strong regioselectivity. These synthetic developments create structurally varied libraries of pyrazole compounds that are suitable to screen against neurodegenerative targets, especially for generation of hybrid or multitarget-directed compounds (El-Gaby et al., 2020).

6.4 PHARMACOLOGICAL POTENTIAL OF PYRAZOLE DERIVATIVES

Pyrazole derivatives display a broad range of pharmacological characteristics pertinent to neurodegenerative illnesses due to its capacity to address oxidative stress, inflammation, protein aggregation, neurotransmitter imbalances and mitochondrial dysfunction. By functioning as multitargeted medicines, pyrazole compounds provide a major advantage in disorders like Parkinson's and Alzheimer's in which several pathogenic pathways overlap (El Alaouy et al., 2024). Depending on substitution pattern and chemical structure, they can function as neuromodulators, mitochondrial stabilizers, antioxidants, enzyme inhibitors or anti-aggregatory compounds (Ashtekar et al., 2023). Hereby, some pyrazole-based marketed drug candidates are displayed in Figure 6.4, i.e., being utilized in different diseases.

6.4.1 Antioxidant and Free Radical Scavenging Activity

Neuronal damage is mostly caused by oxidative stress while pyrazole derivatives have shown potent antioxidant properties in a number of experiments. By activating transcriptional regulators like Nrf2, certain pyrazole derivatives enhance natural antioxidant defense systems and boost expression of detoxification enzymes (Murugan et al., 2024). Pyrazole-based molecules can prevent oxidative chain reactions in addition to directly scavenging radicals *via* chelating redox-active metal ions like iron or copper. Pyrazole derivatives provide attractive options to decrease oxidative processes linked to neurological disorders because of their multifaceted antioxidant properties (Michalska et al., 2020).

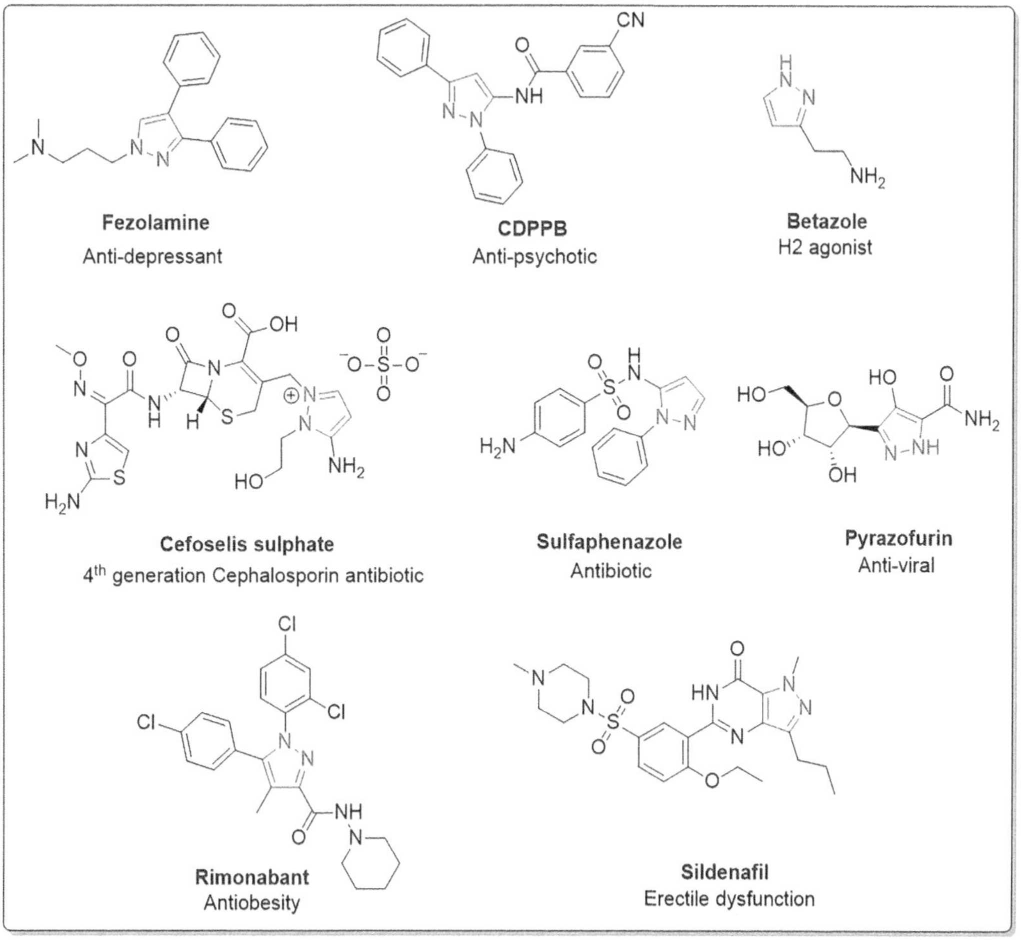

FIGURE 6.4 Pyrazole-based marketed drugs with various pharmacological applications.

6.4.2 Anti-Inflammatory and Microglial Modulatory Effects

Pyrazole derivatives have significant anti-inflammatory effects through regulation of microglial activation, cytokine production and chronic neuroinflammation. They can lower the expression of cytokines like TNF-α, IL-1β and IL-6 by inhibition of pro-inflammatory signaling pathways including NF-κB (Srour et al., 2022). Certain pyrazoles reduce synthesis of nitric oxide and neurotoxic prostaglandins through cyclooxygenase or inducible nitric oxide synthase inhibition. Their capacity to regulate glial reactivity stops long-term inflammatory cycles that worsen neuronal death and synaptic impairment (Ayman et al., 2023). In addition, pyrazole compounds may enhance release of neuroprotective mediators and boost tissue repair by encouraging a shift toward anti-inflammatory microglial phenotypes (Murugan et al., 2024).

6.4.3 Enzyme Inhibition: AChE, BuChE and MAO Inhibitory Activities

Important brain enzymes linked to neurodegenerative pathology are strongly inhibited by pyrazole derivatives. In order to improve acetylcholine levels and maintain cognitive function, cholinesterase (AChE, BuChE) inhibition is crucial in Alzheimer's treatment (Abdelwahab et al., 2023). Herein, strong interactions with cholinesterase active sites are made possible *via* structural changes of pyrazoles through π–π stacking, hydrophobic interactions and H-bonding (Sbai et al., 2026). Furthermore, several pyrazole derivatives function as MAO inhibitors to control metabolism of dopamine, serotonin and norepinephrine. In Parkinson's disease, MAO-B inhibition is especially important because dopamine preservation is required to counteract motor impairment. The versatility of pyrazoles to address neurotransmitter dysregulation is highlighted by these two enzymatic actions (Zugno et al., 2025).

6.4.4 Anti-Amyloidogenic and Anti-Tau Phosphorylation Effects

One of the main pathogenic characteristics of Alzheimer's and associated illnesses is protein misfolding and aggregation. Pyrazole derivatives showed capacity to disrupt amyloid-β aggregation through attachment to oligomeric or fibrillar species. Their planar and aromatic substituents promote π–π interactions that reduce plaque deposition by destabilizing β-sheet formation (Wang et al., 2025). In similar manner, pyrazole drugs can alter tau phosphorylation and prevent neurofibrillary tangles *via* blocking kinases like GSK-3β (Gameiro et al., 2017) or CDK5 (Yerramsetty et al., 2025). Due to encouraging autophagic activity or blocking oxidative modification that causes aggregation, some derivatives also improve removal of misfolded proteins (Potapov et al., 2025).

6.4.5 Metal Chelation and Mitochondrial Protective Roles

In dementia, oxidative damage and protein aggregation are exacerbated by disturbed metal homeostasis. Pyrazole rings can mitigate metal-induced toxicity through coordinating metal ions such as Fe^{2+}, Cu^{2+} and Zn^{2+} with their N-atoms and variable substituents. These chelating interactions stop metal-mediated acceleration of amyloid aggregation and inhibit metal-catalyzed oxidative processes (Singh et al., 2024). Pyrazole derivatives have protective effects on structure as well as function of mitochondria in addition to regulating metals. They can decrease release of pro-apoptotic proteins, increase ATP generation and stabilize potential of mitochondrial membrane. By affecting transcriptional regulators, several molecules also enhance mitochondrial biogenesis (Jiang et al., 2024; Manikandan et al., 2025).

6.5 ROLE OF PYRAZOLE DERIVATIVES IN SPECIFIC NEURODEGENERATIVE DISORDERS

6.5.1 Pyrazole-Based Compounds in Alzheimer's Disease

As pyrazole-containing drugs can disrupt several pathogenic processes that are essential to cognitive decline, they have become highly promising candidates for Alzheimer's disease therapy. These compounds exhibit potent inhibitory effect against butyryl and acetylcholinesterase while re-establishment of cholinergic neurotransmission, i.e., essential for memory formation (Messaad et al., 2022). Several pyrazole analogues also lessen amyloid-β aggregation through attachment toward hydrophobic residues in amyloidogenic sequences, limiting fibrillization and significant disintegration of preexisting aggregates (Vidali et al., 2022). Their ability to influence tau kinases like GSK-3β helps lower tau hyperphosphorylation which is a key factor in development of neurofibrillary tangles. Furthermore, pyrazoles decrease microglial-mediated inflammation while also increases endogenous enzymatic defenses and lowers ROS buildup to produce antioxidant benefits. Their multifunctional therapeutic efficacy in Alzheimer's pathogenesis is reinforced through their metal-chelating activity which further protects against oxidative stress and amyloid aggregation caused by copper and iron (Ali et al., 2020; Bonastre-Sabater et al., 2024).

6.5.2 Role in Parkinson's Disease Therapy

By stabilizing dopaminergic signaling and lowering molecular triggers of neuronal death in substantia nigra, pyrazole analogues offer neuroprotection in Parkinson's disease. By limiting dopamine metabolism and minimizing detrimental oxidative byproducts, their inhibitory effect against MAO-B increases dopamine availability (Zugno et al., 2025). By interfering with β-sheet stacking, a number of derivatives interact with α-synuclein aggregates and reduce cytotoxicity linked to Lewy body development. In addition, pyrazole drugs also affect mitochondrial function *via* maintaining complex I activity, stopping production of ROS and sustaining ATP levels necessary for neuronal survival. Their anti-inflammatory properties further shield dopaminergic neurons against long-term neuroinflammatory damage, particularly by reducing microglial activation and suppressing cytokines (TNF-α, IL-6) (Wang et al., 2025).

6.5.3 Potential in Huntington's Disease and ALS

As excitotoxicity, protein aggregation and mitochondrial dysfunction remains as major contributor to neuronal degeneration in Huntington's disease and amyotrophic lateral sclerosis, pyrazole drugs have therapeutic value in these conditions. Pyrazole-containing molecules lessen mutant huntingtin aggregation in Huntington's disease *via* tangling to polyglutamine-rich areas followed by boosted proteostasis processes including autophagic clearance (Pogoda et al., 2021). Also, their antioxidant activity prevents excessive ROS production leading to speeds up death of striatal neurons. Pyrazoles aid preserves mitochondrial membrane potential, lessen glial-mediated inflammation and decrease motor neurons excitotoxicity in ALS models, i.e., phenomena strongly linked to SOD1 mutations and TDP-43 disease (Rahman et al., 2023).

6.5.4 Cognitive Enhancement and Neurotrophic Modulation

In addition to its impact on specific diseases, pyrazole derivatives also improve neurotransmitter balance, enhance synaptic plasticity and modulate neurotrophic signaling pathways. Numerous pyrazole structures improve synaptic connections and encourage production of dendritic spines *via* upregulation of BDNF and TrkB receptor activation. Further, learning and memory retrieval are improved by their capacity to restore normal glutamatergic and cholinergic activity (Dhapola et al., 2024). Pyrazoles support higher-order cognitive functioning through preservation of synaptic environment, i.e., favorable to long-term potentiation by lowering neuroinflammation and oxidative stress. These characteristics indicate substantial promise of pyrazole derivatives in situations involving cognitive impairment such as moderate cognitive decline and cognitive deficiencies associated with aging (Birmann et al., 2020).

6.6 MOLECULAR MECHANISMS ASSOCIATED WITH NEUROPROTECTION

6.6.1 Regulation of Oxidative Stress Pathways (Nrf2, HO-1)

By triggering Nrf2/HO-1 signaling pathway, i.e., a key regulator of cellular redox balance, pyrazole derivatives shield neurons from oxidative damage (shown in Table 6.2). When Nrf2 is activated, antioxidant enzymes like HO-1, NQO1 and GSH-related proteins can be transcribed. These enzymes work together to neutralize reactive oxygen species and lower lipid peroxidation. In order to stabilize mitochondrial function and stop ROS-driven apoptosis, pyrazole compounds frequently contain EDGs that increase radical scavenging activity. Their high neuroprotective efficacy in a variety of neurodegenerative disease models can be explained by their consistent capacity to enhance endogenous antioxidant defense mechanisms (Alqudah et al., 2025; Murugan et al., 2024).

TABLE 6.2
Molecular Targets, Associated Mechanisms and Potent Pyrazole-Containing Ligands

Entry	Molecular Target	Pathological Process in Neurodegenerative Diseases	Mechanistic Role of Pyrazole Analogues
1.	AChE/BuChE	Cholinergic deficit in Alzheimer's disease	• Competitive binding to catalytic anionic site and peripheral anionic site • Reduces ACh breakdown • In some cases, inhibits Aβ aggregation at PAS
2.	MAO-A/B	Excess monoamine catabolism leads to oxidative stress	• Bind flavin adenine dinucleotide region • Inhibit dopamine/serotonin breakdown
3.	GSK-3β kinase	Tau phosphorylation and synaptic instability	• Occupies ATP-binding pocket • Inhibits aberrant tau phosphorylation
4.	BACE-1	Aβ plaque formation	• Interacts with catalytic Asp residues • Disrupts APP cleavage
5.	Nrf2/ARE pathway	Oxidative stress dysregulation	• Pyrazoles bearing electrophilic centers activate Nrf2 and induces HO-1, NQO1
6.	NF-κB signaling	Microglial activation and neuroinflammation	• Prevent p65 nuclear translocation - Reduce IL-1β, TNF-α, IL-6
7.	α-synuclein aggregation	Parkinson's pathology	• Hydrophobic pyrazoles interfere with fibril growth and seed formation
8.	Mitochondrial Complex I and III	• ROS overproduction • ATP deficit	• Protect against ETC dysfunction - Stabilize mitochondrial membrane potential

6.6.2 Inhibition of Neuroinflammatory Cascades (NF-κB, COX-2, TNF-α)

Pyrazole-based drugs reduce neuroinflammation through disruption of inflammatory transcription factors and cytokine pathways that cause long-term brain damage. They decrease expression of pro-inflammatory mediators such as COX-2, iNOS, TNF-α and IL-1β by blocking NF-κB translocation to nucleus. Also, pyrazole analogues stop prolonged inflammatory milieu that speeds up synaptic dysfunction in Alzheimer's disease, Parkinson's disease and ALS by suppressing microglial activation. Their usefulness as multi-pathway neuroprotective drugs is highlighted by their ability to effectively control several inflammatory nodes (Fadaly et al., 2025; Rahman et al., 2023).

6.6.3 Apoptotic and Autophagy Signaling Modulation

In neuronal cells, pyrazole derivatives affect apoptosis through re-establishment of equilibrium between pro- and anti-apoptotic proteins. They increase production of Bcl-2, suppress Bax activation and stop release of cytochrome c from mitochondria followed by inhibition of caspase initiation. In addition, pyrazole-containing molecules also control autophagy *via* alteration of AMPK and mTOR pathways which facilitates effective removal of misfolded proteins and malfunctioning organelles. These two functions help neurons to survive over long term in stressful situations that are common in neurodegenerative contexts and shield them from programmed cell death (Hua et al., 2024).

6.6.4 Interaction with Mitochondrial Dynamics and Bioenergetics

Molecules based on pyrazole scaffold have a major impact on mitochondrial integrity which is essential for stability of neural energy. They avoid bioenergetic crises in susceptible neuronal populations by sustaining ATP synthesis, stabilizing electron transport chain complexes and maintaining mitochondrial membrane potential. They maintain balanced mitochondrial dynamics, i.e., necessary for synaptic transmission and neural repair by virtue of regulating fission and fusion proteins including Drpl and OPA1. Pyrazole compounds maintain metabolic health of neurons subjected to prolonged degenerative stress through inhibition of ROS production and mitochondrial fragmentation (Jiang et al., 2022).

6.6.5 Molecular Docking and Computational Insights

According to computational studies, pyrazole scaffolds have a substantial binding affinity for several neurodegeneration-related targets such as tau kinases, MAO isoforms, amyloid-β sections, cholinesterases and inflammatory mediators. In this concern, *in silico* molecular docking studies suggested that pyrazole ring contributes to π–π stacking, H-bonding and hydrophobic interactions that improve target selectivity (Tok et al., 2025). Herein, strategies for structural alteration intended to increase potency, selectivity and BBB permeability are influenced by these observations. Additionally, many optimized pyrazole compounds have favorable ADMET features predicted by *in silico* models which supports their potential for additional preclinical development (El Alaouy et al., 2021).

6.7 DESIGN, SYNTHESIS AND OPTIMIZATION STRATEGIES

6.7.1 Hybrid Molecule Design and Multitarget-Directed Ligand Approach

The multitarget-directed ligand method is being utilized to design pyrazole-based neuroprotective medicines that combines pyrazole core with other pharmacophores to create hybrid compounds for combating several pathogenic pathways at the same time. This tactic makes it possible to reduce neuroinflammation, prevent amyloid as well as tau aggregation, modulate oxidative stress and inhibit cholinesterase enzymes. In complex neurodegenerative disorders, such hybrid designs increase possibility of disease change, lower drug load and improve therapeutic efficacy (Jangid et al., 2024).

6.7.2 Synthetic Pathways and Catalytic Techniques

Regioselective cyclocondensation reactions between hydrazines and 1,3-dicarbonyl compounds are commonly used in production of pyrazole derivatives. These reactions are frequently aided by acid or metal catalysts to ensure good yields and structural accuracy. To increase reaction rates and enhance environmental sustainability, modern synthetic methods use transition-metal catalysis, microwave irradiation and solvent-free environments (Zhang et al., 2022). Moreover, variety of substituents that can be effectively added to pyrazole ring is further increased by methods like click chemistry, Suzuki coupling and oxidative heteroannulation to provide more chances for pharmacological optimization (Liu et al., 2024).

6.7.3 Bio-Isosteric Replacement and Lead Optimization

By facilitating structural changes that increase receptor affinity, decrease off-target interactions or improve metabolic stability, bio-isosteric replacement is essential for optimization of pyrazole-based leads. It is possible to customize physicochemical qualities without sacrificing important pharmacophoric characteristics by switching electrically or sterically comparable moieties for functional groups. The therapeutic profile of pyrazole derivatives meant for CNS uses is strengthened by these changes which frequently increase lipophilicity, reduce susceptibility to oxidative metabolism and improve BBB penetration (Sanad & Mekky, 2022).

6.7.4 Quantitative Structure-Activity Relationship Studies

The biological potential of pyrazole analogues is influenced by molecular characteristics like electronic distribution, steric parameters, hydrophobicity and substituent orientation which can be predicted using QSAR models. Enzyme inhibition, antioxidant potential and anti-amyloidogenic effects are certain examples of desired activities that can be correlated with structural characteristics with the help of these computational techniques. Moreover, QSAR analysis expedites optimization of pyrazole-based scaffolds for neurotherapeutic purposes and greatly decreases experimental burden by directing logical design of novel analogues (Nguyen & Kim, 2024).

6.7.5 *In Silico* ADMET and BBB Permeability Predictions

To choose pyrazole compounds with best pharmacokinetic and safety properties appropriate for CNS treatments, *in silico* ADMET assessment is crucial. Early indicators of translational feasibility of molecules are provided by computational models that assess factors like absorption, metabolic stability, hepatotoxicity risk and BBB permeability. These predictions inform structural changes that maximize lipophilicity, molecular weight and polar surface area which are essential for systemic toxicity minimization and assuring effective CNS penetration. These prediction methods increase probability of successful *in vivo* performance and expedite lead selection (Niharika et al., 2025).

6.8 RECENT ARTICLES DEDICATED TOWARD NEUROPROTECTIVE POTENTIAL OF PYRAZOLE DERIVATIVES

Rani et al. (2025) design and developed sulfonated pyrazoles as inhibitors of AChE and BuChE with $A\beta_{1-42}$ aggregation inhibitor in Alzheimer's disease. Moreover, all compounds were screened for enzyme inhibitory assay against AChE and BuChE. Herein, results revealed that **1b** (Figure 6.5) was topmost effective inhibitor of both tested enzymes with IC_{50} values of 1.634 and 0.0285 μM, respectively, in comparison to standard drug (IC_{50} = 2.019 and 17.38 μm against AChE and BuChE, respectively). Furthermore, *in vivo* $A\beta_{1-42}$ inhibition was evaluated by y-maze test in which compound **1b**-increased memory. In addition, analogue **1b** was tested for neurotoxicity and found to be non-toxic against SHSY-5Y cell line (Rani et al., 2025).

Kaya and their co-workers (2025) worked on design, synthesis and *in silico* evaluation of novel pyrazoline–thiazole derivatives as AChE and BChE inhibitors to treat Alzheimer's disease. In this research work, enzyme inhibitory assay was performed of all synthesized molecules on AChE and BChE with comparison to reference drug Tacrine. Resulted indicated that compounds **2g** (Figure 6.5) was most potent among developed series having IC_{50} value of 0.338 and 2.087 μM toward AChE and BChE, respectively, as compared to Tacrine (IC_{50} = 0.15 and 0.215 μM, respectively). Moreover, molecular docking analysis showed that **2g** have good binding affinity toward active site of receptors (Kaya et al., 2025).

In another article, design and synthesis of pyrazolyl triazole hybrid analogues was carried out by Dogan et al. (2025) to assess neurotoxic potential in 6-OHDA induced neurotoxicity model. Among 12 synthesized molecules, some molecules exhibited high efficacy against Bax expression level and caspase-3 of neurotoxicity level. Moreover, AChE assessment was also performed on all molecules which revealed that compound **3l** (Figure 6.5) possessed 30.2% inhibition at 30 μM while 86.2% inhibition at 300 μM, as compared to reference drug Galantamine (% inhibition = 80.7% and 95.9% at 30 and 300 μM, respectively) (DOĞAN et al., 2025).

In year 2024, Pourtaher and their team disclosed one pot sequential synthesis of spiroindolinone-pyrazole as anti-Alzheimer analogues. In this, all generated compounds were evaluated against AChE and BChE enzymes for their inhibitory assay with comparison to standard drug Donepezil. Herein, inhibitory assay resulted that compound **4f** (Figure 6.5) was potent against AChE with IC_{50} value of 6.02 μM while **4p** was most effective against BChE (IC_{50} value = 9.13 μM), in comparison to Donepezil (IC_{50} value = 0.079 and 10.6 μM, respectively). Furthermore, *in silico* docking study demonstrated **4f** as promising ligand toward target site. It was further tested on Neuroblastoma cell line (SH-SY5Y) which showed that **4f** exhibited 64.81% viability at 50 μM dose (Pourtaher et al., 2024).

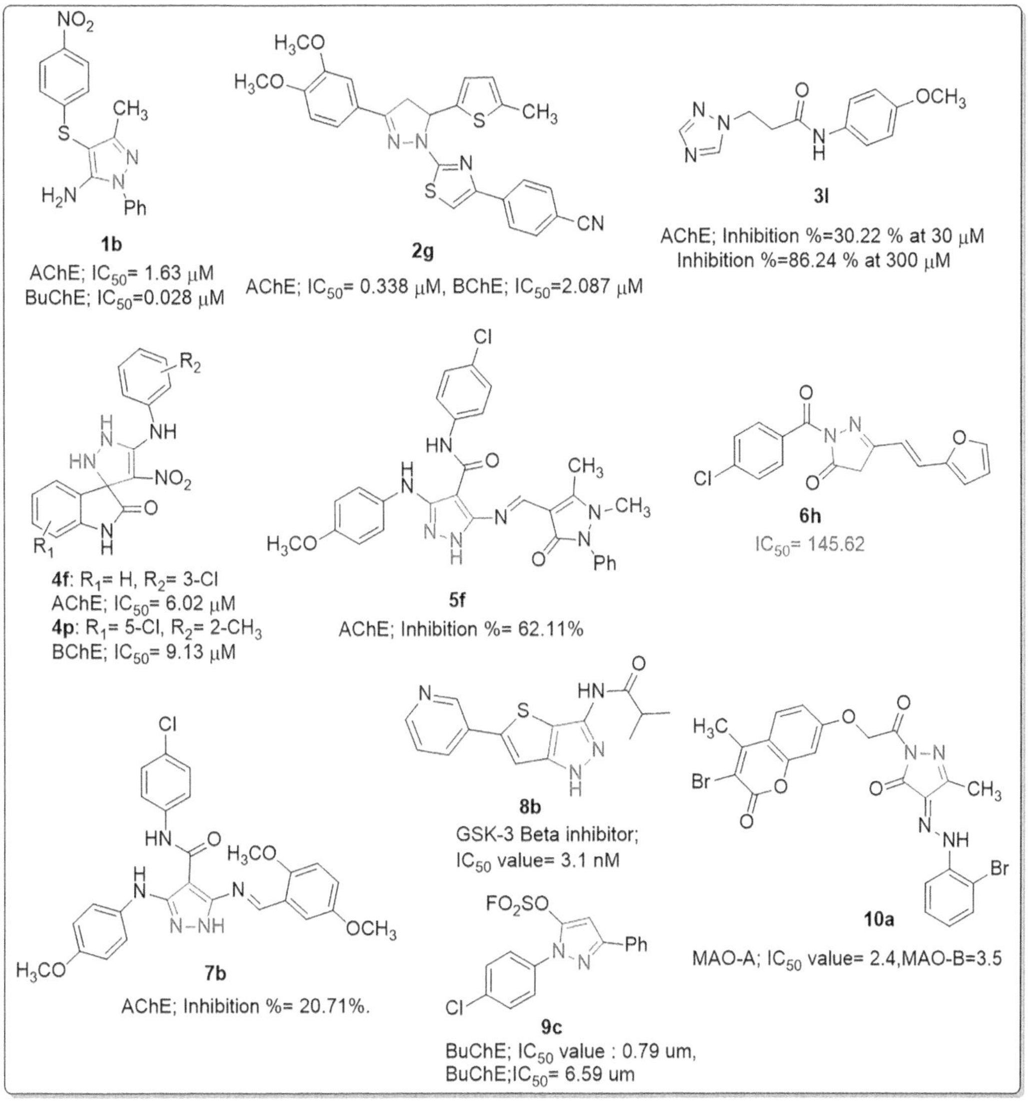

FIGURE 6.5 Structures of most potent pyrazole-based molecules in neurodegenerative diseases.

Naglah and their co-worker (2024) disclosed pyrazole-based Schiff bases as potent anti-diabetic, anti-inflammatory and anti-Alzheimer through *in vivo* and *in silico* investigations. In this work, all compounds were tested for antioxidant evaluation for total antioxidant capacity, iron reducing power and results were compared to reference ascorbic acid. Later on, anti-Alzheimer's disease activity described that **5f** (Figure 6.5) have superior AChE inhibitory profile with % inhibition of 62.11%, as than that of standard Donepezil (% inhibitory value = 70.32%). Moreover, anti-inflammatory, anti-diabetic and cytotoxicity properties were also determined (Naglah et al., 2024).

Manikandan and their team (2024) worked on synthesis of 2,5-substituted pyrazolones and evaluated on human neuroblastoma cells (SH-SY5Y) for neuroprotective potential. They developed ten molecules in which **6h** (Figure 6.5) inhibited Aβ-25–35-induced apoptosis in SH-SY5Y cells with % cell viability of 98.46%. Further, molecular docking study was conducted which revealed that phenyl-ring of **6h** provides strong H-bonding toward amino acids present in active site of target (Docking score = −8.4 kcal/mol) (Manikandan et al., 2025).

In 2023, Alkahtani *et al.* disclosed *in vitro* evaluation of pyrazole bearing Schiff bases as potent antioxidant, anti-diabetic, anti-arthritic and anti-Alzheimer's disease agents. *In vitro* antioxidant study demonstrated that **7b** (Figure 6.5) exhibited strongest antioxidant property compared to ascorbic acid. Later on, anti-Alzheimer activity on AChE enzyme showed that analogue **7b** was most advantageous with 20.71% AChE inhibitory profile. In addition, compound **7b-**elicited promising anti-diabetic and anti-arthritic potential (Alkahtani et al., 2023).

Yan and their team (2022) generated thieno–pyrazole-fused derivatives and assessed them as potent glycogen synthase kinase 3β inhibitors for Alzheimer's disease. They prepared 23 pyrazole-based fused analogues among which **8b** (Figure 6.5) was found to be most effective against GSK-3β with IC_{50} value of 3.1 nM. Further, MTT assay on SH-SY5Y cells showed no significant cytotoxicity at concentration up to 50 μM. In addition, molecular dynamic study also validated that **8b** have higher affinity toward active site of this receptor (Yan et al., 2022).

In the same year, synthesis and biological evaluation of fluorosulfate-containing pyrazole heterocycles was performed by Li and colleagues as selective BuChE inhibitors for treatment of Alzheimer's disease. Among all tested compounds, **9c** (Figure 6.5) experienced most promising AChE and BuChE inhibitory effect with IC_{50} values of 0.79 and 6.59 μM, respectively. It was also highlighted that **9c** was found to be more potent than standard treatments (Rivastigmine, Donepezil) at 20 μM dose. Moreover, PAMPA-BBB penetration study of **9c** showed positive CNS permeability (Li et al., 2022).

Narayanan et al. (2021) reported design, synthesis and biological evaluation of substituted pyrazoles endowed with coumarin pharmacophore as new hybrid molecules against Alzheimer's disease. In this work, *in vitro* MAO inhibitory study revealed that compound **10a** (Figure 6.5) was most potent MAO-A and MAO-B inhibitors by having IC_{50} values of 2.4 and 3.5 μM, respectively, as than that of reference inhibitor (IC_{50} = 3.9 and 4.4 μM, respectively). Later on, antioxidant activity of **10a** elicited IC_{50} value of 23.2 μg/ml which was more potent than reference ascorbic acid (IC_{50} = 29.2 μg/ml). In addition, AChE inhibitory assay showed significant inhibition of **10a** toward specific enzyme (Narayanan et al., 2021).

6.9 FUTURE PERSPECTIVES AND CHALLENGES

It is very crucial to develop pyrazole-based neurotherapeutics in future in order to overcome obstacles related to medication distribution to central nervous system. Despite favorable lipophilicity of several pyrazole derivatives, active efflux mechanisms and enzymatic breakdown at blood–brain barrier renders it difficult to achieve stable and therapeutically relevant concentrations in brain. It will be crucial to optimize structure in order to minimize contact with efflux proteins like P-glycoprotein, maximize carrier-mediated transport and reduce polar surface area. Targeted formulations and sophisticated prodrug techniques can further increase brain absorption, reduce peripheral exposure and improve therapeutic efficacy (Amaradhi et al., 2022). These developments are essential for complete utilization of multitarget pharmacology related to pyrazole derivatives. There is great potential for improvement in stability, bioavailability and targeted distribution if pyrazole-based therapy incorporates nanotechnology. Pyrazole derivatives can be encapsulated in nanocarriers including polymeric nanoparticles, liposomes, solid lipid nanoparticles and dendrimers to prevent premature metabolism and to allow regulated release inside brain microenvironment (Alfei et al., 2021). Further, receptor-mediated transport across BBB is made possible through surface functionalization of nanoparticles with targeting ligands as transferrin, lactoferrin or peptides. This allows for targeted delivery to disease-affected areas. In addition, these nanosystems facilitate co-administration of pyrazoles with synergistic drugs which enhances therapeutic results and lowers systemic toxicity. This combination represents a significant advancement to create neuroprotective formulations of future (Sharma et al., 2022).

Combination approaches utilizing pyrazole derivatives have ability to treat several diseased pathways simultaneously since neurodegenerative illnesses are complex. To produce additive or synergistic neuroprotection, pyrazole analogues can be used with cholinesterase inhibitors, NMDA receptor modulators, antioxidants or anti-inflammatory medications. Herein, combination regimens are rationally constructed through a little dose of each medication to decrease side effects while increasing therapeutic efficacy (Abdelazeem et al., 2024). Moreover, incorporation of pyrazole pharmacophores with other bioactive scaffolds in single hybrid compounds may also result in synergistic interactions in order to afford a more efficient method of multitarget intervention. As neurodegenerative diseases are complicated, combination approaches by employing pyrazole derivatives can treat multiple pathological pathways. Pyrazole analogues can be used with cholinesterase inhibitors, NMDA receptor modulators, antioxidants or anti-inflammatory drugs to create additive or synergistic neuroprotection. Furthermore, pyrazole pharmacophores may interact synergistically with various bioactive scaffolds in single hybrid molecules to provide a more effective multitarget intervention strategy (El Fadili et al., 2022). Future drug development workflows are anticipated to depend extensively on these strategies.

6.10 CONCLUSIONS

Pyrazole-based molecules are a highly adaptable and scientifically proven category of chemicals with substantial therapeutic potential for a variety of neurological diseases. Their wide range of mechanisms is demonstrated by their capacity to control oxidative stress, reduce neuroinflammation, inhibit important enzymes, stop protein aggregation and maintain mitochondrial function. Their promise as multitarget-directed medicines capable to tackle complex pathophysiology of diseases including Alzheimer's, Parkinson's, Huntington's disease and ALS is validated through preclinical evidence that consistently supports their neuroprotective effects in both cellular as well as animal models. Further, pyrazole derivatives offer hope for disease transformation in neuropharmacology due to their varied pharmacological profile. Despite tremendous advancements, there are still a number of issues that need to be resolved before pyrazole derivatives can be used in clinical applications. Therapeutic optimization is hindered by their varied metabolic stability, inadequate characterization of off-target interactions and poor knowledge of their long-term safety. Also, clinical advancement is hampered through translational obstacles such as uneven brain penetration, species-specific variations in pharmacodynamics and intricacy of human neurodegenerative disease biology. Furthermore, more research is needed to determine exact relationship between structural changes and multitarget effects in order to reduce side effects and improve treatment specificity. In order to ensure successful clinical application, these restrictions must be addressed. By acting as modular building blocks for development of highly selective or multitarget-directed neuroprotective treatments, pyrazole molecules are expected to play a crucial role in next-generation neuropharmacology. Additionally, development of optimized derivatives with greater efficacy and safety will be made feasible by their susceptibility to structural alteration as well as developments in targeted delivery technologies and *in silico* drug design. Pyrazole-based medicines are anticipated to contribute to a new era of customized treatments for neurodegenerative illnesses as research continues to incorporate insights from molecular biology, systems neuroscience and medicinal chemistry.

REFERENCES

Abdelazeem, N. M., Aboulthana, W. M., Elshahid, Z. A., El-Hussieny, M., & Al-Ashmawy, A. A. (2024). Synthesis and biological (in vitro and in silico) screening of the 4-aryl-fused pyranopyrazole derivatives as enzyme (α-amylase, α-glucosidase, acetylcholinesterase & proteinase) inhibitors with anti-oxidant and cytotoxic activities. *Journal of Molecular Structure, 1310*, 138224.

Abdel-Maksoud, M. S., Eitah, H. E., Hassan, R. M., & Abd-Allah, W. H. (2025). Design and synthesis of novel pyrimidine-pyrazole hybrids with dual anticancer and anti-inflammatory effects targeting BRAFV600E and JNK. *Molecular Diversity*, 1–21.

Abdelwahab, H. E., Ibrahim, H. Z., & Omar, A. Z. (2023). Design, synthesis, DFT, molecular docking, and biological evalution of pyrazole derivatives as potent acetyl cholinestrease inhibitors. *Journal of Molecular Structure, 1271*, 134137.

AbuSalim, D. I., & Lash, T. D. (2023). Aromatic character and relative stability of pyrazoloporphyrin tautomers and related protonated species: Insights into how pyrazole changes the properties of carbaporphyrinoid systems. *Molecules*, *28*(6), 2854.

Alfei, S., Brullo, C., Caviglia, D., & Zuccari, G. (2021). Preparation and physicochemical characterization of water-soluble pyrazole-based nanoparticles by dendrimer encapsulation of an insoluble bioactive pyrazole derivative. *Nanomaterials*, *11*(10), 2662.

Ali, S. A., Awad, S. M., Said, A. M., Mahgoub, S., Taha, H., & Ahmed, N. M. (2020). Design, synthesis, molecular modelling and biological evaluation of novel 3-(2-naphthyl)-1-phenyl-1H-pyrazole derivatives as potent antioxidants and 15-lipoxygenase inhibitors. *Journal of Enzyme Inhibition and Medicinal Chemistry*, *35*(1), 847–863.

Alkahtani, H. M., Almehizia, A. A., Al-Omar, M. A., Obaidullah, A. J., Zen, A. A., Hassan, A. S., & Aboulthana, W. M. (2023). In vitro evaluation and bioinformatics analysis of Schiff bases bearing pyrazole scaffold as bioactive agents: Antioxidant, anti-diabetic, anti-Alzheimer, and anti-arthritic. *Molecules*, *28*(20), 7125.

Alqudah, A., Qnais, E., Bseiso, Y., Shilbayeh, S. A. R., Aljabali, A. A., & Gammoh, O. (2025). Activation of the Nrf2 pathway by a novel bipyrazole compound mitigates doxorubicin-induced cardiotoxicity in Wistar albino rats. *European Journal of Pharmacology*, 177731.

Amaradhi, R., Mohammed, S., Banik, A., Franklin, R., Dingledine, R., & Ganesh, T. (2022). Second-generation prostaglandin receptor EP2 antagonist, TG8-260, with high potency, selectivity, oral bioavailability, and anti-inflammatory properties. *ACS Pharmacology & Translational Science*, *5*(2), 118–133.

Annes, S. B., Saritha, R., Chandru, K., Mandali, P. K., & Ramesh, S. (2021). Metal-and solvent-free cascade reaction for the synthesis of amino pyrazole thioether derivatives. *The Journal of Organic Chemistry*, *86*(23), 16473–16484.

Ashraf, A., Jeandriens, J., Parkes, H. G., & So, P. W. (2020). Iron dyshomeostasis, lipid peroxidation and perturbed expression of cystine/glutamate antiporter in Alzheimer's disease: Evidence of ferroptosis. *Redox Biology*, *32*, 101494.

Ashtekar, H., Dwivedi, P. S., Aggarwal, N., Fernandes, Z., & Varghese, N. (2023). In-silico validation of pyrazolone derivatives as the potent scaffold for modulating protein abnormalities associated with Parkinson's disease. *Journal of Drug Delivery Technology*, *13*(1), 1–9.

Ayman, R., Abusaif, M. S., Radwan, A. M., Elmetwally, A. M., & Ragab, A. (2023). Development of novel pyrazole, imidazo [1, 2-b] pyrazole, and pyrazolo [1, 5-a] pyrimidine derivatives as a new class of COX-2 inhibitors with immunomodulatory potential. *European Journal of Medicinal Chemistry*, *249*, 115138.

Bajad, N. G., Jangra, J., Kumar, A., Krishnamurthy, S., & Singh, S. K. (2025). Discovery of pyrazoline analogs as multi-targeting cholinesterase, β-secretase and Aβ aggregation inhibitors through lead optimization strategy. *International Journal of Biological Macromolecules*, *301*, 140436.

Benatar, M., Wuu, J., McHutchison, C., Postuma, R. B., Boeve, B. F., Petersen, R., …, & Turner, M. R. (2022). Preventing amyotrophic lateral sclerosis: Insights from pre-symptomatic neurodegenerative diseases. *Brain*, *145*(1), 27–44.

Betcke, I., Götzinger, A. C., Kornet, M. M., & Müller, T. J. (2024). Multicomponent syntheses of pyrazoles via (3+ 2)-cyclocondensation and (3+ 2)-cycloaddition key steps. *Beilstein Journal of Organic Chemistry*, *20*(1).

Birmann, P. T., Casaril, A. M., Hartwig, D., Jacob, R. G., Seixas, F. K., Collares, T., & Savegnago, L. (2020). A novel pyrazole-containing selenium compound modulates the oxidative and nitrergic pathways to reverse the depression-pain syndrome in mice. *Brain Research*, *1741*, 146880.

Bonastre-Sabater, I., Lopera, A., Martínez-Camarena, Á., Blasco, S., Doménech-Carbó, A., Jiménez, H. R., …, & Clares, M. P. (2024). Exo-or endo-1 H-pyrazole metal coordination modulated by the polyamine chain length in [1+ 1] condensation azamacrocycles. Binuclear complexes with remarkable SOD activity. *Dalton Transactions*, *53*(40), 16480–16494.

Chen, Y., Qin, Q., Zhao, W., Luo, D., Huang, Y., Liu, G., …, & Chen, Y. (2022). Carnosol reduced pathogenic protein aggregation and cognitive impairment in neurodegenerative diseases models via improving proteostasis and ameliorating mitochondrial disorders. *Journal of Agricultural and Food Chemistry*, *70*(34), 10490–10505.

Cortes-Flores, H., Torrandell-Haro, G., & Brinton, R. D. (2024). Association between CNS-active drugs and risk of Alzheimer's and age-related neurodegenerative diseases. *Frontiers in Psychiatry*, *15*, 1358568.

Desai, N. C., Bhatt, K., Monapara, J., Pandit, U., & Khedkar, V. M. (2021). Conventional and microwave-assisted synthesis, antitubercular activity, and molecular docking studies of pyrazole and oxadiazole hybrids. *ACS Omega*, *6*(42), 28270–28284.

Dhapola, R., Sharma, P., Kumari, S., Bhatti, J. S., & HariKrishnaReddy, D. (2024). Environmental toxins and Alzheimer's disease: A comprehensive analysis of pathogenic mechanisms and therapeutic modulation. *Molecular Neurobiology, 61*(6), 3657–3677.

Doğan, A. H. T., Saylam, M., YILMAZ, S., Parlar, S., Ballar, P., & Alptüzün, V. (2025). 3-(1H-pyrazole-1-yl/1H-1, 2, 4-triazole-1-yl)-N-propananilide derivatives: Design, synthesis and neuroprotectivity potential against 6-OHDA induced neurotoxicity model. *Turkish Journal of Pharmaceutical Sciences, 22*(1), 1.

El Alaouy, M. A., Alaqarbeh, M., Ouabane, M., Zaki, H., ElBouhi, M., Badaoui, H., …, & Bouachrine, M. (2024). Computational prediction of 3, 5-diaryl-1H-pyrazole and spiropyrazolines derivatives as potential acetylcholinesterase inhibitors for Alzheimer disease treatment by 3D-QSAR, molecular docking, molecular dynamics simulation, and ADME-Tox. *Journal of Biomolecular Structure and Dynamics, 42*(18), 9547–9560.

El Alaouy, M. A., Youness, M., Boutalaka, M., Elbouhi, M. H., Elmernissi, R., Sbai, A., …, & Bouachrine, M. (2021). The 2D-QSAR study, drug likeness and in-silico ADMET prediction of about 3, 5-diaryl-1H-pyrazole derivatives as multifunctional agents for the treatment of Alzheimer's disease. *RHAZES: Green and Applied Chemistry, 13*, 09–28.

El Bakri, Y., Siddique, S. A., Mohamed, S. K., Sarfraz, M., Abuelizz, H. A., Al-Salahi, R., …, & Ahmed, E. A. (2024). Synthesis, investigation of the crystal structure, DFT calculations, and in silico medicinal potential of hydrazono-and aminomethylene substituted pyrazolidine-3, 5-diones as potential anticancer scaffolds. *New Journal of Chemistry, 48*(28), 12591–12608.

Eldehna, W. M., Alkabbani, M. A., Elsayed, Z. M., Ibrahim, K. M., Majrashi, T. A., Elagawany, M., …, & Tawfik, H. O. (2025). Identification of novel Triazole–Pyrazole conjugates as potential anticonvulsant agents: Synthesis and biological evaluations. *ACS Chemical Neuroscience, 16*(15), 3038–3050.

El Fadili, M., Er-rajy, M., Imtara, H., Kara, M., Zarougui, S., Altwaijry, N., …, & Elhallaoui, M. (2022). 3D-QSAR, ADME-tox in silico prediction and molecular docking studies for modeling the analgesic activity against neuropathic pain of novel NR2B-selective NMDA receptor antagonists. *Processes, 10*(8), 1462.

El Fadili, M., Ez-Zoubi, A., Aloui, M., Mujwar, S., Abuelizz, H. A., Elhalaoui, M., & Amin, A. (2025). Design of novel pyrazole and benzofuran-based derivatives as potent acetylcholinesterase inhibitors for Alzheimer's disease management. *Frontiers in Chemistry, 13*, 1614462.

El-Gaby, M., Ammar, Y., IH El-Qaliei, M., Hussein, F. M., & Faraghally, A. F. (2020). Sulfonamides: Synthesis and the recent applications in medicinal chemistry. *Egyptian Journal of Chemistry, 63*(12), 5289–5327.

Fadaly, W. A., Elshewy, A., Abusabaa, A. H., Amin, D. M., Abdelhady, H. K., Haredy, H. H., …, & Nemr, M. T. (2025). Novel pyrazole carboxylate derivatives as lonazolac bioisosteres with selective COX-2 inhibition: Design, synthesis and anti-inflammatory activity. *Molecular Diversity*, 1–26.

Feng, A., Zeng, Q., Wang, J., Li, H., Fang, X., Geng, Y., …, & Dong, J. (2025). Synthesis of novel pyrazole derivatives and neuroprotective effect investigation. *Journal of Enzyme Inhibition and Medicinal Chemistry, 40*(1), 2583820.

Gameiro, I., Michalska, P., Tenti, G., Cores, Á., Buendia, I., Rojo, A. I., …, & León, R. (2017). Discovery of the first dual GSK3β inhibitor/Nrf2 inducer. A new multitarget therapeutic strategy for Alzheimer's disease. *Scientific Reports, 7*(1), 45701.

Gao, M., Qu, K., Zhang, W., & Wang, X. (2021). Pharmacological activity of pyrazole derivatives as an anticonvulsant for benefit against epilepsy. *Neuroimmunomodulation, 28*(2), 90–98.

Haghighijoo, Z., Akrami, S., Saeedi, M., Zonouzi, A., Iraji, A., Larijani, B., …, & Edraki, N. (2020). N-cyclohexylimidazo [1, 2-a] pyridine derivatives as multi-target-directed ligands for treatment of Alzheimer's disease. *Bioorganic Chemistry, 103*, 104146.

Hassan, A. S., & Aboulthana, W. M. (2023). Synthesis, in vitro biological investigation, and in silico analysis of pyrazole-based derivatives as multi-target agents. *Egyptian Journal of Chemistry, 66*(6), 441–455.

Hassan, R. A., & Al-Sabawi, A. H. (2023). Synthesis and spectral study of some new α, β-unsaturated carbonyl compounds and pyrazole derivatives. *Rafidain Journal of Science, 32*(1), 138–148.

Hua, D., Huang, W., Xie, Q., Xu, W., Tang, L., Liu, M., …, & Sheng, Y. (2024). Targeting GPR39 in structure-based drug discovery reduces Ang II-induced hypertension. *Communications Biology, 7*(1), 1441.

Hussain, F., Tahir, A., Rehman, H. M., Wu, Y., Shah, M., & Rashid, U. (2025). Promising thiazolidinedione-thiazole based multi-target and neuroprotective hybrids for Alzheimer's disease: Design, synthesis, in-vitro, in-vivo and in-silico studies. *European Journal of Medicinal Chemistry*, 117327.

Jangid, K., Devi, B., Sahoo, A., Kumar, V., Dwivedi, A. R., Thareja, S., …, & Kumar, V. (2024). Virtual screening and molecular dynamics simulation approach for the identification of potential multi-target directed ligands for the treatment of Alzheimer's disease. *Journal of Biomolecular Structure and Dynamics, 42*(1), 509–527.

Jeong, S., & Joo, J. M. (2021). Transition-metal-catalyzed divergent C–H functionalization of five-membered heteroarenes. *Accounts of Chemical Research, 54*(24), 4518–4529.

Jiang, M., Liu, Z., Shao, J., Zhou, J., Wang, H., Song, C., …, & Zhang, R. (2022). Estrogen receptor α regulates phenotypic switching and proliferation of vascular smooth muscle cells through the NRF1-OMI-mitophagy signaling pathway under simulated microgravity. *Frontiers in Physiology, 13*, 1039913.

Jiang, Y., Wang, Y., Zhao, L., Yang, W., Pan, L., Bai, Y., …, & Li, Y. (2024). P129, a pyrazole ring-containing isolongifolanone-derivate: Synthesis and investigation of anti-glioma action mechanism. *Discover Oncology, 15*(1), 6.

Kanwal, M., Sarwar, S., Nadeem, H., Alghamdi, S. A., Alamro, A. A., Malik, S., …, & Muskan, A. (2025). Pyrazolone-nicotinic acid derivative (4Z)-4-(2-hydroxybenzylidine)-5-methyl-2-(pyridine-3-ylcarbonyl)-2, 4-dihydro-3H-pyrazole-3-one (IIc) as multitarget inhibitor of neurodegeneration and behavioural impairment in dementia. *Journal of Pharmacy and Pharmacology, 77*(2), 275–290.

Kaya, B., Çevik, U. A., Çiftçi, B., Necip, A., Isik, M., Ay, E. N., …, & Kaplancıklı, Z. A. (2025). Design, synthesis, and molecular docking studies of novel pyrazoline-thiazoles as cholinesterase dual-target inhibitors for the treatment of Alzheimer's disease. *ACS Omega, 10*(34), 38427–38439.

Kesari, K. K., Dhasmana, A., Shandilya, S., Prabhakar, N., Shaukat, A., Dou, J., …, & Ruokolainen, J. (2020). Plant-derived natural biomolecule picein attenuates menadione induced oxidative stress on neuroblastoma cell mitochondria. *Antioxidants, 9*(6), 552.

Khan, A. F., & Iturria-Medina, Y. (2024). Beyond the usual suspects: Multi-factorial computational models in the search for neurodegenerative disease mechanisms. *Translational Psychiatry, 14*(1), 386.

Kharde, A. K., Kadu, V. R., & Gholap, S. S. (2025). Aqueous extract of Areca nut as a sustainable and highly efficient green catalyst for the synthesis of pyrazole and isoxazolone derivatives. *Research on Chemical Intermediates*, 1–21.

Kim, K. R., Park, S. E., Hong, J. Y., Koh, J. Y., Cho, D. H., Hwang, J. J., & Kim, Y. H. (2022). Zinc enhances autophagic flux and lysosomal function through transcription factor EB activation and V-ATPase assembly. *Frontiers in Cellular Neuroscience, 16*, 895750.

Krátký, M., Vu, Q. A., Štěpánková, Š., Maruca, A., Silva, T. B., Ambrož, M., …, & Vinšová, J. (2021). Novel propargylamine-based inhibitors of cholinesterases and monoamine oxidases: Synthesis, biological evaluation and docking study. *Bioorganic Chemistry, 116*, 105301.

Leuzy, A., Smith, R., Ossenkoppele, R., Santillo, A., Borroni, E., Klein, G., …, & Hansson, O. (2020). Diagnostic performance of RO948 F 18 tau positron emission tomography in the differentiation of Alzheimer disease from other neurodegenerative disorders. *JAMA Neurology, 77*(8), 955–965.

Li, H. H., Wu, C., Zhang, S. L., Yang, J. G., Qin, H. L., & Tang, W. (2022). Fluorosulfate-containing pyrazole heterocycles as selective BuChE inhibitors: Structure-activity relationship and biological evaluation for the treatment of Alzheimer's disease. *Journal of Enzyme Inhibition and Medicinal Chemistry, 37*(1), 2099–2111.

Liao, L., Jiang, C., Chen, J., Shi, J., Li, X., Wang, Y., …, & Zhang, J. (2020). Synthesis and biological evaluation of 1, 2, 4-triazole derivatives as potential neuroprotectant against ischemic brain injury. *European Journal of Medicinal Chemistry, 190*, 112114.

Liu, Y., Tian, Q., Ge, J., Wu, X., Li, Z., & Cheng, G. (2024). Recent advances in the synthesis of trifluoromethyl-containing heterocyclic compounds via trifluoromethyl building blocks. *Organic & Biomolecular Chemistry.*

Machiela, E., Jeloka, R., Caron, N. S., Mehta, S., Schmidt, M. E., Baddeley, H. J., …, & Southwell, A. L. (2020). The interaction of aging and cellular stress contributes to pathogenesis in mouse and human huntington disease neurons. *Frontiers in Aging Neuroscience, 12*, 524369.

Manikandan, R., Devi, K., & Rajalingam, D. (2025). Design, synthesis, in-silico, and in-vitro assessment of pyrazolone-merged chalcone derivatives as anti-Alzheimer agents. *Brazilian Journal of Pharmaceutical Sciences, 61*, e24291.

Manikandan, R., Devi, K., & Rajalingam, D. (2025b). Synthesis and evaluation of 2, 5-substituted pyrazolone for neuroprotective potential in SH-SY5Y human neuroblastoma cells. *Indian Journal of Pharmaceutical Education and Research, 59*(1s), s323–s332.

Maqbool, M., Rajvansh, R., Srividya, K., & Hoda, N. (2020). Deciphering the robustness of pyrazolo-pyridine carboxylate core structure-based compounds for inhibiting α-synuclein in transgenic C. elegans model of Synucleinopathy. *Bioorganic & Medicinal Chemistry, 28*(17), 115640.

Messaad, M., Dhouib, I., Abdelhedi, M., & Khemakhem, B. (2022). Synthesis, bioassay and molecular docking of novel pyrazole and pyrazolone derivatives as acetylcholinesterase inhibitors. *Journal of Molecular Structure, 1263*, 133105.

Michalska, P., Mayo, P., Fernández-Mendívil, C., Tenti, G., Duarte, P., Buendia, I., ..., & León, R. (2020). Antioxidant, anti-inflammatory and neuroprotective profiles of novel 1, 4-dihydropyridine derivatives for the treatment of Alzheimer's disease. *Antioxidants*, *9*(8), 650.

Montagne, A., Nikolakopoulou, A. M., Huuskonen, M. T., Sagare, A. P., Lawson, E. J., Lazic, D., ..., & Zlokovic, B. V. (2021). APOE4 accelerates advanced-stage vascular and neurodegenerative disorder in old Alzheimer's mice via cyclophilin a independently of amyloid-β. *Nature Aging*, *1*(6), 506–520.

Murugan, R., Nayak, S. R. R., Haridevamuthu, B., Priya, D., Chitra, V., Almutairi, B. O., ..., & Arockiaraj, J. (2024a). Neuroprotective potential of pyrazole benzenesulfonamide derivative T1 in targeted intervention against PTZ-induced epilepsy-like condition in in vivo zebrafish model. *International Immunopharmacology*, *131*, 111859.

Murugan, R., Nayak, S. R. R., Haridevamuthu, B., Priya, D., Rajagopal, R., Pasupuleti, M., ..., & Arockiaraj, J. (2024b). Multifaceted evaluation of pyrazole derivative (T4)-chitosan (CS) nanoparticles: Morphology, drug release, and anti-tumor efficacy in a rat model. *International Journal of Biological Macromolecules*, *283*, 137702.

Naglah, A. M., Almehizia, A. A., Al-Wasidi, A. S., Alharbi, A. S., Alqarni, M. H., Hassan, A. S., & Aboulthana, W. M. (2024). Exploring the potential biological activities of pyrazole-based Schiff bases as anti-diabetic, anti-Alzheimer's, anti-inflammatory, and cytotoxic agents: In vitro studies with computational predictions. *Pharmaceuticals*, *17*(5), 655.

Narayanan, S. E., Narayanan, H., Mukundan, M., Balan, S., Vishnupriya, C. P., Gopinathan, A., ..., & Marathakam, A. (2021). Design, synthesis and biological evaluation of substituted pyrazoles endowed with brominated 4-methyl 7-hydroxy coumarin as new scaffolds against Alzheimer's disease. *Future Journal of Pharmaceutical Sciences*, *7*(1), 161.

Nguyen, H. D., & Kim, M. S. (2024). In silico exploration of promising heterocyclic molecules against both acetylcholinesterase and butyrylcholinesterase enzymes. *Journal of Biomolecular Structure and Dynamics*, *42*(14), 7128–7149.

Niharika, D. G., Salaria, P., & Reddy, M. A. (2025). Integrated computational approaches for identification of potent pyrazole-based glycogen synthase kinase-3β (GSK-3β) inhibitors: 3D-QSAR, virtual screening, docking, MM/GBSA, EC, MD simulation studies. *Molecular Diversity*, *29*(5), 4177–4201.

Nikiforov, E. A., Vaskina, N. F., Moseev, T. D., Varaksin, M. V., Butorin, I. I., Melekhin, V. V., ..., & Chupakhin, O. N. (2023). Indolyl-derived 4 H-imidazoles: PASE synthesis, molecular docking and in vitro cytotoxicity assay. *Processes*, *11*(3), 846.

Norton, D., Bonnette, W. G., Callahan, J. F., Carr, M. G., Griffiths-Jones, C. M., Heightman, T. D., ..., & Davies, T. G. (2021). Fragment-guided discovery of pyrazole carboxylic acid inhibitors of the Kelch-like ECH-associated protein 1: Nuclear factor erythroid 2 related factor 2 (KEAP1: NRF2) protein–protein interaction. *Journal of Medicinal Chemistry*, *64*(21), 15949–15972.

Paciello, F., Pisani, A., Rolesi, R., Montuoro, R., Mohamed-Hizam, V., Boni, G., ..., & Grassi, C. (2024). Oxidative stress and inflammation cause auditory system damage via glial cell activation and dysregulated expression of gap junction proteins in an experimental model of styrene-induced oto/neurotoxicity. *Journal of Neuroinflammation*, *21*(1), 4.

Parambi, D. G. T., Saleem, U., Shah, M. A., Anwar, F., Ahmad, B., Manzar, A., ..., & Mathew, B. (2020). Exploring the therapeutic potentials of highly selective oxygenated chalcone based MAO-B inhibitors in a haloperidol-induced murine model of Parkinson's disease. *Neurochemical Research*, *45*(11), 2786–2799.

Pogoda, A., Chmielewska, N., Maciejak, P., & Szyndler, J. (2021). Transcriptional dysregulation in Huntington's disease: The role in pathogenesis and potency for pharmacological targeting. *Current Medicinal Chemistry*, *28*(14), 2783–2806.

Potapov, K. V., Platonov, D. N., Belyy, A. Y., Novikov, M. A., Tomilov, Y. V., Anashkina, A. A., ..., & Mitkevich, V. A. (2025). Improved synthesis of effective 3-(Indolin-6-yl)-4-(N-pyrazole-sulfonamide)-1 H-pyrrolo [2, 3-b] pyridine-based inhibitors of NADPH oxidase 2. *International Journal of Molecular Sciences*, *26*(8), 3647.

Pourtaher, H., Mohammadi, Y., Hasaninejad, A., & Iraji, A. (2024). Highly efficient, catalyst-free, one-pot sequential four-component synthesis of novel spiroindolinone-pyrazole scaffolds as anti-Alzheimer agents: In silico study and biological screening. *RSC Medicinal Chemistry*, *15*(1), 207–222.

Rahman, A., Saikia, B., & Baruah, A. (2023). In silico analysis of SOD1 aggregation inhibition modes of tertiary amine pyrazolone and pyrano coumarin ferulate as ALS drug candidates. *Physical Chemistry Chemical Physics*, *25*(39), 26833–26846.

Rani, P., Chahal, S., Ranolia, A., Kumar, D., Kataria, R., Kumar, P., …, & Sindhu, J. (2025). Design and development of sulfenylated 5-aminopyrazoles as inhibitors of acetylcholinesterase and butyrylcholinesterase: Exploring the implication for Aβ 1–42 aggregation inhibition in Alzheimer's disease. *RSC Medicinal Chemistry*.

Risen, S. J., Boland, S. W., Sharma, S., Weisman, G. M., Shirley, P. M., Latham, A. S., …, & Moreno, J. A. (2024). Targeting neuroinflammation by pharmacologic downregulation of inflammatory pathways is neuroprotective in protein misfolding disorders. *ACS Chemical Neuroscience*, *15*(7), 1533–1547.

Roecker, A. J., Schirripa, K. M., Loughran, H. M., Tong, L., Liang, T., Fillgrove, K. L., Kuo, Y., Bleasby, K., Collier, H., Altman, M. D., & Ford, M. C. (2023, January 12). Pyrazole ureas as low dose, CNS penetrant glucosylceramide synthase inhibitors for the treatment of Parkinson's disease. *ACS Medicinal Chemistry Letters*, *14*(2), 146–155.

Ryan, P., Xu, M., Jahan, K., Davey, A. K., Bharatam, P. V., Anoopkumar-Dukie, S., …, & Rudrawar, S. (2020). Novel furan-2-yl-1 H-pyrazoles possess inhibitory activity against α-synuclein aggregation. *ACS Chemical Neuroscience*, *11*(15), 2303–2315.

Sanad, S. M., & Mekky, A. E. (2022). [3 + 2] cycloaddition synthesis of new (nicotinonitrile-chromene) hybrids linked to pyrazole units as potential acetylcholinesterase inhibitors. *Synthetic Communications*, *52*(16), 1672–1684.

Satish, D. J., & Shinde, P. (2025). Design, synthesis, and biological evaluation of novel heterocyclic scaffolds targeting neurodegenerative disorders. *International Journal of Advanced Research and Multidisciplinary Trends (IJARMT)*, *2*(1), 823–839.

Sbai, A., El Alaouy, M. A., Maghat, H., Lakhlifi, T., & Bouachrine, M. (2026). Exploring pyrazole and spiropyrazoline derivatives as acetylcholinesterase inhibitors for Alzheimer's treatment: An in-silico study combining Monte Carlo optimization, HQSAR, molecular docking, ADME/Tox, and molecular dynamics. *Moroccan Journal of Chemistry*, *14*(1), J-Chem.

Schrecker, L., Dickhaut, J., Holtze, C., Staehle, P., Vranceanu, M., Hellgardt, K., & Hii, K. K. M. (2023). Discovery of unexpectedly complex reaction pathways for the Knorr pyrazole synthesis via transient flow. *Reaction Chemistry & Engineering*, *8*(1), 41–46.

Sharma, S., Advani, D., Das, A., Malhotra, N., Khosla, A., Arora, V., … & Kumar, P. (2022). Pharmacological intervention in oxidative stress as a therapeutic target in neurological disorders. *Journal of Pharmacy and Pharmacology*, *74*(4), 461–484.

Singh, R., Panghal, A., Jadhav, K., Thakur, A., Verma, R. K., Singh, C., … & Namdeo, A. G. (2024). Recent advances in targeting transition metals (copper, iron, and zinc) in Alzheimer's disease. *Molecular Neurobiology*, *61*(12), 10916–10940.

Smith, R., Capotosti, F., Schain, M., Ohlsson, T., Vokali, E., Molette, J., …, & Hansson, O. (2023). The α-synuclein PET tracer [18F] ACI-12589 distinguishes multiple system atrophy from other neurodegenerative diseases. *Nature Communications*, *14*(1), 6750.

Smolen, P., Dash, P. K., & Redell, J. B. (2023). Traumatic brain injury-associated epigenetic changes and the risk for neurodegenerative diseases. *Frontiers in Neuroscience*, *17*, 1259405.

Srour, A. M., Fahmy, H. H., Khater, M. A., Zarie, E. S., Mohamed, S. S., & Abdelhameed, M. F. (2022). Synthesis, anti-inflammatory properties, molecular modelling and potential COX-2, TNF-α, PGE2 and IL1β inhibitors of pyrazole-based scaffolds. *Journal of Molecular Structure*, *1266*, 133499.

Sun, J., Zhang, Y., Kong, Y., Ye, T., Yu, Q., Satyanarayanan, S. K., …, & Liu, J. (2022). Microbiota-derived metabolite indoles induced aryl hydrocarbon receptor activation and inhibited neuroinflammation in APP/PS1 mice. *Brain, Behavior, and Immunity*, *106*, 76–88.

Tok, F., Özer, F. A., Kuzu, Z. S., Sıcak, Y., Kemaloğlu, F., Baday, S., …, & Öztürk, M. (2025). Synthesis, biological evaluation and molecular docking of novel 2-pyrazoline-1-carboxamides as anti-Alzheimer agents. *Future Medicinal Chemistry*, 1–17.

Vidali, V. P., Nigianni, G., Athanassopoulou, G. D., Canko, A., Mavroidi, B., Matiadis, D., …, & Sagnou, M. (2022). Synthesis of novel pyrazolo [3, 4-b] pyridines with affinity for β-amyloid plaques. *Molbank*, *2022*(1), M1343.

Wang, H., Wu, Y., Liu, A., Li, S., Zhu, P., Zuo, J., …, & Jiang, X. (2025). Design, synthesis and biological evaluation of novel pyrazolinone derivatives as multifunctional ligands for the treatment of Alzheimer's disease. *Bioorganic Chemistry*, *154*, 108052.

Yan, N., Shi, X. L., Tang, L. Q., Wang, D. F., Li, X., Liu, C., & Liu, Z. P. (2022). Synthesis and biological evaluation of thieno [3, 2-c] pyrazol-3-amine derivatives as potent glycogen synthase kinase 3β inhibitors for Alzheimer's disease. *Journal of Enzyme Inhibition and Medicinal Chemistry*, *37*(1), 1724–1736.

Yerramsetty, S., Ayyagari, L., Makani, V. K. K., Mendonza, J. J., Ajjarapu, S. M., Kamal, A., …, & Bhadra, M. P. (2025). Neuroprotective and anti-apoptotic activity of anthranilamide pyrazolo [1, 5-a] pyrimidine derivative against Parkinson's disease model in rotenone-induced SH-SY5Y cells. *International Journal of Pharmaceutical Investigation*, *15*(2).

Zhang, D., Ren, L., Liu, A., Li, W., Liu, Y., & Gu, Q. (2022). One-pot solvent-free synthesis of 1, 3, 5-trisubstituted 1 H-pyrazoles catalyzed by H3 [PW12O40]/SiO2 under microwave irradiation. *Monatshefte für Chemie-Chemical Monthly*, *153*(3), 257–266.

Zheng, C., Stehouwer, J. S., Ummenthala, G. R., Munot, Y. S., & Vasdev, N. (2025). Multi-fused S, N-heterocyclic compounds for targeting α-synuclein aggregates. *Cells*, *14*(19), 1531.

Zugno, G. P., Birmann, P. T., Sinott, A., Fetter, J., de Oliveira, R. L., Valente, L. C., …, & Savegnago, L. (2025). Modulating monoaminergic systems: A study on 3, 5-dimethyl-1-phenyl-4-(phenylselenyl)-1H-pyrazole in lipopolysaccharide-induced depressive-like symptoms in mice. *European Journal of Pharmacology*, 178176.

7 Quinoline and Isoquinoline Analogs in Neuroprotection

Moharir Shreyash Rajendra, Kosana Sai Chaitanya, Rafael Balaña Fouce, Estela Melcón Fernández, Murugesan Sankaranarayanan, and Kondapalli Venkata Gowri Chandra Sekhar

7.1 INTRODUCTION

Neurodegenerative disorders are conditions that affect the body's nervous system, the brain, spinal cord, and the network of peripheral nerves. More than one in three people globally live with a neurodegenerative condition, making these disorders the leading cause of illness and disability, as reported by the WHO. They have a wide range of effects because the nervous system controls nearly all body functions. They can develop due to many reasons, including inherited genetic changes, infections, injuries, autoimmune reactions, neurodegeneration, or problems with blood flow to the brain (Figure 7.1). Some well-known neurodegenerative disorders include Parkinson's disease (PD), Alzheimer's disease (AD), Huntington's disease (HD), multiple sclerosis (MS), stroke, epilepsy, and various forms of peripheral neuropathy (Figure 7.2). Under these conditions, individuals may experience muscle weakness, tremors, loss of balance or coordination, changes in sensation, seizures, chronic pain, memory problems, or difficulties with thinking and concentration. In many cases,

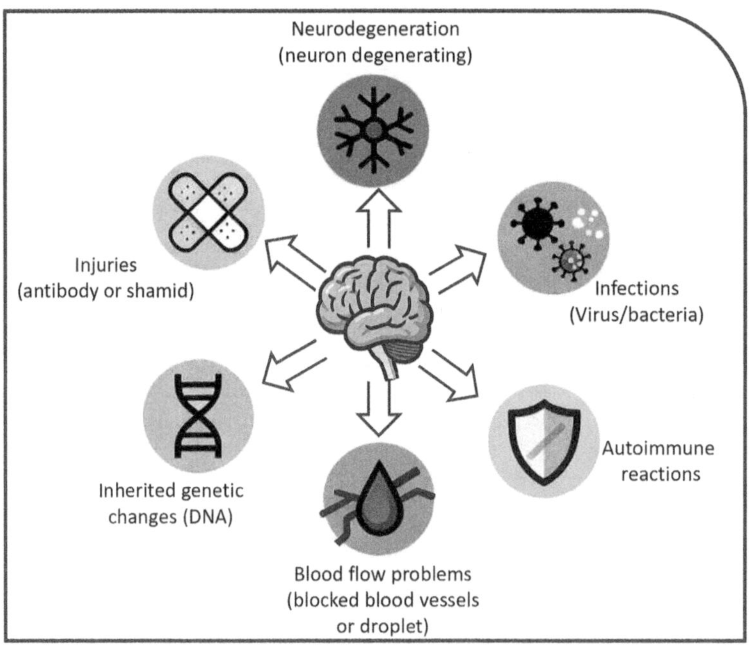

FIGURE 7.1 Causes of neurodegenerative disorders.

DOI: 10.1201/9781003772019-7

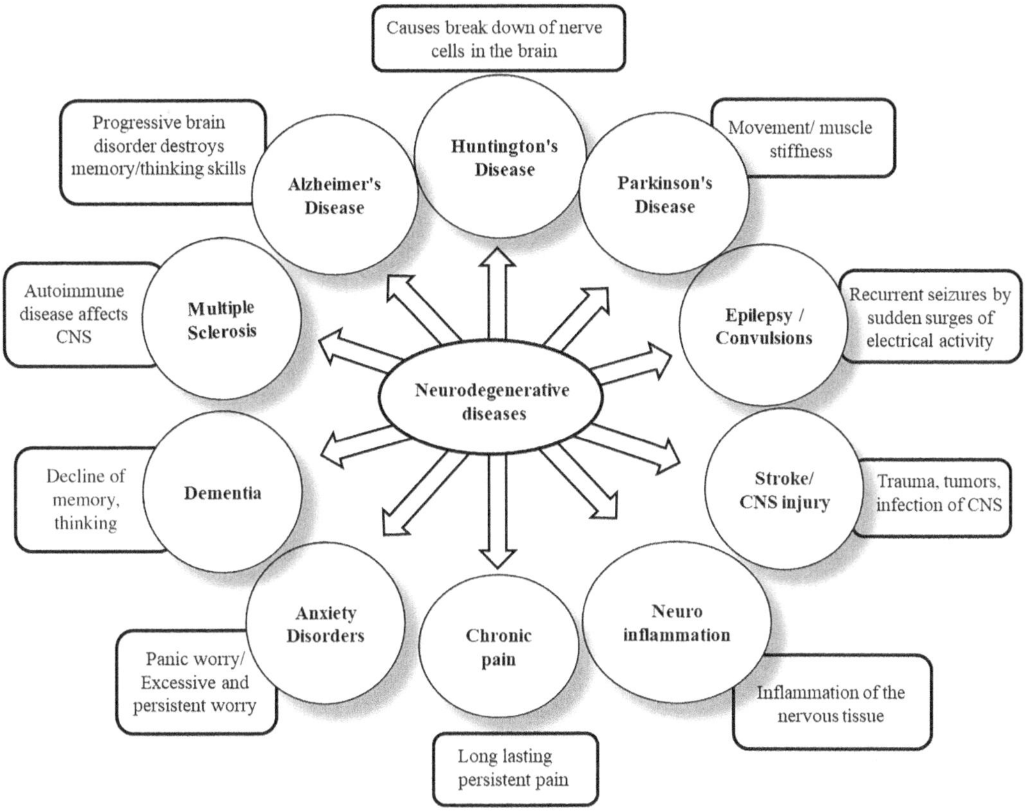

FIGURE 7.2 Types of neurodegenerative disorders.

emotional regulation and daily activities are also affected, making neurodegenerative disorders challenging both for patients and caregivers (Gadhave et al., 2024).

Heterocyclic rings are fundamental in medicinal chemistry, serving as privileged scaffolds that modulate key properties such as polarity, hydrogen bonding, and lipophilicity to optimize drug behavior (Jampilek, 2019). Among these, quinolines have proven to be extraordinarily versatile, used in anticancer, antimicrobial, and neuroprotective agents, thanks to their rigid bicyclic nitrogen structure (Yadav & Shah, 2021). Meanwhile, isoquinolines, especially from natural alkaloids, exhibit strong multitarget potential in many neurodegenerative disorders (Plazas & Muñoz, 2022). Over the past two decades, quinolines and isoquinolines have consistently attracted strong research interest, reflected by a continuous rise in scientific publications. Their versatile heterocyclic structures enable them to interact with multiple neurobiological targets, making them promising scaffolds for developing effective therapies against several neurodegenerative disorders (Figure 7.3).

7.2 QUINOLINES

Quinoline, a bicyclic nitrogen-containing heterocycle composed of a fused benzene and pyridine ring, has long served as a preferred scaffold in medicinal chemistry. Naturally, quinoline was first discovered through the degradation of quinine, an alkaloid from Cinchona bark, which marked a foundational moment in early medicinal chemistry. Quinoline and its derivatives, which were initially known for their antimalarial and antimicrobial properties (quinine, chloroquine, and 8-hydroxyquinoline), have sparked significant interest in recent decades due to their potential neuroprotective applications in the treatment of various neurodegenerative disorders (Senerovic et al., 2019). The significance of

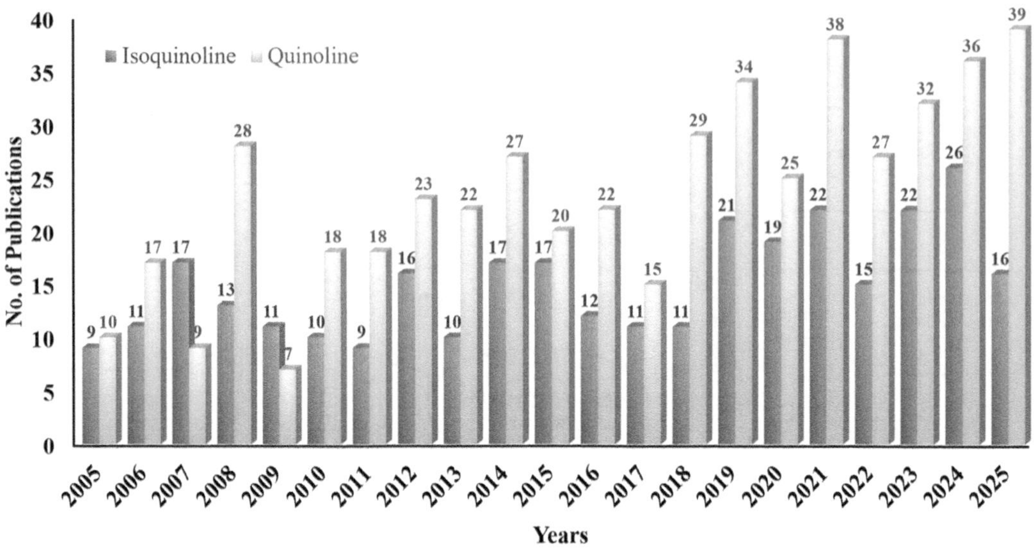

FIGURE 7.3 List of publications on quinoline and isoquinoline scaffolds on neurodegenerative disorders.

quinoline-based compounds comes from their synthetic accessibility, emerging pharmacology, and structural modifiability. There are various quinoline-based USFDA-approved drugs currently in the pharmaceutical market for treating various diseases such as cancer, malaria, tuberculosis, and several drugs act as antibiotics for treating bacterial infections (Figure 7.4) (Chandra et al., 2025).

Quinine and chloroquine are two of the earliest antimalarial medications; quinine (derived from Cinchona bark) and chloroquine (a synthetic analogue) work as blood schizonticides by preventing *Plasmodium* parasites from detoxifying their heme. Quinine also serves as an off-label agent for relieving nocturnal leg cramps (Kingston & Cassera, 2022). Hydroxychloroquine, an antimalarial and disease-modifying antirheumatic drug (DMARD) used to treat disorders such as rheumatoid arthritis and lupus erythematosus, is believed to exert its immunomodulating action through its impact on Toll-like receptor (TLR) signaling and lysosomal acidification. Mefloquine is an oral medication used to prevent malaria that is effective against *Plasmodium* species resistant to chloroquine (Schrezenmeier and Dörner). Primaquine is well-known for eliminating latent liver stages, but it is also distinctive in inhibiting the further transmission of *Plasmodium.* Bedaquiline, which targets mycobacterial ATP synthase and has been approved for multidrug-resistant TB, offers a breakthrough for chronic infections. The scaffold's adaptability beyond antimalarial and anti-inflammatory indications, as demonstrated by its quinoline chemical core. Irinotecan, which is a semisynthetic Camptothecin derivative, is used to treat lung, colon, and other malignancies, via targeting topoisomerase-I enzyme (Kciuk et al., 2020). Yet there are no quinoline-based USFDA-approved drugs for the treatment of neurodegenerative disorders. However, their emergence as multitarget-directed ligands for neuroprotection has only recently been recognized.

Quinolines have recently gained interest as potential neuroprotective agents due to their chemical versatility and ability to affect multiple neuropathological pathways. The complicated aetiologies of neurodegenerative disorders, including AD, PD, HD, and others, include oxidative stress (OS), neuroinflammation, protein aggregation, and neuronal death (Hernández-Ayala et al., 2023). As multitarget molecules, quinoline derivatives are ideally suited to treat these multifactorial processes involved in neurodegenerative disorders.

This section defines the structural characteristics, pharmacological mechanisms, structure–activity relationships (SARs), and therapeutic potential i.e., detailed medicinal chemistry of quinoline analogues in the context of neuroprotection.

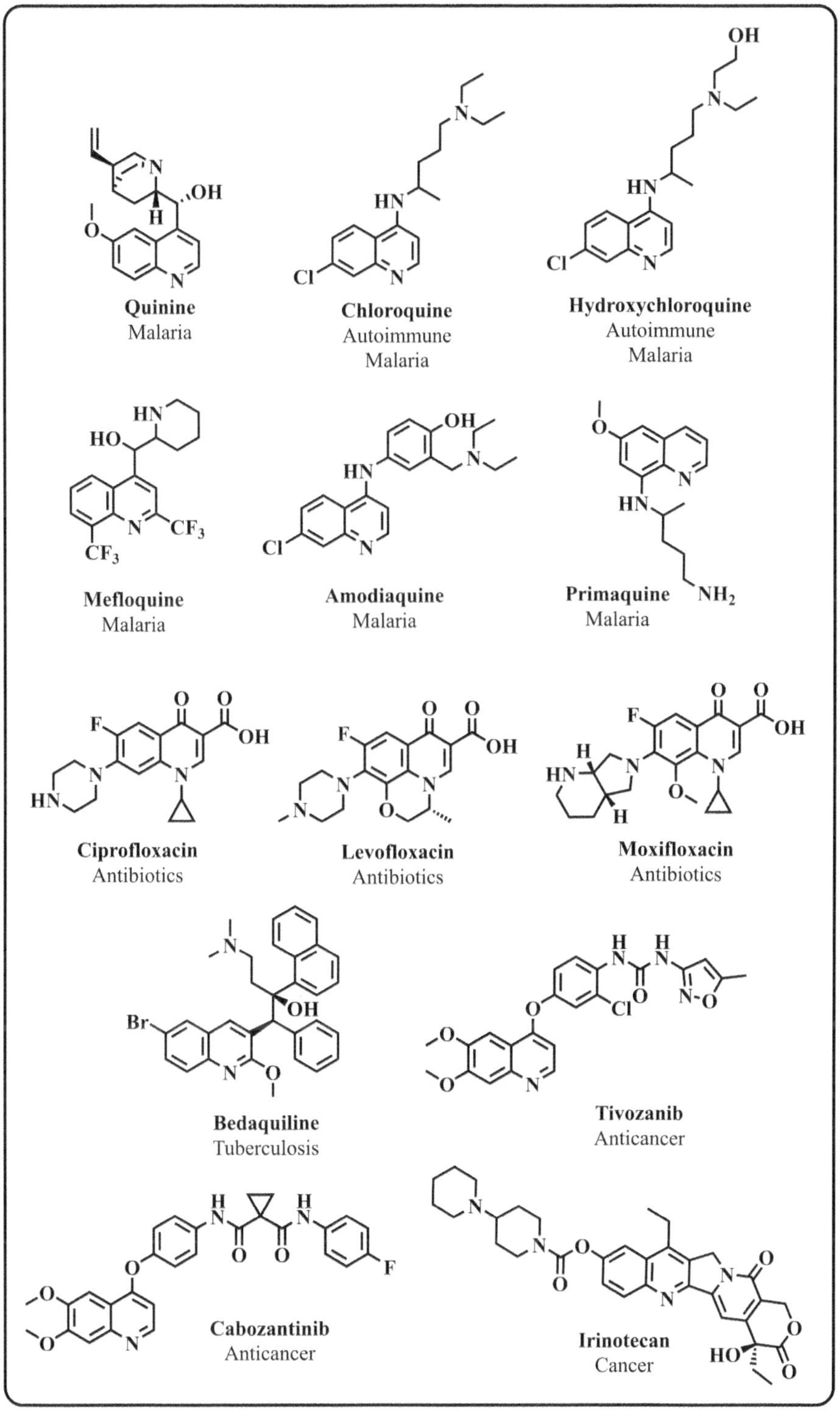

FIGURE 7.4 Structure of quinoline-based USFDA-approved drugs with corresponding disorder.

7.2.1 Chemical Structure and Origin of Quinoline Analogs

Quinoline (C_9H_7N, colorless liquid) is characterized by a planar, aromatic framework consisting of a benzene ring fused to a pyridine ring with nitrogen positioned at atom 1 (i.e., 1-azanaphthalene) (Figure 7.32). The nitrogen in the ring confers electronic properties (heterocyclic basicity, hydrogen bonding, and substitution at multiple ring positions such as 2, 4, 6, and 8) that make it a privileged scaffold in medicinal chemistry. Substituted quinolines have the potential to significantly modify lipophilicity, membrane permeability, and binding affinity toward a range of broad biological targets. Derivatives can have a variety of substituents (such as alkyl, alkoxy, halogen, amine, hydroxy, and carboxy) at different positions, which significantly impact both biological activity and pharmacokinetics. Different substitution patterns were made possible by the evolution of its derivatives via both synthetic and natural biosynthetic routes (Sahu et al., 2024).

In the 19th century, Friedlieb Ferdinand Runge extracted quinoline from coal tar. Later, it was derived shortly after by Charles Frederic Gerhart from the decomposing process of the Cinchona alkaloids, Quinine and Cinchonidine. Quinoline is abundantly found in coal tar, oil shale, and petroleum, and it may also be obtained by the combustion of a variety of substances, including tobacco. However, the major source of commercial quinoline is coal tar. Synthetic approaches such as the Skraup, Doebner-Miller, Combes, Conrad–Limpach, (in acidic conditions) Friedlander, and Pfitzinger (in basic conditions) synthesis make it easier to generate quinoline analogue libraries for biological screening. All of them include a cyclo-addition of aniline (or an aniline derivative) to a carbonyl compound, followed by aromatization via dehydration/oxidation processes (Figure 7.5). The starting materials and reaction conditions are used (basic or acidic) where these procedures differ from one another (Patel et al., 2022). The synthetic approaches are described below:

- *Skraup Synthesis*: Condensation of aniline with glycerol in the presence of an oxidizing agent and strong acid, results in unsubstituted or 2-substituted quinolines.
- *Doebner-Miller Synthesis*: Condensation of aniline with an α,β-unsaturated carbonyl compound under acidic conditions to yield 2,3- or 2,4-disubstituted quinolines.

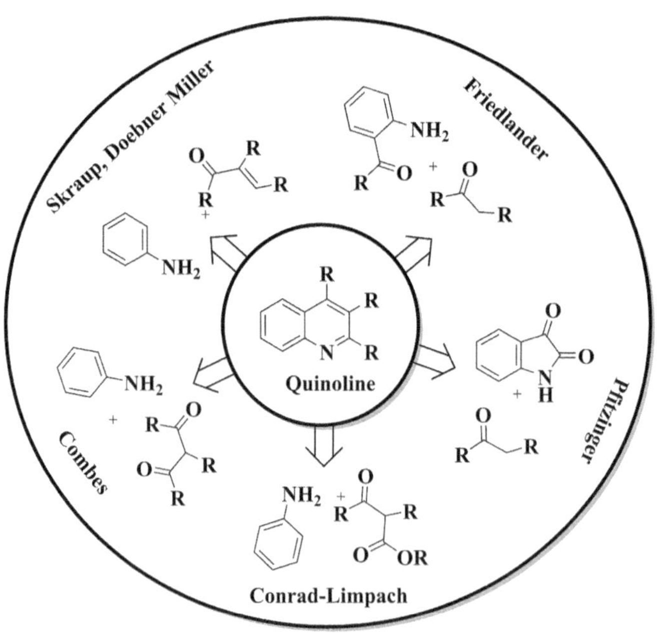

FIGURE 7.5 Conventional synthesis strategies toward substituted quinolines.

- *Combes Synthesis*: Aniline reacts with a β-diketone under acidic conditions, giving substituted quinolines, depending on the diketone and aniline used.
- *Conrad–Limpach Synthesis*: Aniline (or substituted anilines) with β-ketoesters, under acid catalysis, followed by cyclization, and dehydration, gives 4-hydroxyquinolines and derivatives.
- *Friedlander Synthesis*: 2-aminoaryl ketone or 2-aminobenzaldehyde with a carbonyl compound (such as a ketone) in the presence of base, results in 2-substituted quinolines.
- *Pfitzinger Synthesis*: Isatin (an indole derivative) reacts with methylene ketones in basic conditions to give quinolone and its derivatives.

Although many of these traditional procedures are highly successful and remain widely used today, they have numerous significant environmental downsides. Many of them require extreme conditions, such as strong acids or bases at high temperatures and prolonged durations. Additionally, they generate a significant amount of waste and can require costly or hazardous metal catalysts (Weyesa & Mulugeta, 2020). Therefore, for both synthetic and medicinal chemists, creating novel quinolines using effective and environmentally acceptable methods, such as microwave-irradiated synthesis, photocatalytic synthesis, ultrasound-assisted synthesis, and various green chemistry methods (Figure 7.6), has become both a challenge and an essential aim in recent years.

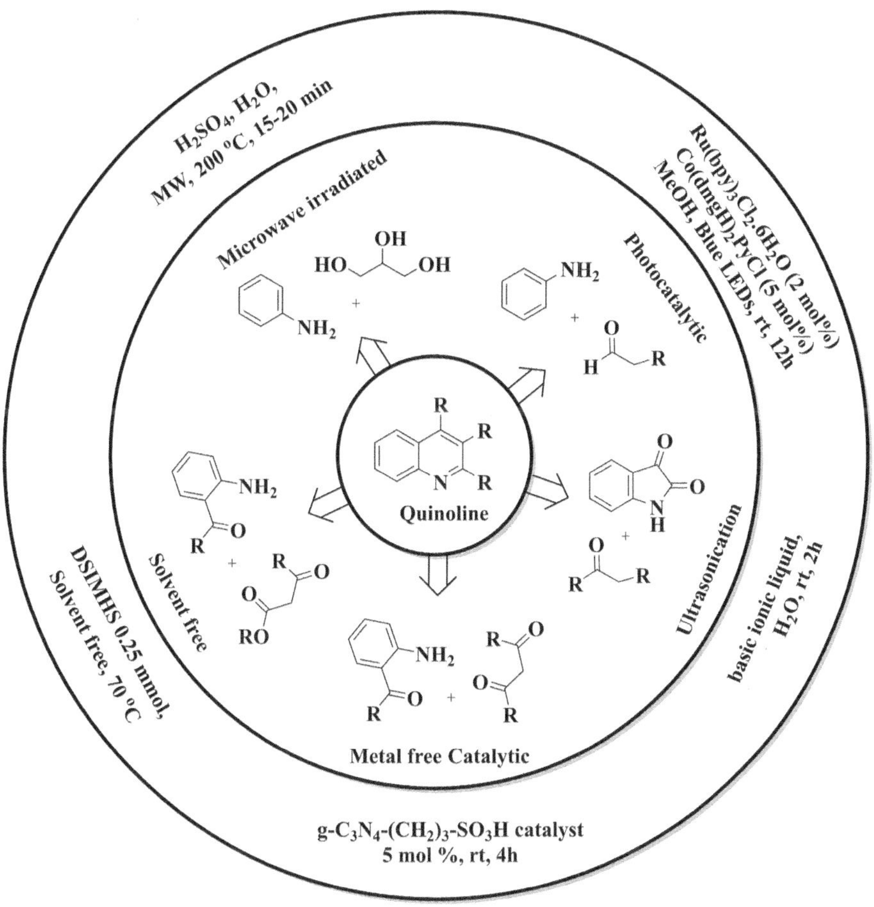

FIGURE 7.6 Green synthesis strategies toward substituted quinolines.

7.2.2 Neuroprotective Role of Quinoline Analogs

Quinoline derivatives affect neurodegenerative disorders through several interconnected pathways in the following ways.

7.2.2.1 Antioxidant and Metal Chelation

The pathophysiological basis of neurodegenerative disorders is OS, which arises from an imbalance between the generation of reactive oxygen species (ROS) and antioxidant defenses. Through their ability to scavenge radicals and chelate metals, quinoline analogues inhibit ROS, particularly when they are replaced with groups that donate electrons (such as hydroxy or methoxy). For instance, 8-hydroxyquinoline derivatives effectively chelate redox-active metals, Fe(II), Cu(II), and Zn(II), which prevents neuronal damage and inhibits Fenton-type reactions that produce ROS. Such multifunctional antioxidants protect neurons against OS-induced apoptosis (Hernández-Ayala et al., 2023).

7.2.2.2 Enzyme Inhibition and Multitargeting Effects

A vital element in treating symptoms of AD is the use of cholinesterase inhibitors. Quinoline analogues have shown no Pargyline-like inhibitory effects on butyrylcholinesterase (BuChE) and acetylcholinesterase (AChE). Molecular docking studies show hydrogen-bonding interactions and π–π stacking in the catalytic anionic site of AChE. Additionally, certain derivatives inhibit monoamine oxidase B (MAO-B), an enzyme involved in oxidative damage and dopamine metabolism in PD. Quinoline pharmacophore-based multitarget-directed ligands (MTDLs) have the potential to concurrently target multiple enzymes implicated in central nervous system (CNS) disorders (Zhou & Huang, 2021).

7.2.2.3 Amyloidogenesis and Tauopathy Modulation

Tau hyperphosphorylation and amyloid-beta (Aβ) aggregation are key components of AD pathogenesis. Amyloid and tau fibrils can be bound by quinoline-based probes and inhibitors, which can alter aggregation and encourage the removal of hazardous oligomers. Metal-chelating quinolines, including clioquinol, have dual neuroprotective effects by reducing aggregation and facilitating the elimination of metal-ion-linked aggregates. Certain quinoline analogues promote non-amyloidogenic processing of amyloid precursor protein (APP) via activating α-secretase (Paclibar et al., 2025).

7.2.2.4 Anti-Inflammatory Effects and Receptor Modulation

By activating microglia and astrocytes and generating pro-inflammatory cytokines and ROS, neuroinflammation worsens neurodegenerative processes. Quinoline analogues can inhibit the generation of TNF-α and IL-1β, decrease inflammatory cascades, and alter important signaling pathways, including NF-κB and Nrf2. Additionally, novel quinolines activate ERK1/2 and Akt pro-survival pathways and inhibit neuronal death by binding to CB2 cannabinoid receptors (Dhapola et al., 2021).

7.2.2.5 Mitochondrial Function and Mitophagy

Neuronal loss in HD and PD is caused by impaired mitochondrial dynamics and poor mitophagy (the removal of damaged mitochondria). In models of cerebral ischemia and dementia, quinoline analogues with resveratrol-like alterations, such as RV02, activate Parkin-mediated mitophagy and maintain mitochondrial function, preventing cell death (Chen & Chan, 2009).

7.2.3 Quinoline Analogs for Major Neurodegenerative Disorders

Assis de Oliveira and coworkers in 2025 (de Oliveira et al., 2025) synthesized 4-substituted quinoline derivatives. The study found that sulfonamide derivatives, specifically **QN1, QN2,** and **QN3**, were the most potent (Figure 7.7). Interestingly, compound **QN1** showed expressive results as an inhibitor for all target enzymes *in-vitro* (iGSK3B 73.4%, iBACE 81.2%, and iAChE 94.6%) and

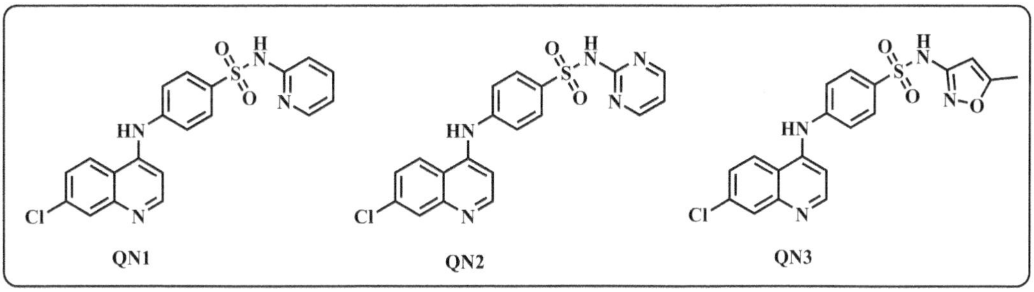

FIGURE 7.7 4-Substituted quinoline-based sulfonamide derivatives.

in-silico (HsBACE1 −9.5, HsGSK3B −9.7, HsAChE −11.9, and EeAChE −11.5 kcal/mol). Therefore, this study presents a molecule with multitarget properties, suggested as a potential inhibitor of AChET activity.

S. Jena and coworkers in 2024 (Jena et al., 2024) designed and synthesized novel set of 5,6-dihydropyrido [2′,1′:2,3]imidazo [4,5-*c*]quinolines-based compounds through microwave-assisted one-pot telescopic approach. A total of 18 derivatives were synthesised and evaluated for their neuroprotective activity. The findings showed that **QN4, QN5,** and **QN6** compounds are just as effective as ebselen at shielding neuron-like cells against MPP+-induced cell death and total cellular superoxide generation. Subsequent studies revealed involvement with two well-established mechanisms, CB2 signaling and acetylcholine metabolism (Figure 7.8), that are pertinent to the treatment of neurodegeneration. Such dual-acting compounds can provide both symptom alleviation and neuroprotection.

Wang et al. (2024) synthesised a novel set of melatonin hydroxyquinoline–based hybrid molecules that target both metal-chelating and antioxidation at the same time. The compounds demonstrated high blood–brain barrier (BBB) permeability, substantial oxygen radical absorbance capacity (**QN7** = 2.76±0.01 and **QN8** = 3.23±0.02, clioquinol= 0.54±0.17), and suppression of Aβ_{1-42} aggregation (**QN7** = 63.24±2.02% and **QN8** = 40.33±2.09%, clioquinol = 30.76±1.08%). Among these, **QN7** and **QN8** (Figure 7.9) show metal-chelating characteristics with a chelation ratio of 2:1 and are effective at reducing OS caused by hydrogen peroxide. Additionally, **QN7** can considerably reduce Aβ aggregation caused by metal ions.

B. Çiftci et al. (2024) synthesized quinoline derivatives and evaluated them for their anti-epileptic, anti-AD, and antimicrobial activities. Significant differences in enzyme inhibition were found by replacing methoxy groups with nitro or bromine groups at specific positions in the quinoline and tetrahydroquinoline rings. With IC$_{50}$ values of 1.54 and 0.13 μM, the test findings show that **QN11, QN13,** and **QN14** are efficient inhibitors of the hCA I and hCA II enzymes, respectively (Figure 7.10).

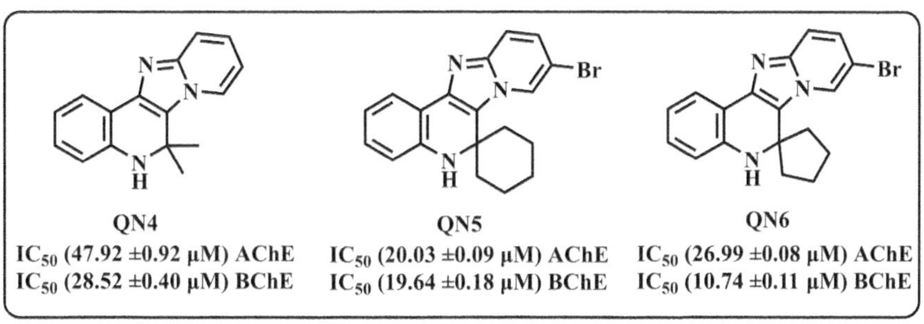

FIGURE 7.8 Imidazo[4,5-*c*]quinolines derivatives.

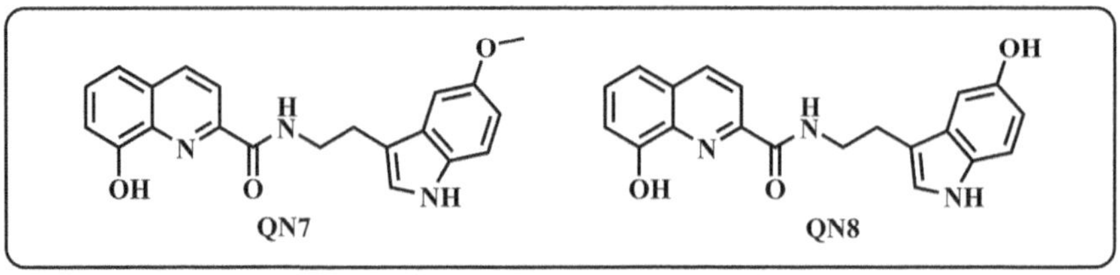

FIGURE 7.9 Melatonin hydroxyquinoline–based hybrids.

QN9 and **QN12** were found to be the most potent against AChE, with IC_{50} values of 3.15 and 4.96 μM, respectively, compared to Tacrine at 53.31 μM. The experimental results were validated by molecular docking investigations.

Kilic and Dogruer (2024) designed eight novel N′-(quinolin-4-ylmethylene)propanehydrazide derivatives (Figure 7.11) as hits for targeting AD and investigated them for their anti-ChE, antioxidant potential. Five compounds showed AChE inhibition (IC_{50} ranging from 0.69 to 11.62 μM). Furthermore, all of the active compounds showed considerable BChE inhibition (IC_{50} ranging from 16.06 to 26.0 μM), except **QN18** (Table 7.1). The most active **QN16** (AChE) and **QN19** (BChE) showed mixed-type inhibitory behavior, according to enzyme kinetic tests. Additionally, they were found to be chelators for copper, iron, and zinc ions, as determined by metal binding experiments.

M. Noori et al. (2023) developed phenyl-quinoline-based derivatives. Results showed that derivatives were more potent against BChE than AChE, despite their promising effectiveness against both AChE and BChE. The most potent BChE derivative was **QN20**, which outperformed donepezil by almost 75 folds. With Ki and Kis values of 4.44 and 13.50 μM, respectively, **QN21**, the most active anti-AChE drug, demonstrated mix-type inhibition. Whereas **QN20**, a strong BChE inhibitor, demonstrated a competitive mode of inhibition with Ki of 0.14 μM (Figure 7.12). Molecular docking studies of these derivatives reveal that both derivatives interact critically with the enzyme's PAS and CAS pockets, impeding the hydrolysis of the ACh neurotransmitter. Overall, this investigation shows that the proposed inhibitors may be excellent options for further development of anti-AD agents.

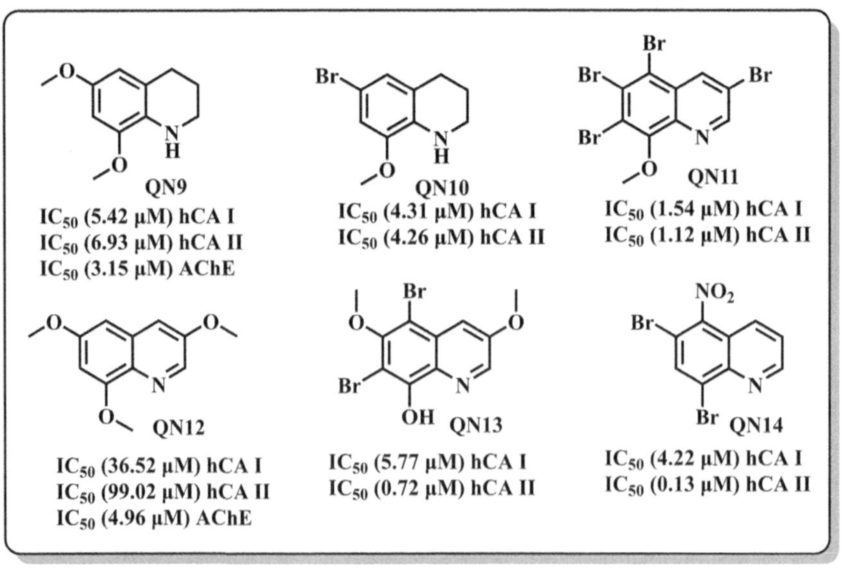

FIGURE 7.10 Quinoline derivatives.

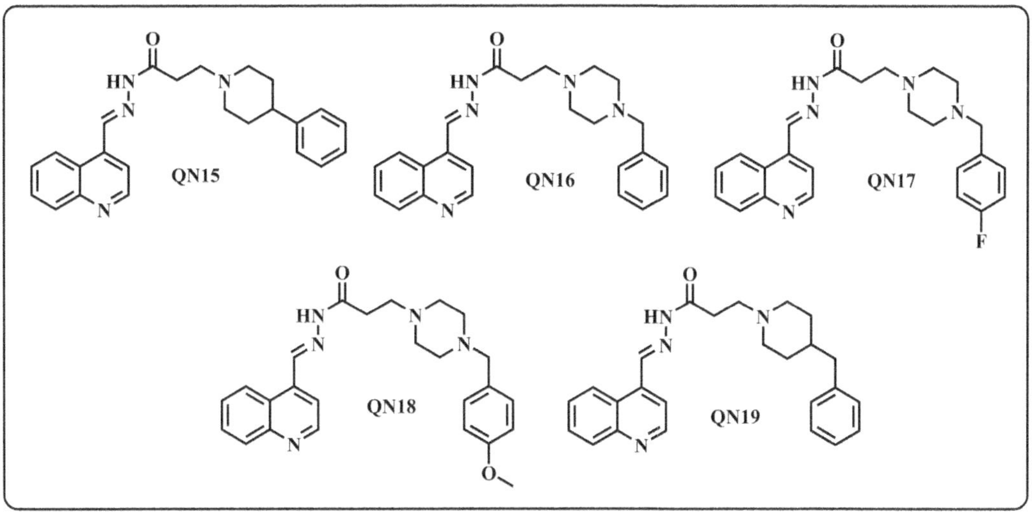

FIGURE 7.11 N′-(quinolin-4-ylmethylene)propanehydrazide derivatives.

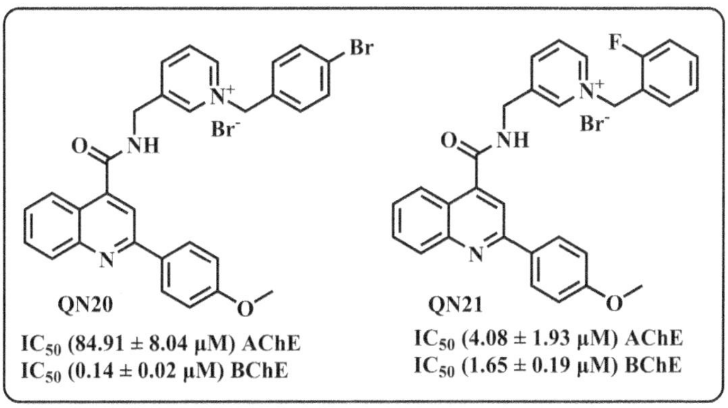

FIGURE 7.12 Phenyl-quinoline-based derivatives.

TABLE 7.1
Anticholinesterase Activity of N′-(quinolin-4-ylmethylene)propanehydrazide Derivatives

	eeAChE			eqBChE		
Compound Code	% inh 10 μM	% inh 100 μM	IC_{50} (μM)	% inh 10 μM	% inh 100 μM	IC_{50} (μM)
QN15	57.9±2.1	86.6±0.5	11.62±0.04	36.7±3.5	91.7±2.1	17.36±0.2
QN16	85.7±0.9	94.6±2.9	0.69±0.19	19.8±1.4	77.9±4.8	26.00±0.03
QN17	83.8±4.5	95.0±0.1	2.46±0.08	48.4±0.1	91.0±0.3	21.09±0.05
QN18	30.4±1.2	64.4±2.2	8.16±0.07	≤10	33.0±5.7	≥100
QN19	61.5±3.4	82.0±11.3	7.04±0.05	36.1±4.9	88.3±1.3	16.06±0.02
Donepezil	98.5±0.1	–	0.062±0.002	90.1±1.1	–	3.55±0.07
Galantamine	92.1±1.1	–	–	31.5±1.8	–	–

All values are expressed as the mean±SD (n = three experiments).

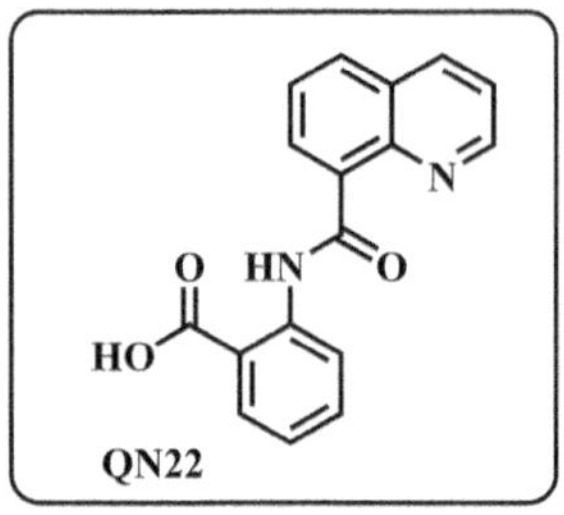

FIGURE 7.13 Quinoline carboxamidobenzoate alkaloid.

T. Y. Lee et al. (2023) synthesized a naturally occurring quinoline carboxamidobenzoate alkaloid that was extracted from a fungus that originated in the deep sea. Their synthesis involved amidation of 8-quinolinecarboxylic acid with methyl anthranilate, which on further hydrolysis resulted in **QN22** (Figure 7.13). The substantial neuroprotective potential of **QN22** against MPP_3 and α-syn-mediated DA neurotoxicity was demonstrated by pharmacological investigations utilizing a C. elegans PD model. With these studies, **QN22** could be further explored and used as a potentially effective therapy option for PD.

R. Cai et al. (2019) developed novel 4-N-phenylaminoquinoline derivatives and evaluated their potential on AD. Three derivatives among the synthesized ones were considerable for their activities. **QN23, QN24,** and **QN25** (Figure 7.14) had comparable inhibition activities to galantamine in both AChE and BChE. **QN23** was the most potent, with IC_{50} values of 1.20±0.18 and 18.52±1.21 μM against eeAChE and eqBChE, respectively, when compared to the standard reference galantamine (IC_{50} = 1.28±0.01 and 24.41±2.01 μM). **QN23** was found to be a reversible inhibitor and showed a mixed type of inhibition on AChE through enzyme kinetic studies. Additionally, **QN23** was inferred to interact with the CAS and PAS of both AChE and BChE by molecular docking experiments. **QN24** showed 25.80% of propidium iodide displacement from the PAS of AChE. **QN25** EC_{50} value was 0.328 μM, nearly as high as that of ascorbic acid (EC_{50} = 0.095 μM), and its DPPH radical scavenging activity reached 84% (at 1 mg/mL). Furthermore, **QN23, QN24,** and **QN25** showed significant metal-chelating capabilities toward ions of Al^{3+}, Fe^{2+}, Cu^{2+}, and Zn^{2+}.

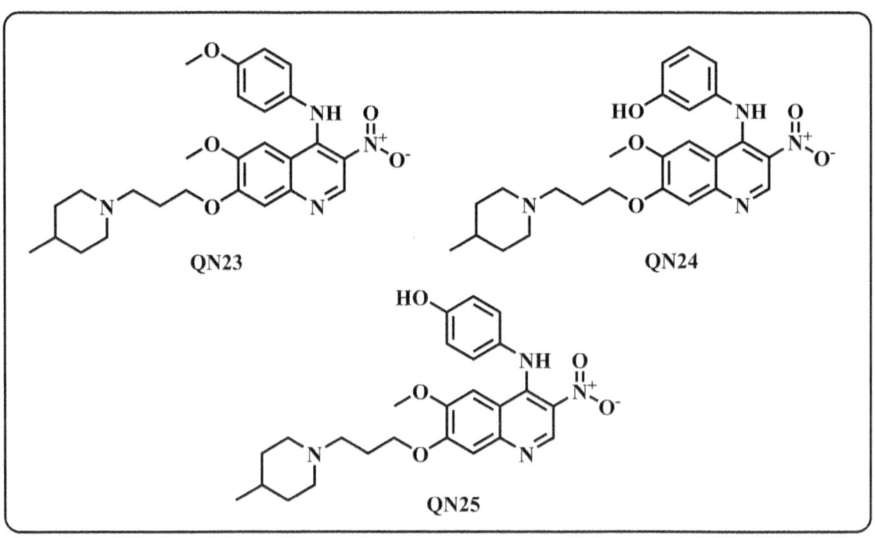

FIGURE 7.14 4-N-phenylaminoquinoline derivatives.

7.3 ISOQUINOLINES

Isoquinolines are an essential class of nitrogen-containing heterocycles with a core structure similar to that of benzene and pyridine rings in a 2,3-system. Although isoquinoline is an isomer of quinoline, the different position of the nitrogen atom gives it distinct electronic properties and reactivity. This unique arrangement allows isoquinoline-based scaffolds to undergo diverse chemical transformations and interact with a wide range of biological targets (Mahadeviah et al., 2024). Isoquinoline-based frameworks are widely regarded as privileged scaffolds in modern medicinal chemistry, as their rigid yet adaptable structure provides an excellent foundation for designing and optimizing new drug candidates (Vijayakumar et al., 2023). Their capacity to interact with numerous biological targets has led to the discovery of isoquinoline derivatives with an extensive range of pharmacological activities. These compounds have been reported to exhibit anticancer, antioxidant, antiplatelet, anti-inflammatory, antihyperglycemic, antiplasmodial, antifungal, antibacterial, antidepressant, and anti-TB activities (Figure 7.15) (Mahadeviah et al., 2024).

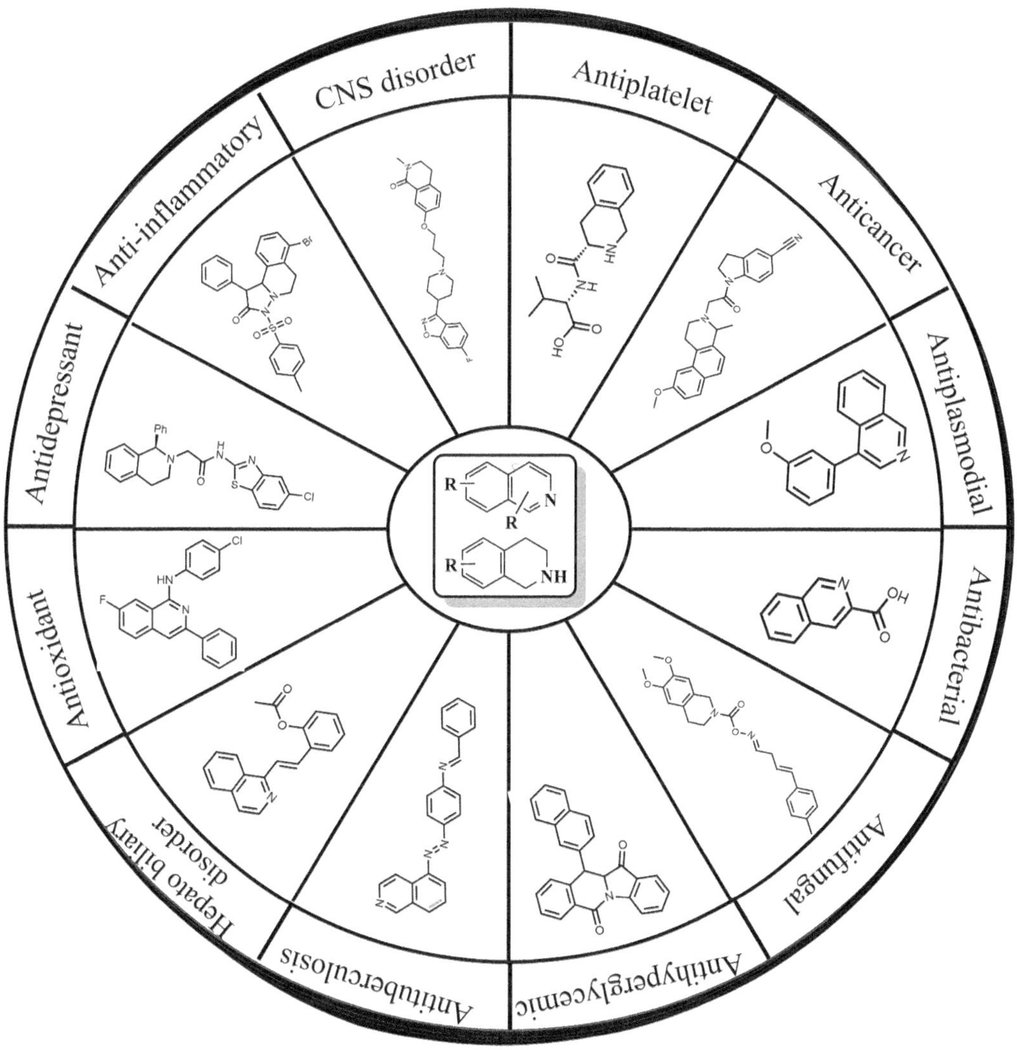

FIGURE 7.15 Examples of reported isoquinolines in various medicinal applications.

This remarkable pharmacological versatility highlights the enduring importance of the isoquinoline framework in past decades (Figure 7.3). Noting its interaction with various biological targets, this provides numerous opportunities for structural modification, which is crucial in the fields of medicinal chemistry and drug discovery (Plazas & Muñoz, 2022). Some of the clinically approved drugs containing the isoquinoline framework are depicted in Figure 7.16.

Isoquinolines are obtained from both natural and synthetic sources. They exist in different types based on structural modifications and substitution patterns (Figure 7.17). These can be broadly classified into simple isoquinolines, hydrogenated derivatives, substituted isoquinolines, condensed/fused isoquinoline systems, and isoquinoline alkaloids (Qing et al., 2017).

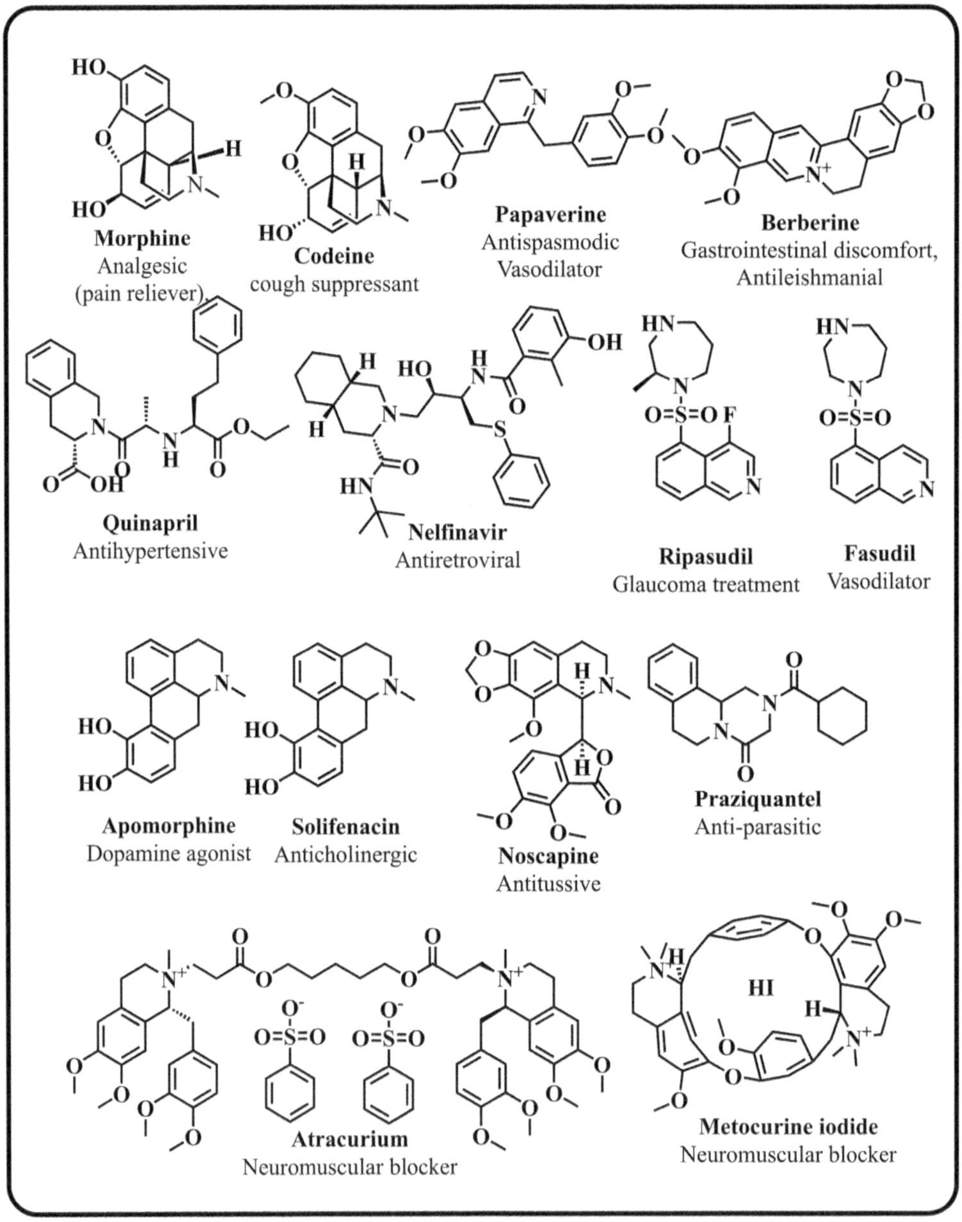

FIGURE 7.16 List of isoquinoline scaffold containing USFDA-approved drugs.

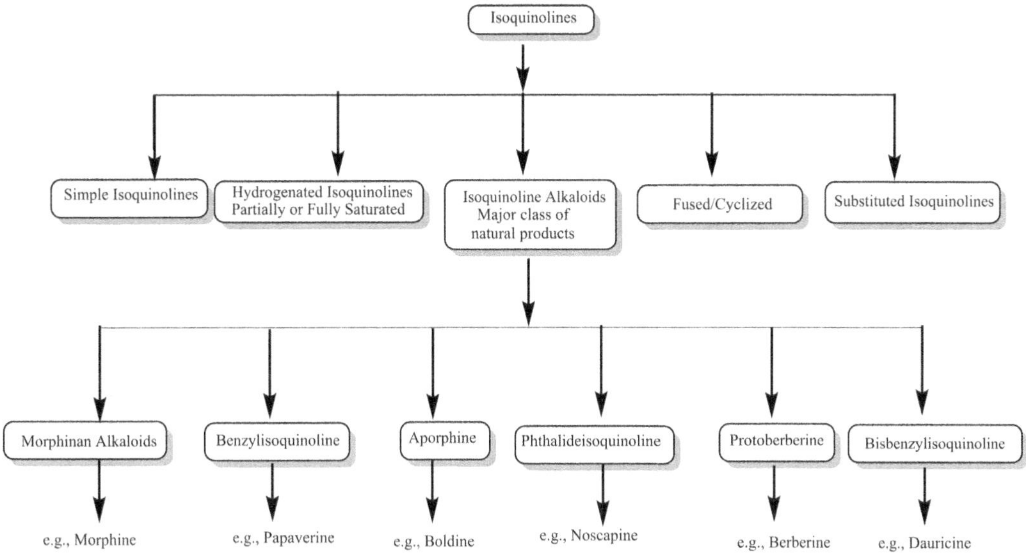

FIGURE 7.17 Classifications of isoquinolines based on their structural modifications and substitution patterns.

7.3.1 Natural Isoquinolines

Natural isoquinolines, extracted from plants, exhibit a range of pharmacological activities (Diamond & Desgagné-Penix, 2016). These compounds are primarily sourced from plant families such as Papaveraceae, Berberidaceae, Ranunculaceae, Menispermaceae, and Fumariaceae (Yang et al., 2024). Some of the natural isoquinoline alkaloids with neuroprotective effects are mentioned in Figure 7.18.

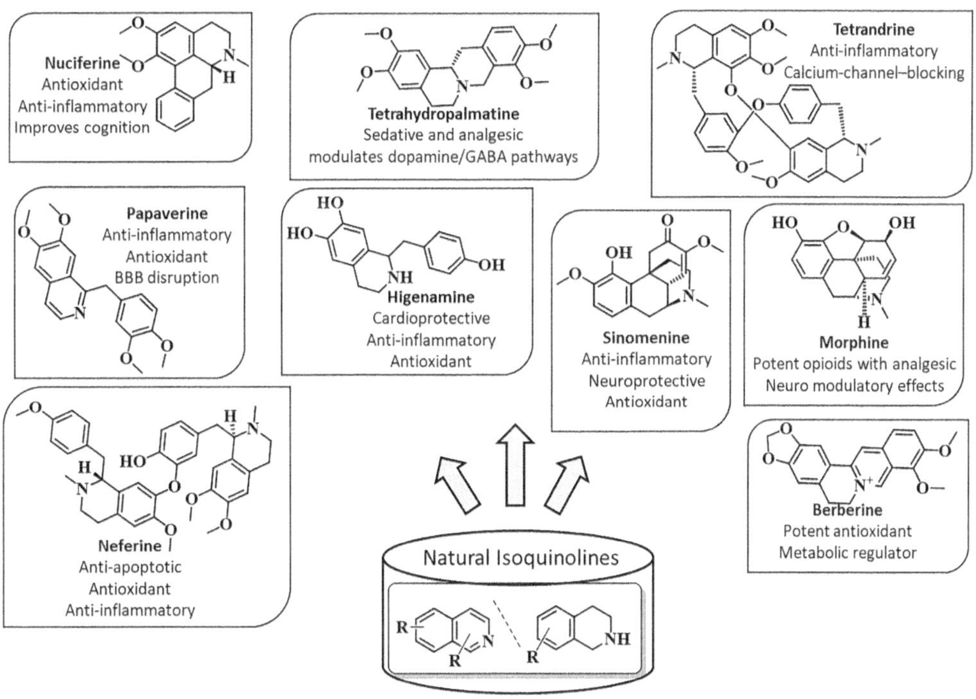

FIGURE 7.18 Some of the natural isoquinolines in neurodegenerative disorders.

7.3.2 Synthesis of Isoquinolines

Several classical and modern methods are used for the synthesis of isoquinolines. Over time, classical annulation pathways are surpassed by modern catalytic approaches in various applications (Figure 7.19).

Some of the classical methods to synthesize the isoquinolines are the following:

- Bischler–Napieralski reaction, initially reported in 1893 by Bischler and Napieralski (Mandrekar et al., 2019).
- Pomeranz–Fritsch reaction, reported by C. Pomeranz and P. Fritsch in 1893 (Wang et al., 2019).
- Pictet–Spengler reaction, discovered in 1911 by Amé Pictet and Theodor Spengler (Stöckigt et al., 2011).
- Pictet–Gams reaction, a modified Bischler–Napieralski reaction (Li, 2014).

These reactions involve intramolecular cyclization via an electrophilic aromatic substitution pathway using strong dehydrating agents, such as phosphorus oxychloride ($POCl_3$) and phosphorus pentoxide (P_2O_5), among others.

The modern and green synthesis approaches include the following:

- Transition metal-mediated C–H activation and annulation strategies involve Rh(III), Pd, and Co catalysts for cyclization with alkynes, often necessitating a directing group to achieve regioselectivity (Li, 2014).
- Microwave-assisted synthesis facilitates by rapidly heating reaction mixtures, significantly shortening reaction time to minutes by enhancing yields. This method typically employs catalysts such as palladium or copper for coupling and annulation (Yang et al., 2012).
- Visible-light photocatalysis employs photocatalysts like Ru- and Ir-complexes or organic dyes, utilizing visible light (Wei et al., 2019).

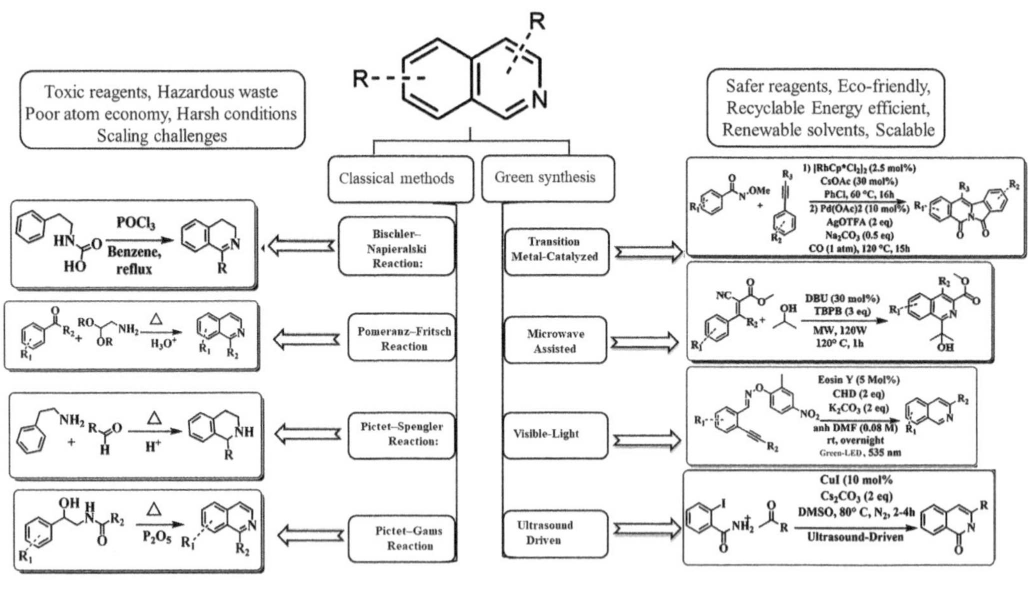

FIGURE 7.19 Classical and modern synthetic methods of isoquinolines.

- Ultrasound-assisted synthesis utilizes ultrasonic waves to generate localized cavitation zones, thereby promoting green and energy-efficient processes (Wei et al., 2019).
- Both visible-light and ultrasound methods facilitate single-electron transfer and radical cyclizations or annulations under mild, eco-friendly conditions (Jiang et al., 2014).

7.3.3 Mechanisms of Biological Activity Isoquinoline in Neuroprotective Actions

Isoquinoline alkaloids, naturally found in many medicinal plants, show broad pharmacological benefits, especially in protecting nerve cells from damage associated with CNS disorders. They not only prevent neuronal injury but also support the brain's repair and regeneration processes. Their neuroprotective actions include antioxidant, anti-inflammatory, and anti-apoptotic effects, inhibition of toxic protein aggregation, modulation of neurotransmitters, activation of neurotrophic pathways, and improved cerebral blood flow. By acting on multiple targets simultaneously, these compounds offer strong potential for treating complex neurodegenerative disorders where conventional single-target drugs often fall short (Plazas & Muñoz, 2022). The neuroprotective activity of isoquinoline alkaloids arises from multiple interrelated mechanisms (Figure 7.20).

7.3.3.1 Anti-Inflammatory Effects

Neurodegenerative disorders involve chronic brain inflammation from neuronal stress, leading to cell death and harmful immune responses. This excessive inflammation damages brain tissue and disrupts the BBB, thereby exacerbating the disease (Yang et al., 2019). Effective treatments should focus on reducing this inflammatory damage. Natural isoquinoline alkaloids, such as berberine, tetrandrine, and nuciferine, possess anti-inflammatory properties. Tetrandrine inhibits NLRP3 inflammasome activation and reduces inflammatory mediators related to ischemia/reperfusion. It also enhances Sirt-3, thereby decreasing CHOP, IRE1α, and p-JNK, while reversing inflammatory gene transcription and suppressing cytokine secretion. Berberine reduces neuroinflammation and preserves the BBB by inhibiting TNF-α, COX-2, and iNOS, while activating the PI3K-AKT pathway and suppressing NF-κB signaling. Nuciferine inhibits NF-κB activation and reduces pro-inflammatory mediators, thereby mitigating Lipopolysaccharide (LPS)-induced inflammatory responses in BV2 cells (Kong et al., 2021).

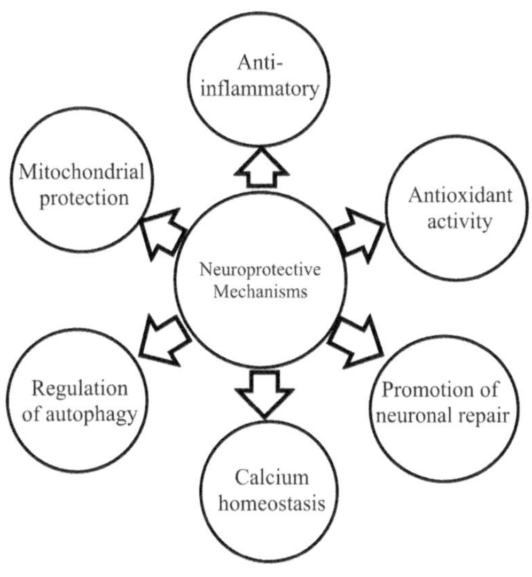

FIGURE 7.20 Various mechanisms of isoquinolines in neuroprotection.

7.3.3.2 Antioxidant Activity

OS arises from an imbalance between ROS and antioxidants, such as superoxide dismutase (SOD) and glutathione, which is linked to neurodegeneration. It can be intensified by conditions that produce excessive free radicals, resulting in inflammation and cellular damage. This activates transcription factors that influence inflammation-related gene expression, potentially exacerbating the progression of diseases (Chaudhary et al., 2023).

Berberine protects neural stem cells by lowering ROS levels through the NRF1/2–NQO1–HO-1 pathway, reducing pro-apoptotic factors while increasing protective proteins and antioxidant enzymes. It further boosts cell survival via the ERK/p-ERK and WNT/β-catenin pathways. Tetrahydropalmatine also enhances antioxidant enzyme levels and reduces oxidative damage, contributing to neuroprotection. Nuciferine counters OS in nerve cells by restoring antioxidant enzyme activity impaired by high ROS, showing effectiveness in diabetic rats and potential relevance to AD's disease (Uttara et al., 2009).

7.3.3.3 Promotion of Neuronal Repair

Isoquinoline alkaloids promote endothelial cell proliferation and neuronal regeneration, resulting in enhanced memory and cognitive function following ischemic stroke. Tetrahydropalmatine has been shown to bind to VEGFR2 and promote angiogenesis, while tetrandrine increases VEGF-A synthesis, boosting endothelial activity and recovery from ischemic injury, which underscores their potential in treating neurodegenerative disorders (Wang et al., 2021).

7.3.3.4 Regulation of Autophagy

Autophagy is a cellular process that degrades proteins and organelles, and excessive activation can lead to autophagic cell death. Traditionally, it was thought that neuronal injury activates apoptosis followed by autophagy; however, recent findings indicate that prolonged autophagy can actually worsen neuronal injury, highlighting the need for careful regulation. Additionally, tetrandrine has been shown to reduce malondialdehyde (MDA) levels and increase glutathione (GSH) content following traumatic brain injury (TBI) (Liu et al., 2022).

7.3.3.5 Mitochondrial Protection

Energy metabolism, driven by mitochondria, is crucial for cellular activities. Dysfunction in mitochondria can lead to neuronal injury and is associated with disorders such as AD and PD. Tetrandrine improves mitochondrial function and protects neurons, while morphine reduces mitochondrial dysfunction. Berberine may offer neuroprotection in PD by enhancing mitochondrial function and inhibiting harmful processes in neurons (Spinelli & Haigis, 2018).

7.3.3.6 Calcium Homeostasis

Calcium (Ca^{2+}) is essential for cellular functions, with imbalances associated with various diseases, particularly neurodegenerative disorders. Disruption causes neurotoxicity, inflammation, and mitochondrial dysfunction. Methamphetamine worsens this by interfering with mitochondrial calcium ATPase, leading to excess calcium and cell death. Low-dose morphine helps mitigate calcium overload, enhancing cell survival by decreasing cytotoxicity and inflammation. Berberine also protects oligodendrocytes from excitotoxicity by blocking intracellular calcium accumulation during ischemic conditions in a dose-dependent manner (Sun et al., 2025).

This chapter reviews isoquinolines from natural and synthetic sources (2004–2025), focusing on their roles in neurodegenerative disorders. It details structural classes, synthetic methods, and SARs, highlighting the therapeutic potential of isoquinoline derivatives as neuroprotective agents for treating conditions like AD, PD, HD, and other neuroinflammatory disorders.

7.3.4 Isoquinoline Analogs for Major Neurodegenerative Disorders

In 2024, Saltanat Aghayeva and colleagues (Ahmad et al., 2024) reported novel isoquinoline-based derivatives as potential cholinesterase inhibitors for AD, synthesized from isoquinoline-2-carboxylate

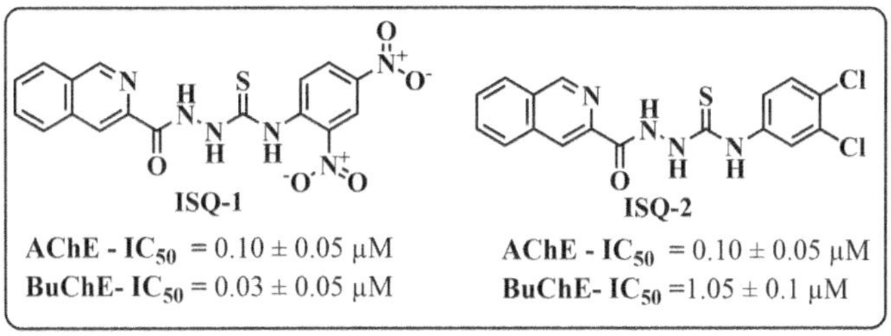

FIGURE 7.21 Isoquinoline derivatives.

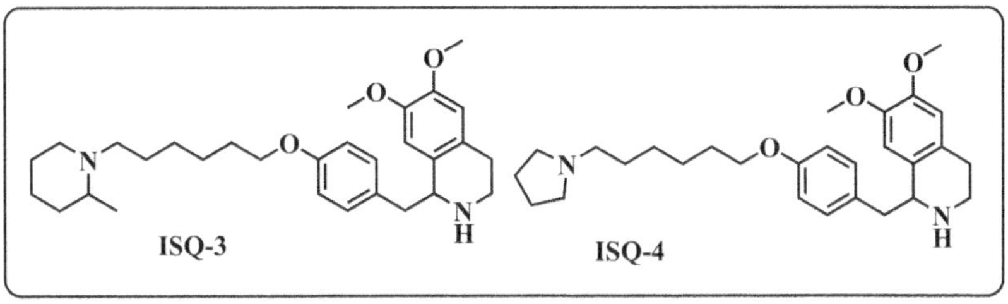

FIGURE 7.22 Benzylisoquinolines derivatives.

and were fully characterized before biological testing. All compounds demonstrated micromolar inhibitory activity against both AChE and BuChE. Derivatives **ISQ-1** and **ISQ-2** (Figure 7.21) showed the greatest potency, with **ISQ-1** being more potent than the positive control donepezil with IC_{50} of 0.16 and 4.5 μM for AChE and BuChE, respectively. SAR analysis indicated that EWGs at the ortho/para positions enhanced activity, whereas electron donating groups (EDGs) weakened it. Molecular docking supported these findings by stable interactions within the active sites of both enzymes.

Xu et al. (2014) developed benzylisoquinoline derivatives by linking various amine groups to a core structure, targeting cholinesterase activity and β-amyloid ($A\beta_{42}$) aggregation for AD's treatment. The compounds, synthesized through Bischler–Napieralski cyclization and alkylation, exhibited strong biological activity (Figure 7.22). **ISQ-3** emerged as the most effective AChE inhibitor, while **ISQ-4** was best for BuChE inhibition, both reducing $A\beta_{42}$ aggregation significantly (Table 7.2). Docking and kinetic studies confirmed dual-site interactions within AChE, and cytotoxicity tests indicated low toxicity.

TABLE 7.2
Inhibition of ChEs Activity, Human ChEs Activity, and Self-Induced $A\beta_{42}$ Aggregation

	IC_{50} (μM)		IC_{50} (μM)		
Compound Code	**AChE[a]**	**BuChE[a]**	**h-AChE[a]**	**h-BuChE[a]**	**$A\beta_{42}$ aggregation**
ISQ-3	**0.95 ± 0.02**	**3.13 ± 0.16**	**2.20 ± 0.12**	**5.83 ± 1.43**	**78.4 ± 0.6**
ISQ-4	**1.28 ± 0.13**	**2.34 ± 0.28**	**2.98 ± 0.20**	**2.27 ± 0.04**	**69.0 ± 1.3**
Benzylisoquinoline	**>100**	**39.7 ± 0.54**	**>100**	**46.77 ± 0.36**	**61.52± 1.76**
Galanthamine	**2.67 ± 0.15**	**12.7 ± 0.62**	–	–	**36.0 ± 5.7**
Resveratrol	–	–	–	–	**61.5 ± 2.1**

[a] Inhibitor concentration (mean ± SEM of three experiments) required for 50% inactivation.

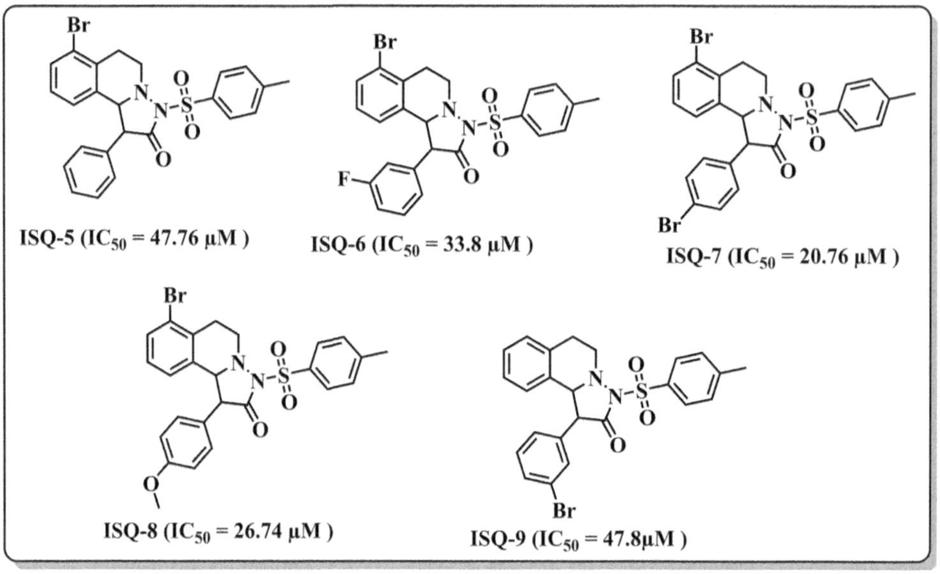

FIGURE 7.23 Pyrazolo-isoquinoline derivatives.

Akhtar et al. (2023) developed chiral pyrazolo–isoquinoline derivatives, which exhibited over 90% cell viability in LPS-stimulated RAW 264.7 macrophages at doses up to 100 μM. The derivatives, **ISQ-5** and **ISQ-9,** showed significant effects (Figure 7.23); notably, **ISQ-7** demonstrated dose-dependent inhibition of LPS-induced nitric oxide (NO) production and effectively reduced inducible nitric oxide synthase (iNOS) expression as determined by western blot studies, indicating its anti-inflammatory action through the suppression of iNOS-mediated NO production.

Dactyllactone A (**ISQ-10**), a new isoquinoline alkaloid isolated from Dactylicapnos scandens by Xiao-Dong Luo and colleagues in 2018 (Wang et al., 2018), is identified as a novel subtype in the aporphine family. Its structure was elucidated using spectroscopic techniques and confirmed by single-crystal X-ray diffraction. Dactyllactone A (Figure 7.24) demonstrates significant anti-inflammatory activity, with IC_{50} values between 2.4 and 12 μM, effectively suppressing IL-1β and PGE_2 production in LPS-stimulated RAW 264.7 macrophages, outperforming dexamethasone. Importantly, (3-(4,5-dimethylthiazol-2-yl)-2,5-diphenyltetrazolium bromide) MTT assays indicate the absence of cytotoxicity at tested concentrations, suggesting that the anti-inflammatory effects are not due to cell death. This research enhances the structural diversity of aporphine-type alkaloids and positions Dactyllactone A as a potential candidate for future anti-inflammatory drug development.

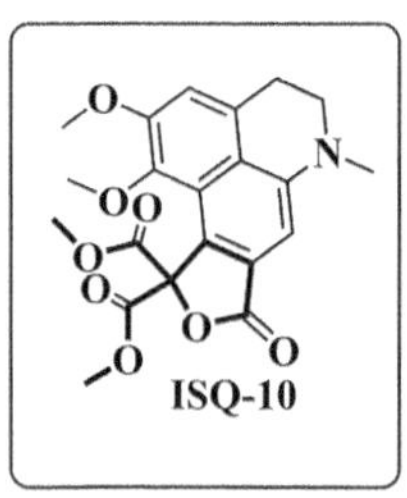

FIGURE 7.24 Bis-seco-aporphine scaffold.

Liu et al. (2022) developed benzothiazoles and tetrahydroisoquinolines as MTDLs to alleviate depressive symptoms linked with neurodegenerative disorders. *In vitro* studies showed **ISQ-11 to ISQ-16** (Figure 7.25) inhibited MAOs compared to Clorgyline (84.21%) for MOA-A, and cholinesterases (ChEs), with **ISQ-15** and **ISQ-16** showing the strongest MAO-B inhibition compared to Pargyline (93.47%) in MOA-B. *In vivo* tests indicated **ISQ-15** provided significant antidepressant-like effects by reducing immobility time in the forced swim test (FST). Molecular docking studies confirmed stable interactions of **ISQ-16** with MAO-B and BuChE, suggesting its potential as an effective treatment for neurodegenerative conditions with depressive symptoms.

Gao et al. (2019) developed N^1-isoquinolin-5-yl-N^2-phenylpyrrolidine-1,2-dicarboxamide derivatives as TRPV1 antagonists for chronic pain management. TRPV1 mediates responses to capsaicin, heat, and acid, linking it to neuropathy and inflammatory pain. **ISQ-17 to ISQ-20** (Figure 7.26) showed higher activity than the reference antagonist BCTC, with **ISQ-17** being the most potent, and it did not cause hyperthermia, unlike other TRPV1 inhibitors. SAR analysis indicated that para-substitutions are crucial for potency. **ISQ-17** reduced pain in capsaicin- and heat-induced mouse models more effectively than BCTC (*in vivo*), although it was less effective against acid-induced pain. Where hTRPV1(CAP) and hTRPV1(pH) are human TRPV1 receptor activated by capsaicin and human TRPV1 receptor activated by low pH, respectively.

In 2016, Zhe-Shan Quan and colleagues (Zhang et al., 2016) synthesized two series of 3,4-dihydroisoquinoline derivatives and assessed their anticonvulsant effects in mice through MES and PTZ seizure models at a dose of 100 mg/kg. **ISQ-21** and **ISQ-22** exhibited significant seizure protection, with **ISQ-21** outpacing valproate (ED_{50} MES, (mg/kg) = 272) in efficacy, yet remaining less effective than carbamazepine (ED_{50} MES, (mg/kg) = 11.8). **ISQ-22** had a higher ED_{50} but

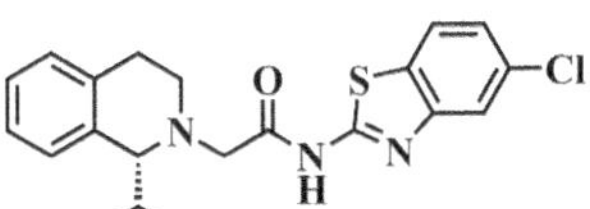

ISQ-11

IC_{50} (μM)	Inhibition (%)
MOA = 38.82	MOA-A = 39.27
AChE = >150	MOA-B = 39.27
BuChE = 17.59	AChE = 4.42
	BuChE = 69.44

Duration of Immobility (s) = 85.7

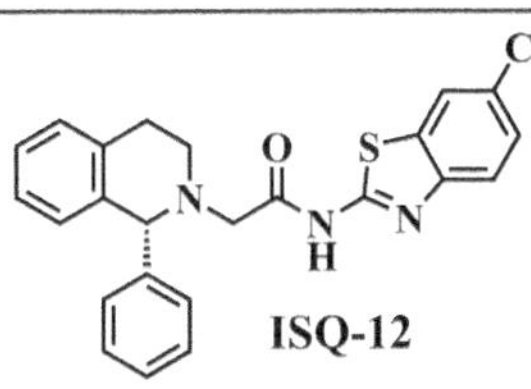

ISQ-12

IC_{50} (μM)	Inhibition (%)
MOA = 45.9	MOA-A = 21.59
AChE = >150	MOA-B = 21.59
BuChE = >150	AChE = 1.10
	BuChE = 68.8

Duration of Immobility (s) = 77.0

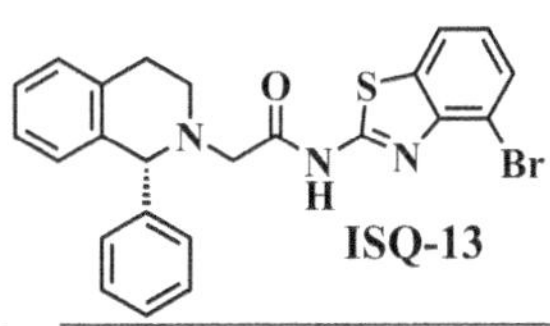

ISQ-13

IC_{50} (μM)	Inhibition (%)
MOA = 64.83	MOA-A = 43.37
AChE = >150	MOA-B = 43.37
BuChE = 14.61	AChE = 0.55
	BuChE = 77.76

Duration of Immobility (s) =62.3

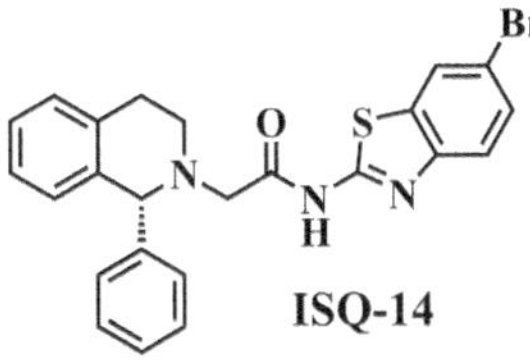

ISQ-14

IC_{50} (μM)	Inhibition (%)
MOA = 41.78	MOA-A = 24.15
AChE = >150	MOA-B = 24.15
BuChE = >150	AChE = 1.10
	BuChE = 32.32

Duration of Immobility (s) = 71.7

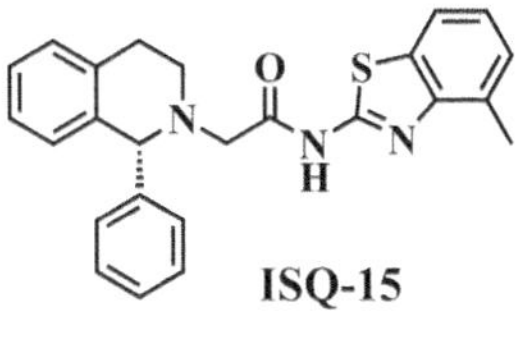

ISQ-15

IC_{50} (μM)	Inhibition (%)
MOA = 14.8	MOA-A = 57.11
AChE = >150	MOA-B = 57.11
BuChE = 30.35	AChE = 1.10
	BuChE = 72.64

Duration of Immobility (s) = 69.0

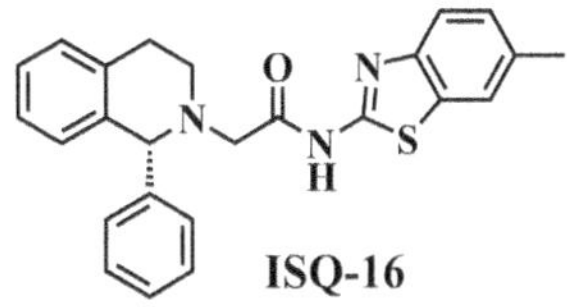

ISQ-16

IC_{50} (μM)	Inhibition (%)
MOA = 18.53	MOA-A = 26.46
AChE = >150	MOA-B = 26.46
BuChE = 19.17	AChE = 2.21
	BuChE =47.68

Duration of Immobility (s) = 83.8

FIGURE 7.25 Benzothiazole–isoquinolines derivatives.

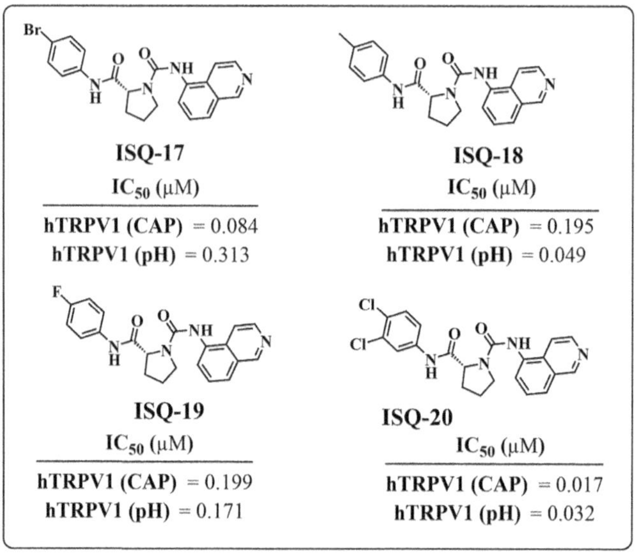

FIGURE 7.26 Phenylpyrrolidine–isoquinoline derivatives.

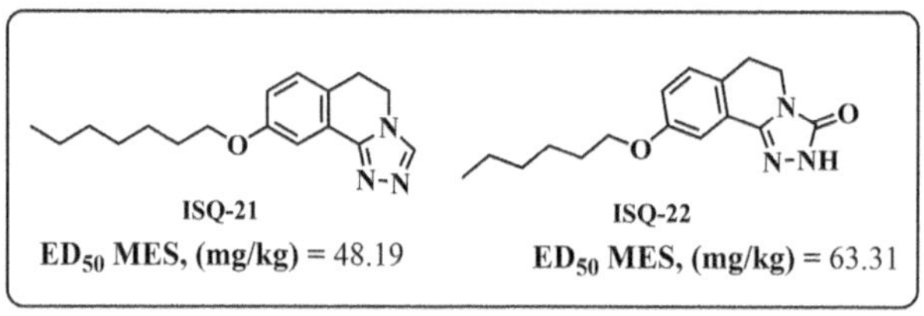

FIGURE 7.27 3,4-dihydroisoquinolines derivatives.

maintained a broader safety margin compared to valproate (Figure 7.27) (Table 7.3). **ISQ-21** is believed to exert its effects by acting as a benzodiazepine-receptor agonist at the GABA*A* receptor.

Rosaria Gitto et al. (2004) developed new 1-aryl-6,7-dimethoxy-1,2,3,4-tetrahydroisoquinoline derivatives as potential non-competitive AMPA receptor antagonists for anticonvulsant therapy. They synthesized N-acetylated and N-unsubstituted compounds using Pictet–Spengler cyclization and tested their activity in DBA/2 mice against established AMPA antagonists (Figure 7.28). Results indicated that the N-acetylated series generally outperformed the references, with **ISQ-25** being the most effective, though some N-unsubstituted **ISQ-23** to **ISQ-27** also showed significant activity, suggesting the acetyl group enhances but is not essential for efficacy.

In 2014, Gilson Zeni and colleagues (Mantovani et al., 2014) developed an efficient synthetic route for 1-amino-isoquinoline derivatives using AgOTf-catalyzed cyclization of o-alkynylbenzaldoximes with isocyanates under aerobic conditions, achieving high yields. Optimal conditions included DMF

TABLE 7.3
Effect of Induced Seizures in Mice

Compound Code	Doses (mg/kg)	Clonic Seizures (%)	Tonic Seizures (%)	Lethality (%)
ISQ-21	100	100	10	60
Carbamazepine	50	100	0	0
DMSO	–	100	100	90

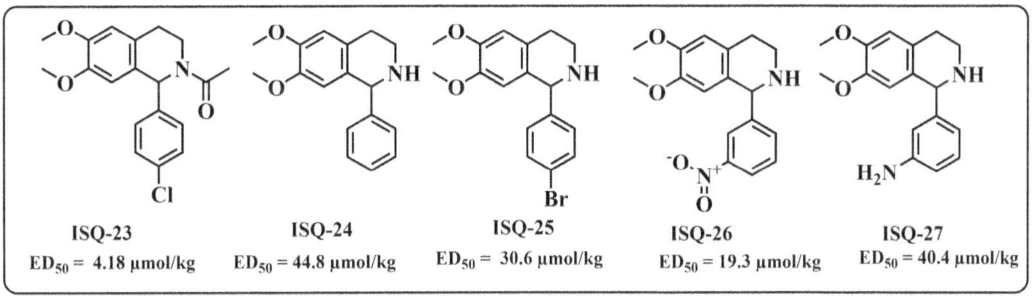

FIGURE 7.28 Tetrahydroisoquinoline derivatives.

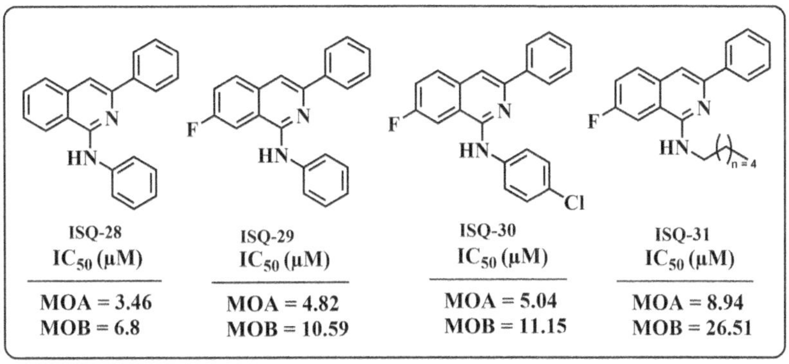

FIGURE 7.29 1-amino-isoquinoline derivatives.

and silver triflate at 140°C. Several derivatives **ISQ-28** to **ISQ-31** demonstrated significant antioxidant activity and effectively inhibited MAO-A and MAO-B at low concentrations. *In vivo* tests revealed **ISQ-29** and **ISQ-30** (Figure 7.29) exhibiting clear antidepressant-like effects at 10 mg/kg without toxicity or locomotor activity impairment, indicating their efficiency as neuroprotective candidates.

Chen et al. (2006) identified isoquinoline-1,3,4-trione (**ISQ-30**) as a potent caspase-3 inhibitor with an IC_{50} of 0.15 μM after screening 8,000 compounds. They synthesized new derivatives for improved activity, with many showing low-nanomolar potency *in vitro* (Figure 7.30). **ISQ-33** provided enhanced protection against camptothecin-induced apoptosis in human Jurkat T cells. Further

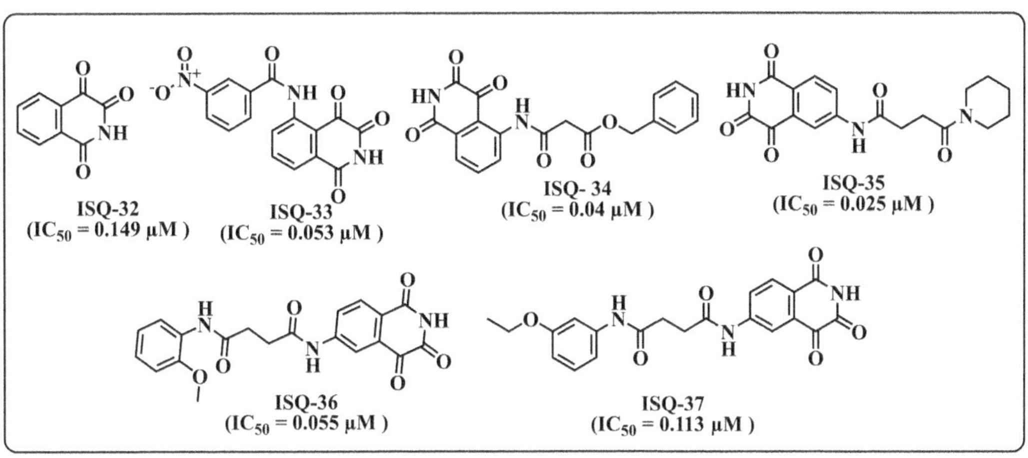

FIGURE 7.30 Isoquinoline-1,3,4-trione derivatives.

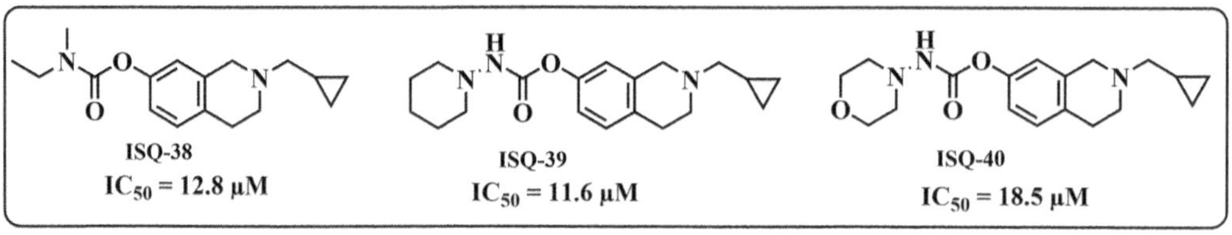

FIGURE 7.31 1,2,3,4-tetrahydroisoquinoline derivatives.

validation in an in vivo MCAO stroke model highlighted **ISQ-37**, which demonstrated significant neuroprotective effects by reducing infarct volume by up to 77% at a 1 mg/kg subcutaneous dose, indicating its potential as a neuroprotective agent.

In a 2008 study by Ming-Kuan Hu et al. (2008), 1,2,3,4-tetrahydroisoquinoline derivatives were investigated for their effects on APP processing, crucial for AD's treatment. These compounds acted as MAO-B inhibitors, enhancing non-amyloidogenic APP processing through PKC and ERK/MAPK pathways, thereby increasing sAPPα production and neuronal survival (Figure 7.31). Notably, cyclopropylmethyl derivatives **ISQ-38** to **ISQ-40** significantly reduced γ-secretase activity by 35%–72% at 10 µM, outperforming rasagiline (IC_{50} = 2.1 µM). While propargyl analogues exhibited weak MAO-B inhibition, ISQ-34 notably activated ERK and boosted sAPPα secretion, suggesting potential neuroprotective properties, and showed no cytotoxicity in HEK293 cells.

7.4 GENERAL STRUCTURE ACTIVITY RELATIONSHIP (SAR) OF QUINOLINE/ISOQUINOLINE NEUROPROTECTIVE AGENTS

Rational drug design requires an understanding of the connection between neuroprotective effectiveness and quinoline/isoquinoline core. Recent developments have shed light on a number of SAR aspects for these scaffolds, and we have summarized the same in Figure 7.32 (Li et al., 2023; Mantoani et al., 2016).

- ❑ **C2:** Alkyl/aryl substitution enhances Lipophilicity & CNS penetration
- ❑ **C4:** Hydroxy/amino/carboxyl groups enhances Antioxidant & chelation capabilities
- ❑ **C6C8:** -OMe/-OH enhances Radical scavenging, MAO-B inhibition
- ❑ **N (position 1):** Changes may affect metabolic stability, receptor affinity, & water solubility

- ❑ **C1:** Chiral center (important in alkaloids)
- ❑ **C3:** Alkyl/aryl substitution enhances CNS penetration
- ❑ **C4:** Antioxidant, chelating groups (Fe^{2+}, Cu^{2+})
- ❑ **C6/C8:** -OMe/-OH boost radical scavenging, MAO-B inhibition.
- ❑ **C7:** Halogen/alkoxy enhances enzyme inhibitor activity.
- ❑ **N (position 2):** Impacts metabolic stability, receptor affinity, solubility

FIGURE 7.32 General SAR of quinoline and isoquinoline scaffold as neuroprotective agents.

7.5 CONCLUSION

In conclusion, quinolines and isoquinolines have gained significant attention as promising building blocks for developing new treatments for major neurodegenerative disorders. Research over the past 20 years, from studying natural products to designing modern synthetic molecules, has shown that these scaffolds are highly flexible and can influence several pathways involved in disease. They have demonstrated the ability to act on ChEs, MAOs, amyloid formation, inflammation, ion channels, and cell-death regulators, making them strong candidates for multitarget drug design, which is crucial for complex disorders like AD, PD, and MS. Although many quinoline/isoquinoline derivatives exhibit strong neuroprotective and anti-inflammatory effects in early studies, translating these findings into safe and effective medicines for humans remains a significant challenge. Continued improvements in pharmacokinetics, brain penetration, synthetic methods, and computational design will be essential. Overall, current progress clearly shows the lasting importance and future promise of quinoline/isoquinoline frameworks in developing disease-modifying therapies for the brain.

REFERENCES

Ahmad, Mushtaq, Imad Uddin, Marukh Murad, Laiba Maryam, Zakia Subhan, Muhammad Saleem Khan, Kainat Nabi, Muhammad Nabi, Fazal Rahim, and Hayat Ullah. 2024. "Cholinesterase inhibitors for the treatment of Alzheimer's disorder: Synthesis, biological analysis and molecular modelling studies of iso-quinoline derivatives." *Results in Chemistry 9*:101661.

Akhtar, Maryam, Jiabin Niu, Yujie Zhu, Zhaoyi Luo, Ting Tian, Yuanliang Dong, Yuan Wang, Muhammad Subaan Fareed, and Li Lin. 2023. "Anti-inflammatory efficacy and relevant SAR investigations of novel chiral pyrazolo isoquinoline derivatives: Design, synthesis, in-vitro, in-vivo, and computational studies targeting iNOS." *European Journal of Medicinal Chemistry 256*:115412.

Cai, Rong, Li-Ning Wang, Jing-Jing Fan, Shang-Qi Geng, and Yu-Ming Liu. 2019. "New 4-N-phenylaminoquinoline derivatives as antioxidant, metal chelating and cholinesterase inhibitors for Alzheimer's disease." *Bioorganic Chemistry 93*:103328.

Chandra, Devesh, Sachin Sachin, Shourabh Rav, and Upendra Sharma. 2025. "Quinoline-based FDA-approved drugs: Synthetic route and clinical uses." *Synlett 36*(16):2537–2552.

Chaudhary, Priya, Pracheta Janmeda, Anca Oana Docea, Balakyz Yeskaliyeva, Ahmad Faizal Abdull Razis, Babagana Modu, Daniela Calina, and Javad Sharifi-Rad. 2023. "Oxidative stress, free radicals and antioxidants: Potential crosstalk in the pathophysiology of human diseases." *Frontiers in Chemistry 11*:1158198.

Chen, Hsiuchen, and David C Chan. 2009. "Mitochondrial dynamics–fusion, fission, movement, and mitophagy–in neurodegenerative diseases." *Human Molecular Genetic 18*(R2):R169–R176.

Chen, Yi-Hua, Ya-Hui Zhang, Hua-Jie Zhang, Da-Zhi Liu, Min Gu, Jing-Ya Li, Fang Wu, Xing-Zu Zhu, Jia Li, and Fa-Jun Nan. 2006. "Design, synthesis, and biological evaluation of isoquinoline-1, 3, 4-trione derivatives as potent caspase-3 inhibitors." *Journal of Medicinal Chemistry 49*(5):1613–1623.

Çiftci, Bilge, Salih Ökten, Ümit Muhammet Koçyiğit, Vildan Enisoğlu Atalay, Mehmet Ataş, and Osman Çakmak. 2024. "Functionalized methoxy quinoline derivatives: Experimental and in silico evaluation as new antiepileptic, anti-Alzheimer, antibacterial and antifungal drug candidates." *European Journal of Medicinal Chemistry Reports 10*:100127.

de Oliveira, Bruno Assis, Filipe Gonçalves de Oliveira, Otavio de Assis Cruz, Pollyana Mendonça de Assis, Nicolas Glanzmann, Adilson David da Silva, Nádia Rezende Barbosa Raposo, and Priscila Vanessa Zabala Capriles Goliatt. 2025. "In silico and in vitro evaluation of quinoline derivatives as potential inhibitors of AChE, BACE1, and GSK3β for neurodegenerative diseases treatment." *Chemistry & Biodiversity 22*(3):e202401629.

Dhapola, Rishika, Subhendu Shekhar Hota, Phulen Sarma, Anusuya Bhattacharyya, Bikash Medhi, and Dibbanti HariKrishna Reddy. 2021. "Recent advances in molecular pathways and therapeutic implications targeting neuroinflammation for Alzheimer's disease." *Inflammopharmacology 29*(6):1669–1681.

Diamond, Andrew, and Isabel Desgagné-Penix. 2016. "Metabolic engineering for the production of plant isoquinoline alkaloids." *Plant Biotechnology Journal 14*(6):1319–1328.

Gadhave, Dnyandev G., Vrashabh V. Sugandhi, Saurav Kumar Jha, Sopan N. Nangare, Gaurav Gupta, Sachin Kumar Singh, Kamal Dua, Hyunah Cho, Philip M. Hansbro, and Keshav Raj Paudel. 2024. "Neurodegenerative disorders: Mechanisms of degeneration and therapeutic approaches with their clinical relevance." *Ageing Research Reviews 99*:102357.

Gao, Mingxiang, Cunbin Nie, Jinyu Li, Beibei Song, Xinru Cheng, Erying Sun, Lin Yan, and Hai Qian. 2019. "Design, synthesis and biological evaluation of N1-(isoquinolin-5-yl)-N2-phenylpyrrolidine-1, 2-dicarboxamide derivatives as potent TRPV1 antagonists." *Bioorganic Chemistry 82*:100–108.

Gitto, Rosaria, Roberta Caruso, Valerie Orlando, Silvana Quartarone, Maria Letizia Barreca, Guido Ferreri, Emilio Russo, Giovambattista De Sarro, and Alba Chimirri. 2004. "Synthesis and anticonvulsant properties of tetrahydroisoquinoline derivatives." *Il farmaco 59*(1):7–12.

Hernández-Ayala, Luis Felipe, Eduardo Gabriel Guzmán-López, and Annia Galano. 2023. "Quinoline derivatives: Promising antioxidants with neuroprotective potential." *Antioxidants 12*(10):1853.

Hu, Ming-Kuan, Yung-Feng Liao, Jung-Fang Chen, Bo-Jeng Wang, Ying-Tsen Tung, Hui-Ching Lin, and Kang-Po Lee. 2008. "New 1, 2, 3, 4-tetrahydroisoquinoline derivatives as modulators of proteolytic cleavage of amyloid precursor proteins." *Bioorganic & Medicinal Chemistry 16*(4):1957–1965.

Jampilek, Josef. 2019. *Heterocycles in Medicinal Chemistry*. MDPI.

Jena, Sushovan, Gabriel Gonzalez, Dominik Vítek, Marie Kvasnicová, Šárka Štěpánková, Miroslav Strnad, Jiří Voller, and Kaushik Chanda. 2024. "Novel neuroprotective 5, 6-dihydropyrido [2′, 1′: 2, 3] imidazo [4, 5-c] quinoline derivatives acting through cholinesterase inhibition and CB2 signaling modulation." *European Journal of Medicinal Chemistry 276*:116592.

Jiang, Heng, Yuanzheng Cheng, Ruzhi Wang, Yan Zhang, and Shouyun Yu. 2014. "Synthesis of isoquinolines via visible light-promoted insertion of vinyl isocyanides with diaryliodonium salts." *Chemical Communications 50*(46):6164–6167.

Kciuk, Mateusz, Beata Marciniak, and Renata Kontek. 2020. "Irinotecan—Still an important player in cancer chemotherapy: A comprehensive overview." *International Journal of Molecular Sciences 21*(14):4919.

Kilic, Burcu, and Deniz S Dogruer. 2024. "Synthesis and investigation of the cholinesterase inhibitory and antioxidant capacities of some novel N′-(quinolin-4-ylmethylene) propanehydrazides against Alzheimer's disease." *Drug Development Research 85*(3):e22183.

Kingston, David G. I., and Maria Belen Cassera. 2022. "Antimalarial natural products" In *Antimalarial Natural Products*, 1–106. Springer.

Kong, Yew Rong, Kai Ching Tay, Yi Xiang Su, Choon Kwang Wong, Wen Nee Tan, and Kooi Yeong Khaw. 2021. "Potential of naturally derived alkaloids as multi-targeted therapeutic agents for neurodegenerative diseases." *Molecules 26*(3):728.

Lee, Tae Yeon, Wooin Yang, Dong Seok Cha, and Young Taek Han. 2023. "Synthesis of a natural quinoline alkaloid isolated from the deep-sea-derived fungus and its potential as a therapeutic for Parkinson's disease." *Journal of Asian Natural Products Research 25*(5):446–455.

Li, Jie Jack. 2014. "Pictet–Gams isoquinoline synthesis" In *Name Reactions: A Collection of Detailed Mechanisms and Synthetic Applications Fifth Edition*, 478–479. Springer.

Li, Zhao-Hui, Li-Quan Yin, Dong-Hai Zhao, Lian-Hai Jin, Ya-Juan Sun, and Cheng Tan. 2023. "SAR studies of quinoline and derivatives as potential treatments for Alzheimer's disease." *Arabian Journal of Chemistry 16*(2):104502.

Liu, Huan, Shiqing He, Jianpeng Wang, Chong Li, Yongshi Liao, Qin Zou, and Rui Chen. 2022. "Tetrandrine ameliorates traumatic brain injury by regulating autophagy to reduce ferroptosis." *Neurochemical Research 47*(6):1574–1587.

Liu, Weihua, Donghai Zhao, Zhiwen He, Yiming Hu, Yuxia Zhu, Lingjian Zhang, Lianhai Jin, Liping Guan, and Sihong Wang. 2022. "Synthesis, characterization and biological evaluation of benzothiazole–isoquinoline derivative." *Molecules 27*(24):9062.

Mahadeviah, Chitrashree, Purushotham Karadigere Nagaraj, Thoppalada Yunus Pasha, and Priyanka Marathi. 2024. "Exploring the pharmacological potential of isoquinoline derivatives: A comprehensive review." *International Journal of Pharmaceutical Investigation 14*(4).

Mandrekar, Ketan S., Hari K. Kadam, and Santosh G. Tilve. 2018. "Domino Bischler–Napieralski–Michael reaction and Oxidation–new route to coumarin-pyrrole-isoquinoline fused pentacycles." *European Journal of Organic Chemistry* (47):6665–6670.

Mantoani, Susimaire P., Talita P. C. Chierrito, Adriana F. L. Vilela, Carmen L. Cardoso, Ana Martínez, and Ivone Carvalho. 2016. "Novel triazole-quinoline derivatives as selective dual binding site acetylcholinesterase inhibitors." *Molecules 21*(2):193.

Mantovani, Anderson C., Ana Paula Pesarico, Tuane B. Sampaio, Cristina W. Nogueira, and Gilson Zeni. 2014. "Synthesis of pharmacologically active 1-amino-isoquinolines prepared via silver triflate-catalyzed cyclization of o-alkynylbenzaldoximes with isocyanates." *European Journal of Pharmaceutical Sciences 51*:196–203.

Noori, Milad, Navid Dastyafteh, Sajedeh Safapoor, Minoo Khalili Ghomi, Romina Tanideh, Kamiar Zomorodian, Haleh Hamedifar, Mahintaj Dara, Shahrokh Zare, and Cambyz Irajie. 2023. "Phenyl-quinoline derivatives as lead structure of cholinesterase inhibitors with potency to reduce the GSK-3β level targeting Alzheimer's disease." *International Journal of Biological Macromolecules 253*:127392.

Paclibar, Cayz G., Deanna M. Schafer, Agnes P. Biju, Fariha Karim, Stephanie A. Sison, Christopher Liang, Shamiha T. Ahmed, and Jogeshwar Mukherjee. 2025. "Towards imaging tau hyperphosphorylation: Is DYRK1A a potential target for imaging hyperphosphorylation of tau? Molecular modeling assessment and synthesis of [125I] radioiodinated DYRK1A inhibitor." *Molecules 30*(5):990.

Patel, Ashish, Stuti Patel, Meshwa Mehta, Yug Patel, Rushi Patel, Drashti Shah, Darshini Patel, Umang Shah, Mehul Patel, and Swayamprakash Patel. 2022. "A review on synthetic investigation for quinoline-recent green approaches." *Green Chemistry Letters and Reviews 15*(2):337–372.

Plazas, Erika, and Diego R Muñoz. 2022. "Natural isoquinoline alkaloids: Pharmacological features and multi-target potential for complex diseases." *Pharmacological Research 177*:106126.

Qing, Zhi-Xing, Peng Yang, Qi Tang, Pi Cheng, Xiu-Bin Liu, Ya-jie Zheng, Yi-Song Liu, and Jian-Guo Zeng. 2017. "Isoquinoline alkaloids and their antiviral, antibacterial, and antifungal activities and structure-activity relationship." *Current Organic Chemistry 21*(18):1920–1934.

Sahu, Adarsh, Shweta Mishra, Pranay Wal, Biplab Debnath, Deepesh Chouhan, Sachinkumar Dnyaneshwar Gunjal, and Arpan Kumar Tripathi. 2024. "Novel quinoline-based RAF inhibitors: A comprehensive review on synthesis, SAR and molecular docking studies." *ChemistrySelect 9*(23):e202400347.

Senerovic, Lidija, Dejan Opsenica, Ivana Moric, Ivana Aleksic, Marta Spasić, and Branka Vasiljevic. 2019. "Quinolines and quinolones as antibacterial, antifungal, anti-virulence, antiviral and anti-parasitic agents." *Advances in Microbiology, Infectious Diseases and Public Health 14*:37–69.

Spinelli, Jessica B., and Marcia C. Haigis. 2018. "The multifaceted contributions of mitochondria to cellular metabolism." *Nature Cell Biology 20*(7):745–754.

Stöckigt, Joachim, Andrey P. Antonchick, Fangrui Wu, and Herbert Waldmann. 2011. "The Pictet–Spengler reaction in nature and in organic chemistry." *Angewandte Chemie International Edition 50*(37):8538–8564.

Sun, Jianxu, Ge Gao, Sitong Wang, Hongmei Liu, and Tie-Shan Tang. 2025. "Decoding the influence of mitochondrial Ca^{2+} regulation on neurodegenerative disease progression." *Mitochondrial Communications 3*:1–5.

Uttara, Bayani, Ajay V. Singh, Paolo Zamboni, and Raghunath Mahajan. 2009. "Oxidative stress and neuro-degenerative diseases: A review of upstream and downstream antioxidant therapeutic options." *Current Neuropharmacology 7*(1):65–74.

Vijayakumar, Archana, M. Manod, R. Bharath Krishna, Abra Mathew, and Chithra Mohan. 2023. "Diversely functionalized isoquinolines and their core-embedded heterocyclic frameworks: A privileged scaffold for medicinal chemistry." *RSC Medicinal Chemistry 14*(12):2509–2534.

Wang, Bei, Zi-Feng Yang, Yun-Li Zhao, Ya-Ping Liu, Jun Deng, Wan-Yi Huang, Xiao-Nian Li, Xin-Hua Wang, and Xiao-Dong Luo. 2018. "Anti-inflammatory isoquinoline with bis-seco-aporphine skeleton from Dactylicapnos scandens." *Organic Letters 20*(6):1647–1650.

Wang, Hong, Lu Chen, Shaoxia Wang, Xiaoxuan Tian, Lusha Zhang, Huiying Li, Chunxiao Li, Yuejin Xue, Qianyi Wang, and Leyu Fang. 2021. "Tetrandrine promotes angiogenesis via transcriptional regulation of VEGF-A." *Vascular Pharmacology 141*:106920.

Wang, Wei, Tingting Pan, Rui Su, Mingbin Chen, Wandi Xiong, Congjun Xu, and Ling Huang. 2024. "Discovery of novel melatonin–mydroxyquinoline hybrids as multitarget strategies for Alzheimer's disease therapy." *Frontiers in Chemistry 12*:1374930.

Wang, Yuanze, Pravin Patil, Katarzyna Kurpiewska, Justyna Kalinowska-Tluscik, and Alexander Dömling. 2019. "Diverse isoquinoline scaffolds by Ugi/Pomeranz–Fritsch and Ugi/Schlittler–Müller reactions." *Organic Letters 21*(10):3533–3537.

Wei, Yun-Long, Jian-Qiang Chen, Bo Sun, and Peng-Fei Xu. 2019. "Synthesis of indolo [2, 1-a]isoquinoline derivatives via visible-light-induced radical cascade cyclization reactions." *Chemical Communications 55*(42):5922–5925.

Weyesa, Abdanne, and Endale Mulugeta. 2020. "Recent advances in the synthesis of biologically and pharmaceutically active quinoline and its analogues: A review." *RSC Advances 10*(35):20784–20793.

Xu, Zi-Chen, Xiao-Bing Wang, Wen-Ying Yu, Sai-Sai Xie, Su-Yi Li, and Ling-Yi Kong. 2014. "Design, synthesis and biological evaluation of benzylisoquinoline derivatives as multifunctional agents against Alzheimer's disease." *Bioorganic & Medicinal Chemistry Letters 24*(10):2368–2373.

Yadav, Pratibha, and Kamal Shah. 2021. "Quinolines, a perpetual, multipurpose scaffold in medicinal chemistry." *Bioorganic Chemistry 109*:104639.

Yang, Changjun, Kimberly E. Hawkins, Sylvain Doré, and Eduardo Candelario-Jalil. 2019. "Neuroinflammatory mechanisms of blood-brain barrier damage in ischemic stroke." *American Journal of Physiology-Cell Physiology 316*(2):C135–C153.

Yang, Dingqiao, Satvika Burugupalli, David Daniel, and Yu Chen. 2012. "Microwave-assisted one-pot synthesis of isoquinolines, furopyridines, and thienopyridines by palladium-catalyzed sequential Coupling–Imination–Annulation of 2-bromoarylaldehydes with terminal acetylenes and ammonium acetate." *The Journal of Organic Chemistry 77*(9):4466–4472.

Yang, Xiaorong, Xiaolou Miao, Lixia Dai, Xiao Guo, Janar Jenis, Jiyu Zhang, and Xiaofei Shang. 2024. "Isolation, biological activity, and synthesis of isoquinoline alkaloids." *Natural Product Reports 41*(11):1652–1722.

Zhang, Hong-Jian, Qing-Kun Shen, Chun-Mei Jin, and Zhe-Shan Quan. 2016. "Synthesis and pharmacological evaluation of new 3, 4-dihydroisoquinolin derivatives containing heterocycle as potential anticonvulsant agents." *Molecules 21*(12):1635.

Zhou, Shiyang, and Gangliang Huang. 2021. "Synthesis and activities of acetylcholinesterase inhibitors." *Chemical Biology & Drug Design 98*(6):997–1006.

8 Indole-Based Compounds for the Treatment of Neurological Disorders

Mohammed Er-rajy, Mohamed El Fadili, Mohammed Zerrouk, Arzine Aziz, Khalil Azzaoui, and Menana El Hallaoui

8.1 INTRODUCTION

Neurodegenerative diseases (NDs), such as Alzheimer's, Parkinson's, Huntington's, and Amyotrophic Lateral Sclerosis (ALS), are progressive conditions that lead to the gradual loss of neuronal function and structure (Gadhave et al., 2024). NDs comprise increasingly widespread diverse neurological disorders: millions of people around the world are affected by these diseases that cause severe dysfunctions especially in older people. The increase in NDs is mainly due to the growth and aging of the global population. If the cases continue to increase, by 2025 more than 65% of people over the age of 65 will be affected by these diseases. Despite a decline in communicable NDs over the past 30 years, the number of deaths has increased by 39% (Feigin et al., 2020). As predicted by the World Health Organization, NDs will overtake cancer within 20 years and become the second leading cause of death, after cardiovascular disease.

Despite considerable research efforts, effective treatments remain scarce, highlighting the need for innovative therapeutic approaches. One promising strategy is the development of compounds based on indoles, a class of heterocyclic aromatic compounds with a bicyclic structure, consisting of a six-membered benzene ring fused with a five-membered nitrogen-containing pyrrole ring. These substances' many biological characteristics, such as their anti-inflammatory, antioxidant, and neuroprotective actions, have been highlighted by recent research, making them desirable options for the treatment of NDs (Marino et al., 2022). Complex processes such as oxidative stress, mitochondrial failure, protein aggregation, and neuroinflammation are often involved in NDs. Indole-based medications appear to interact with these pathways, according to recent study (Mohd Sairazi & Sirajudeen, 2020). One of the main causes of neurodegeneration is oxidative stress, which encourages cell damage and planned neuronal death. Indole derivatives, particularly melatonin and its equivalents, have remarkable antioxidant properties that fight oxidative stress by neutralizing free radicals and enhancing endogenous antioxidant defense mechanisms (Gandhi & Abramov, 2012). Preclinical studies have shown that melatonin lowers oxidative damage in models of Parkinson's and Alzheimer's disease (AD). Protein aggregation and misfolding are common in many neurodegenerative disorders (Ribeiro et al., 2025).

Alpha-synuclein and amyloid-beta (Aβ) peptide, two proteins implicated in Parkinson's and ADs, respectively, are prevented from aggregating by substances like indirubin and its derivatives (Phukan et al., 2023). By interfering with the fibrillation process, these indole derivatives lessen neurotoxicity. Another significant factor in neuronal death is mitochondrial dysfunction. By enhancing respiratory function and lowering mitochondrial oxidative stress, compounds with an indole structure, such as some tryptophan metabolites, protect mitochondrial integrity. For instance, the tryptophan metabolite kynurenic acid acts on mitochondrial pathways to produce neuroprotective effects. NDs also progress more quickly when the nervous system is chronically inflamed. Indole derivatives have strong anti-inflammatory properties, especially those that come from the kynurenine pathway. By interacting with different receptors and preventing the synthesis of pro-inflammatory cytokines, substances like kynurenine control immune responses.

DOI: 10.1201/9781003772019-8

8.2 NEUROLOGICAL DISORDERS

The term "neurological disorders" refers to a broad category of conditions that affect the central and peripheral nervous systems' structural or functional integrity (Brew, 1992). Numerous factors, including vascular anomalies, neurodegenerative processes, genetic mutations, immunological dysregulation, and metabolic problems, may contribute to these illnesses. Cognitive, motor, sensory, or behavioral impairments are some of their clinical manifestations. According to global epidemiological data, these disorders have a significant impact on morbidity and mortality globally and affect over 1 billion people. Public health, society, and the economy are all significantly impacted by the pathophysiological complexity and frequently chronic course of these illnesses.

Neurological diseases are classified into several key groups, including cerebrovascular diseases, neurodegenerative disorders, rare neurological problems, and lower-priority but very widespread disorders such as sleep difficulties (Carroll, 2019). One of the leading causes of neurological disability is cerebrovascular illness, particularly stroke (Losseff et al., 2016). The primary cause of these conditions is aberrant cerebral blood flow, which rapidly destroys neurons. Reducing stroke incidence globally still requires effective prevention through diabetes management, lifestyle changes, and hypertension control (Kuriakose & Xiao, 2020).

Another important group is neurodegenerative disorders, which are distinguished by a gradual loss of neural structure and function. AD affects around 50 million people globally, and as the population ages, this figure is predicted to climb dramatically in the next decades (Zhang et al., 2025a). Parkinson's disease affects around 10 million people and is becoming more common (Ben-Shlomo et al., 2024). Despite the dearth of curative treatments, breakthroughs in biomarker research, medicine, and neurorehabilitation have helped with diagnosis and symptom management. Nonetheless, these illnesses continue to cause serious long-term cognitive, emotional, and physical impairments.

Despite their relatively low incidence, rare neurological disorders affect millions of people. Dystonia affects tens of thousands of people, underlining the difficulties that patients confront when they are often diagnosed late and have limited access to professional care. Because many rare neurological illnesses are caused by genetic or developmental flaws, current advances in molecular diagnostics, gene sequencing, and targeted treatments offer promising possibilities for better clinical results.

Finally, disorders such as chronic headaches and sleep difficulties, which are commonly seen as minor treatment issues, contribute significantly to neurological morbidity. More than 1 billion people suffer from migraines, which are one of the primary causes of disability, and about one-third of the world's population has difficulty sleeping. Although these diseases are rarely deadly, they significantly impair quality of life, productivity, and cognitive function. Better screening procedures, increased awareness, and improved access to multidisciplinary care are all required to reduce their impact on society.

8.3 INDOLE COMPOUNDS

One of the most significant heteroaromatic scaffolds in organic and medicinal chemistry is indole, sometimes referred to as benzopyrrole (Barresi et al., 2023). It is structurally composed of a bicyclic aromatic framework made up of a six-membered benzene ring fused to a five-membered pyrrole ring.

Researcher Adolf von Baeyer developed the first method for synthesizing the indole ring in 1866, marking the start of a period of intensive research into this structure. Its ability to interact with a wide range of biological targets subsequently led to the indole ring becoming a key model for drug development (see Figure 8.1).

Over the years, numerous synthetic methodologies have been developed to access the indole framework. These include classical and modern approaches such as the Fischer indole synthesis (Robinson, 1963), Julia synthesis (Gribble, 2000), Reissert synthesis (Li, 2009b), Baeyer–Emmerling

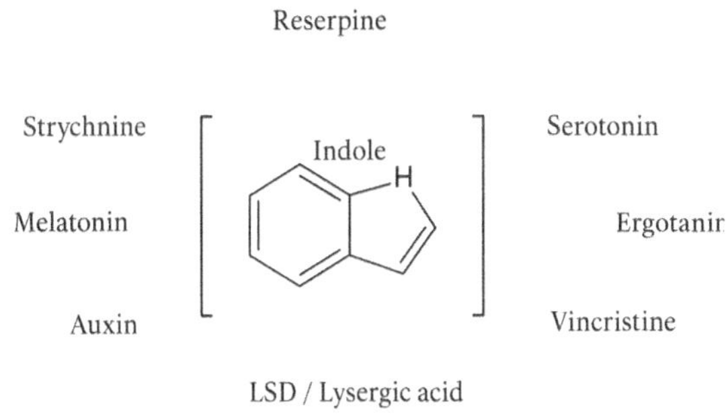

FIGURE 8.1 The chemical structure of indole-rich natural products.

reaction (Sánchez-Viesca, 2018), Larock cyclization (Ma et al., 2001), Bartoli reaction (Li, 2014), Madelung synthesis (Tois et al., 2003), Fukuyama approach (Tang et al., 2019), Bischler synthesis (Benchekroun et al., 2016), Gassman method (Li, 2009a), Hemetsberger reaction (Inman & Moody, 2012), Sundberg oxidation pathway (Sundberg, 2012), and the Leimgruber–Batcho synthesis (Clark, 1984). Recent literature also highlights the role of various catalysts in improving efficiency, selectivity, and sustainability in indole synthesis (Nasri et al., 2022).

8.4 INDOLE-BASED COMPOUNDS AS POTENTIAL ANTI-NEURODEGENERATIVE AGENTS

8.4.1 Naturally Occurring Compounds with Indole Nucleus for the Treatment of Neurological Disorders

Naturally occurring indole-based compounds represent a structurally privileged class of biomolecules with significant therapeutic potential in neurological disorders. The indole nucleus, an aromatic bicyclic framework composed of a benzene ring fused to a pyrrole ring, exhibits high chemical versatility and the capacity to interact with a wide array of molecular targets in the central nervous system. Because many neurological diseases, such as Alzheimer's disease, Parkinson's disease, and major depressive disorder, involve oxidative stress, neuroinflammation, mitochondrial dysfunction, and neurotransmitter imbalance, indole derivatives are increasingly recognized as valuable scaffolds for designing neuroprotective agents with multi-target activities.

One of the most prominent natural indole derivatives is melatonin, a neurohormone biosynthesized from tryptophan in the pineal gland (Singh et al., 2014). In addition to its physiological role in circadian rhythm modulation, melatonin has anti-inflammatory, anti-apoptotic, and antioxidant qualities. According to experimental study, melatonin may help treat Parkinson's and AD by lowering β-amyloid toxicity, repairing mitochondrial function, and controlling inflammatory signaling in the brain. Its ability to effectively pass the blood–brain barrier (BBB) increases its utility in neurological treatments.

Another important class of naturally occurring indole-containing chemicals found in *Peganum harmala* and other plant species are the β-carboline alkaloids, which include harmine, harmaline, and tetrahydroharmine (Herraiz et al., 2010). These substances have potent inhibitory effects on monoamine oxidase (MAO-A) and modulatory effects on central nervous system pathways linked to neuroplasticity. It has been demonstrated that harmine enhances adult neurogenesis, increases brain-derived neurotrophic factor (BDNF) signaling, and reduces neuroinflammatory responses. Because of these properties, β-carboline alkaloids are a prospective lead chemical for the treatment of cognitive decline and neurodegenerative illnesses.

Because of its potential therapeutic use in neuropsychiatric illnesses, naturally occurring tryptamine derivatives such as psilocybin and related indole alkaloids produced by *Psilocybe* mushrooms have attracted scientific attention (Hagel et al., 2025). Clinical research has demonstrated the long-lasting effects of psilocybin on mood, anxiety, and cognitive flexibility. These effects are believed to be caused by strong agonism at the 5-HT2A receptor and the activation of structural and functional neuroplasticity. These results have rekindled interest in investigating indole-based compounds as serotonin receptor modulators in neurological and psychiatric therapeutic approaches.

Additionally, indole-type antioxidants found in many medicinal plants have the ability to prevent neuronal damage and modulate oxidative stress (Ciaglia et al., 2024a). A number of these substances can lower reactive oxygen species, inhibit cholinesterase enzymes, or control calcium homeostasis biological pathways that are crucial for neurodegeneration. Indole-bearing metabolites exhibit a wide range of molecular actions, which implies that they could be used as models to create multi-target drug candidates that can address the intricate pathophysiology of neurological disorders.

When combined, naturally occurring indole-containing compounds provide a diverse chemical reservoir with important neuroprotective, anti-inflammatory, antioxidant, and neuromodulatory properties. Their structural versatility and capacity to interact with many neuronal pathways emphasize their usefulness in the current drug discovery efforts aimed at neurological illnesses.

8.4.1.1 An Illustration of a Natural Indole Molecule Used to Treat Alzheimer's

Indole-based structures, which are present in many natural systems, make up a number of vital endogenous biomolecules, such as melatonin, serotonin, tryptophan, and its metabolite tryptamine. The biological value of cruciferous vegetables is further enhanced by the presence of indole derivatives. Due to their ability to neutralize hydroxyl radicals and reduce mitochondrial electron leakage, two naturally occurring indole compounds, indole-3-propionic acid (IPA) and indole-3-propionamide (IPAM), have attracted more attention. Dietary treatments that can increase IPA levels in the brain have been linked to neuroprotective advantages and may decrease the progression of AD (Liu et al., 2025). These findings demonstrate the importance of the indole pharmacophore in the production of compounds with anticholinesterase characteristics, an essential tactic in the treatment of AD.

Physostigmine, an alkaloid derived from *Physostigma venenosum*, is one of the first examples of a naturally occurring indole with anticholinesterase action (Zhao et al., 2004). The therapeutic potential of indole derivatives is demonstrated by their capacity to inhibit acetylcholinesterase (AChE) and improve cholinergic neurotransmission. Physostigmine's short biological half-life and narrow therapeutic index limit its clinical utility, despite its strong permeability across the BBB (Bernardo-Castro et al., 2020). Later, Zhan et al. identified eight indole alkaloids from *Ervatamia hainanensis*, two of which showed AChE inhibition similar to that of the commercial medication galantamine (Zhan et al., 2010).

Another source of indole is *Rauvolfia reflexa*, which has traditionally been used to treat age-related cognitive impairment (Fadaeinasab et al., 2015). Rauvolfine C, rescinnamine, undulifoline, and isoresrpiline are among the indoles and β-carboline alkaloids found in the bark of this plant that exhibit selective or dual inhibition of AChE and butyrylcholinesterase (BuChE). According to docking studies, methoxy-substituted regions stabilize the ligand within the BuChE active site, whereas the indole core interacts strongly with the peripheral anionic site of AChE (Yusufzai et al., 2018). Electronic and steric characteristics that improve multi-target activity can be captured by incorporating these binding insights into structure-based quantitative structure–activity relationships (QSAR) models.

Two important indole alkaloids from *Rauwolfia serpentina* roots, ajmalicine and reserpine, were shown to have multi-target action associated with AD, including the inhibition of AChE, β-site amyloid-cleaving enzyme (BACE-1), and monoamine oxidase-B (MAO-B) (Kashyap et al., 2020). Computational study revealed strong interactions with key catalytic residues, confirming their potential as multi-target ligands. Additionally, these findings provide a helpful basis for molecular docking studies, which provide an effective assessment of the ligands' stability within the active site. Similar neuroactive indole alkaloids, such as 10-hydroxyinfractopicrine and infractopicrine from *Cortinarius infractus*, have also been shown to inhibit AChE at levels comparable to those of galantamine (ALNasser et al., 2025).

Passos et al. investigated a wide range of β-carboline and monoterpene indole alkaloids (MIAs) from *Psychotria* species in addition to isolated alkaloids, and they found broad inhibitory activities across AChE, BuChE, and MAO-A (Passos et al., 2013).

Among these, prunifoline, 14-oxoprunifoline, and several MIAs exhibited low micromolar IC_{50} values. Their multifunctional behavior confirms the growing interest in natural indole structures as models for multi-target-directed ligand (MTDL) design. Collectively, these studies underscore the importance of indolic alkaloids as promising leads in AD drug development.

8.4.2 Indole Alkaloids as Multi-Target-Directed Ligands (MTDLs)

Indole alkaloids form a broad family of natural compounds distinguished by a compact bicyclic framework that supports rich chemical diversification (Kaushik et al., 2013). This structural adaptability makes them attractive as starting points for designing molecules capable of acting on several biological targets at once-a central requirement for developing MTDLs. Because many diseases involve interconnected pathways, indole-based derivatives provide an opportunity to intervene at multiple levels with a single optimized scaffold (Hossain & Hussain, 2025).

Indole-based MTDLs exhibit a wide range of biological activities relevant to neuroprotection. They can inhibit key enzymes such as AChE, MAO-A, and MAO-B, reduce the breakdown of neurotransmitters, and improve neuronal communication. In addition, these compounds often possess antioxidant properties, scavenging reactive oxygen and nitrogen species (ROS/RNS), and reducing oxidative stress that contributes to neuronal injury. Some indoles also demonstrate anti-inflammatory activity by modulating pro-inflammatory mediators, while others can chelate toxic metal ions, preventing metal-induced neuronal damage.

Protein misfolding and aggregation are central to many neurodegenerative disorders. Indole MTDLs have been shown to inhibit the aggregation of neurotoxic proteins, such as amyloid-beta and tau in AD, while also modulating glutamate-induced excitotoxicity in other conditions. By limiting excessive calcium influx and reducing excitatory stress, indole MTDLs can help maintain neuronal homeostasis and prevent cell death.

Recent studies have reported several indole derivatives with significant multi-target activity (Justino et al., 2025). Hybrid molecules combining indole with known neuroactive scaffolds, such as rivastigmine, have demonstrated enzyme inhibition, metal chelation, antioxidant effects, and anti-aggregation properties simultaneously. Natural indole alkaloids, such as ajmalicine and reserpine, have also shown activity against multiple neurodegenerative targets, supporting their potential as MTDLs for a range of neurological disorders.

The structural flexibility of the indole scaffold enables chemical modification to maximize pharmacokinetic properties, brain penetration, and multi-target effects (Mo et al., 2024). Indole-based MTDLs may be able to slow the progression of the disease more successfully than single-target medications. Achieving selectivity for pathological targets without interfering with healthy pathways, reducing potential toxicity, and converting preclinical results into successful clinical trials are still obstacles.

All things considered, indole alkaloids offer a flexible and promising framework for the creation of MTDLs in neurological conditions. They are promising candidates for the creation of next-generation neuroprotective treatments because of their capacity to regulate a variety of pathological mechanisms, including enzyme activity, oxidative stress, inflammation, metal dysregulation, and protein aggregation.

8.4.3 Neuroprotective Properties of Synthetic-Fused Indole Frameworks

The indole core is naturally incorporated into a number of endogenous molecules, such as tryptophan, melatonin, and serotonin (Paredes et al., 2009). Despite their structural similarities, each has a unique physiological function.

An essential amino acid, tryptophan, contributes to the structure of proteins and generates a number of metabolites with biological activity (Badawy, 2019). Serotonin, also known as 5-hydroxytryptamine, or 5-HT, is a neurotransmitter that plays a role in emotional equilibrium and general health. The human neurohormone melatonin (N-acetyl-5-methoxytryptamine) is a potent antioxidant; its indole ring is especially effective at providing electrons to counteract free radicals (Zhang et al., 2025b). Notably, melatonin levels decline in patients with AD, making it a meaningful chemical template for slowing disease progression. Because of these inherent properties, medicinal chemists have explored a wide range of indole-based hybrid molecules to design MTDLs capable of addressing key pathological elements of AD. The following subsections summarize the main synthetic strategies and biological advances reported most recently.

8.4.3.1 Indole Hybrids Targeting Aβ Aggregation and Metal Dysregulation

8.4.3.1.1 Indole–Quinolone Hybrid Systems

Earlier studies identified 8-hydroxyquinoline derivatives as capable metal chelators. Building on this, Bowroju and co-workers created hybrid molecules combining 8-hydroxyquinoline (8-HQ) with indole to simultaneously disrupt Aβ self-assembly and chelate metal ions implicated in amyloid-mediated toxicity (Bowroju et al., 2020).

Their multistep synthetic route involved generating an iodinated quinoline intermediate, converting it to an azide, and subsequently reducing it to the corresponding amine. Coupling reactions using 1-ethyl-3-(3-dimethylaminopropyl)carbodiimide (EDC) produced a series of indole–8-HQ amide hybrids.

A second structural series used piperazine as a flexible bridge between the two pharmacophores. The presence of piperazine enhanced the inhibition of Aβ1–42 aggregation and significantly improved Cu^{2+}/Zn^{2+} chelation.

Molecular docking revealed that both quinoline and indole motifs participate in important interactions with key residues, while hydroxyl substitution on the quinoline ring enabled the formation of other important bonds.

Overall, these hybrids demonstrated dual action by blocking amyloid aggregation and competing with the metal-binding residues of Aβ.

8.4.3.1.2 N-Fused Indole–Oxazine Frameworks

N-fused indole–oxazine frameworks have attracted attention due to their rigid polycyclic architectures, which support predictable interactions with biological targets. Formal [3 + 2] and [4 + 2] cycloadditions of indoles with 1,2-diaza-1,3-dienes (DDs) have been devised as a Zn(II)-catalyzed divergent synthesis to create functionalized polycyclic indolines (Chen et al., 2025). Access to two different structural motifs under the same catalytic circumstances is made possible by the kind and location of substituents on the substrates acting as a chemical switch, guiding the reaction toward either the [3 + 2] or [4 + 2] pathway.

Each cycloaddition pathway's electrical and steric variables were identified through mechanistic investigations bolstered by density functional theory (DFT) computations. Certain indole-oxazine compounds showed protective effect against Aβ-induced neurotoxicity in SH-SY5Y neuronal cells, according to biological evaluations. The importance of substituent effects for neuroprotective potency was highlighted by the considerable modulation of activity caused by structural factors such as electron-donating or electron-withdrawing groups on the indole ring. These substances showed little protection in oxidative or metabolic stress models and specificity for amyloid-related toxicity, indicating that their main mechanism is interference with amyloidogenic pathways rather than broad-spectrum antioxidant actions. The findings highlight the potential of N-fused indole–oxazine frameworks catalyzed by Zn(II) as scaffolds for the development of MTDLs for neurological disorders.

8.4.3.2 Indole Hybrids as Cholinesterase Inhibitors and Antioxidants

8.4.3.2.1 Indole–Meptazinol Multifunctional Constructs

Using an MTDL strategy, Cheng and colleagues combined the opioid analgesic meptazinol (MEP) with melatonin-inspired indole fragments (Cheng et al., 2015). Since (MEP) binds both the catalytic and peripheral sites of AChE, and melatonin contributes antioxidant and anti-amyloid properties, linking these two motifs created promising multifunctional candidates. Several hybrids were produced using tryptamine derivatives and variable linker lengths. A seven-carbon linker yielded the most potent AChE inhibitors ($IC_{50} \approx 0.3–0.9$ μM), surpassing the parent molecules. The best compound, designated 4, also showed strong free-radical-scavenging ability and effectively reduced Aβ1–40 and Aβ1–42 aggregation.

8.4.3.2.2 Indole–Benzyl Pyridinium Bromide Hybrids

Cheng et al. (2015) synthesized indole–benzyl pyridinium derivatives to exploit the known cholinesterase-binding properties of pyridinium salts. The quaternary ammonium center facilitates interactions with aromatic residues at the AChE gorge. Among the generated compounds, compound 5 (bearing a 3-fluorobenzyl group) exhibited pronounced AChE inhibition ($IC_{50} = 0.11$ μM), moderate BuChE inhibition, and strong antioxidant effects. This hybrid also protected SH-SY5Y cells from oxidative stress without displaying cytotoxicity.

8.4.3.2.3 Indole–Dibenzyl-(N-Methyl)amine Hybrids

López-Iglesias and collaborators fused melatonin-like indole structures with N,N-dibenzyl-(N-methyl)amine to obtain multifunctional derivatives (López-Iglesias et al., 2014). These hybrids demonstrated antioxidant activity, neuroprotection, and cholinesterase inhibition at low micromolar concentrations.

The standout compound, 6, inhibited both hAChE and hBuChE ($IC_{50} = 1.9$ and 3.7 μM) and effectively displaced propidium from the Peripheral Anionic Site (PAS), indicating strong potential to block Aβ aggregation. Favorable CNS penetration and low toxicity further support the relevance of these hybrids as lead candidates.

8.4.3.2.4 Bivalent β-Carboline (Pyrido[3,4-b]indole) Hybrids

Although they are poorly soluble, β-carbolines (Harman alkaloids) naturally inhibit AChE, BuChE, and MAO-B. In order to get around this, Zhao et al. (2018) created bivalent hybrids by joining two β-carboline units with hydrophilic connections like diamines or piperazine (Zhao et al., 2018). Two compounds, 7 and 8, showed favorable binding to the aromatic residues of the enzyme through π–π and cation–π interactions, as well as strong, selective inhibition of BuChE. These compounds are intriguing scaffolds for future research because they also demonstrated a variety of AD-relevant actions.

8.4.3.2.5 Indole–Cinnamic Acid Hybrids

The nuclear factor erythroid 2–related factor 2 (Nrf2) pathway, a critical regulator of antioxidant defense in the brain, has been shown to be induced by derivatives of cinnamic acid (Pourparizi et al., 2023). Using this characteristic, scientists have created hybrid compounds that combine cinnamic acid and indole cores to produce multifunctional neuroprotective drugs. The objective is to treat amyloid pathology and oxidative stress in AD at the same time.

In these hybrids, the cinnamic acid moiety offers extra antioxidant and anti-amyloid activity, while the indole part gives electron-rich π-systems that can scavenge free radicals. Typically, synthetic approaches begin with commercially available tryptamine, which is converted into tetrahydro-β-carboline intermediates via Pictet–Spengler cyclization with aldehydes. The final indole–cinnamic acid hybrids are obtained by combining with activated cinnamic acid derivatives.

According to biological evaluation, these substances shield neuronal cells from various insults such as oxidative stress caused by hydrogen peroxide, toxicity caused by okadaic acid, and cytotoxicity caused by Aβ1–42 in human PC12 and SH-SY5Y cells. In microglial BV2 cells exposed to lipopolysaccharide, the hybrids also decrease the formation of ROS, suggesting that they may have anti-inflammatory properties.

According to docking studies, the cinnamic acid phenyl ring engages in π–π stacking interactions inside the active site of cholinesterase enzymes, whilst the indole core establishes hydrogen bonds with important residues. When compared to their methoxylated counterparts, compounds containing polyhydroxylated cinnamic acid rings showed better suppression of Aβ1–42 aggregation, indicating that electron-rich hydroxyl groups improve target interactions.

These hybrids have favorable pharmacokinetic characteristics, such as high oral absorption and BBB permeability, in addition to strong antioxidant and anti-amyloid activities. This makes them interesting leads for additional structural optimization in the development of anti-AD drugs.

8.4.3.3 Miscellaneous Indole-Based Neuroprotective Scaffolds

8.4.3.3.1 Aryl Indoles via Transition-Metal Catalysis

Parveen et al. (2018) synthesized aryl indole derivatives using microwave-assisted Bischler–Möhlau reactions, followed by palladium-catalyzed Suzuki cross-coupling. Optimization of reaction parameters, including Pd(dppf)Cl_2 catalyst, base, temperature, and solvents, enabled high-yield synthesis.

Most compounds displayed strong AChE inhibition (IC_{50} = 0.27–21.5 μM). Notably, compounds 31 and 32 showed dual inhibitory activity against AChE (IC_{50} = 5.65 and 1.37 μM) and BuChE (IC_{50} = 2.42 and 1.85 μM), likely due to electron-withdrawing 5-fluoro substituents enhancing binding. Molecular docking suggested these compounds adopt multiple binding modes, interacting with amino acids similarly to the reference inhibitor, huprin W.

8.4.3.3.2 Fascaplysin and Derivatives

Fascaplysin, a benzoyl-linked β-carboline alkaloid isolated from the marine sponge *Fascaplysinopsis bergquist*, exhibits AChE inhibition, prevents Aβ aggregation, and protects neurons from Aβ oligomer toxicity. Pan et al., in 2019, synthesized fascaplysin derivatives and evaluated their anti-ChE and neuroprotective activities, including BBB permeability and in vivo cognitive protection in scopolamine-induced models (Pan et al., 2019).

The derivatives demonstrated potent AChE inhibition (IC_{50} = 0.95–9.55 μM) and moderate BuChE inhibition (IC_{50} = 2.79–15.98 μM), with enhanced selectivity for AChE. Compounds 33 and 34 were most effective, interacting with PAS residues (Tyr337, Tyr341, Trp286, Phe338) and CAS (Trp86) via π–π stacking and hydrogen bonding. Compound 34 notably increased SH-SY5Y cell viability at nanomolar concentrations and mitigated oxidative stress by modulating antioxidant enzymes. Molecular docking and molecular dynamic studies confirmed stable interactions with AChE, supporting the observed dual-site inhibition and potential anti-Aβ aggregation effects.

Additionally, 9-methylfascaplysin showed enhanced inhibition of Aβ aggregation at nanomolar concentrations, interacting with negatively charged Aβ42 residues through hydrogen bonds and π–π stacking.

8.4.3.3.3 Neuroprotective Effects of Multifunctional Indole Hybrid Compounds

Multifunctional indole-based hybrid compounds have gained attention as potential therapeutics for NDs such as Alzheimer's and Parkinson's disease. These compounds are specifically designed to interact with multiple molecular targets simultaneously, including the inhibition of β-amyloid aggregation, modulation of cholinesterase enzymes (AChE and BuChE), attenuation of oxidative stress, and reduction of neuroinflammation. Multifunctional indole hybrids have been produced in a number of classes, such as the following:

Indole–quinolone hybrids prevent Aβ aggregation and metal-assisted neurotoxicity by combining the indole nucleus with the metal-chelating characteristics of 8-hydroxyquinoline.

- N-fused indole–oxazine derivatives can prevent Aβ25–35-induced neuronal damage and show stereochemistry-dependent neuroprotective effects.
- In addition to acting as dual cholinesterase inhibitors, indole–meptazinol and indole–benzyl pyridinium hybrids also have antioxidant properties, and they stop Aβ aggregation.
- Derivatives of fascaplysin and β-carboline exhibit multifunctional inhibition of AChE/BuChE, reduce β-amyloid formation, and shield neuronal cells from oxidative stress.

These hybrids have notable neuroprotective efficacy, including dual cholinesterase inhibition, suppression of β-amyloid aggregation, and radical-scavenging activities, according to in vitro investigations, especially on SH-SY5Y human neuroblastoma cell lines. Their pharmacological activity and BBB permeability are improved by structural changes such as the inclusion of bioactive moieties or functional groups.

These indole-based hybrids are very intriguing candidates for additional preclinical testing due to their multifunctional profile (Er-rajy et al., 2022, 2025). Compared to single-target medicines, their capacity to target several pathogenic systems at once offers a strategic advantage that may more successfully halt or stop neurodegenerative processes.

8.4.3.3.4 Current Various Indole-Based Neuroprotective Scaffolds (2023–2024)

Innovative derivatives with multi-target actions and improved pharmacological profiles have been highlighted in recent research that broaden the reach of several indole-based neuroprotective scaffolds.

8.4.3.3.4.1 Indol-3-Yl-Phenyl Allylidene Hydrazine Carboximidamide Derivatives A total of 41 indol-3-yl-phenyl allylidene hydrazine carboximidamide derivatives were synthesized, and their inhibitory efficacy against AChE and BACE1 was assessed by Sharma et al. (2024a). At micromolar doses, one of the strongest substances showed dual enzyme inhibition. While ADME predictions indicated good drug-like characteristics, molecular docking experiments showed favorable binding at both the BACE1 active site and the AChE catalytic site. This work demonstrates that indole hybrids can serve as MTDLs for AD therapy.

8.4.3.3.4.2 Pyrrol-2-Yl-Phenyl Allylidene Hydrazine Carboximidamide Derivatives In a related study, Sharma et al. developed 40 pyrrol-2-yl-based derivatives to target both AChE and BACE1 (Sharma et al., 2024b). These substances demonstrated comparable dual inhibitory characteristics, demonstrating the adaptability of non-classical indole scaffolds in the development of multifunctional neuroprotective drugs.

8.4.3.3.4.3 Indole-3-Carbinol (I3C) and Diindolylmethane (DIM) The neuroprotective properties of natural indole derivatives like I3C and DIM have lately been studied. These compounds prevent Aβ-induced toxicity and regulate oxidative stress through Nrf2 activation (Singh et al., 2024). The potential of natural indole scaffolds in multi-target neuroprotection was highlighted by in vitro and in silico analyses that verified interactions with AChE and proposed additional neuroprotective mechanisms, such as anti-inflammatory and antioxidant properties.

8.4.3.3.4.4 Indole–Phenolic Hybrids In neuroblastoma cell lines, novel indole–phenolic hybrids effectively inhibited Aβ aggregation, chelated copper, and exhibited antioxidant activity (Ciaglia et al., 2024b). The ROS were reduced and cell viability was restored in experimental experiments; interactions with Aβ fibrils were confirmed by molecular docking and biophysical assays. Phenolic substituents on the indole ring improve metal chelation and anti-aggregative activity, according to structure–activity relationship studies, highlighting the significance of structural optimization.

8.4.3.3.4.5 Multi-Target Indole Derivatives with Anti-Inflammatory Potential Recent synthetic indole derivatives have been developed as MTDLs combining **AChE/BuChE inhibition** with anti-inflammatory activity (Lamie et al., 2022). Compounds displaying dual enzyme inhibition also reduced nitric oxide production and modulated pro-inflammatory cytokines (IL-1β, TNF-α), supporting the concept of indole-based scaffolds as multifunctional neuroprotective agents. Docking studies confirmed stable interactions with key residues in both cholinesterase, while structure–activity relationship analyses identified substituent positions that influence dual-target efficacy and anti-inflammatory potency.

We have not been able to list all the efforts made in these complex diseases, but we have mentioned the most important ones. Despite all this research, the majority of current treatments merely alleviate symptoms; they neither halt nor reverse the course of the disease. Numerous overlapping pathological pathways are often involved in NDs, and mounting evidence indicates that single-pathway therapeutic approaches may not be adequate. Consequently, the creation of multi-target approaches is increasingly recognized as a viable tactic for successful intervention. Indolone molecules offer significant potential in the treatment of NDs, due to their simple chemical structure, ability to cross the BBB and affordability compared to more expensive biologic therapies. This is why small molecules continue to be useful in the treatment of NDs.

8.5 CONCLUSION AND FUTURE PERSPECTIVE

Millions of people worldwide suffer from a group of progressive nervous system disorders known as NDs, especially the elderly. The most prevalent of these are Parkinson's disease and AD, but other major causes of neurological disability include Huntington's disease, multiple sclerosis, ALS, and a number of neuropsychiatric disorders. Progressive neuronal loss is a hallmark of NDs, resulting in behavioral, motor, and cognitive impairments. Oxidative stress, aberrant protein aggregation, inflammation, neuronal degeneration, and mitochondrial dysfunction are common pathological mechanisms. Disease onset and progression are further complicated by lifestyle, environmental, and genetic factors.

Despite decades of research, the majority of current treatments only alleviate symptoms; they do not stop or reverse the course of the disease. Numerous overlapping pathological pathways are frequently involved in NDs, according to mounting evidence, indicating that therapeutic approaches focusing on a single pathway might not be adequate. As a result, the creation of multi-targeted approaches is becoming more widely acknowledged as a viable tactic for successful intervention. Because of their straightforward chemical structures, capacity to cross the BBB, and affordability in comparison to larger biological therapeutics, small molecules continue to be useful in ND therapy.

Indole-based medicines can target various molecular targets in NDs, including cholinesterases, β-amyloid, MAO-B, adenosine A2A receptors, PERK signaling, AMPK, and serotonin receptors (5-HT6R). Incorporating indole into hybrid structures allows for the synthesis of MTDLs that inhibit Aβ aggregation, reduce oxidative stress, and restore cholinergic balance simultaneously. In addition, new research using human neuroblastoma SH-SY5Y cell models show that indole-based derivatives are effective in vitro, with promising neuroprotective benefits, antioxidant activity, and multi-target inhibition. A number of these compounds have progressed to preclinical testing, implying that they may be developed into effective ND therapies.

The pharmacokinetic and pharmacodynamic profiles of indole-based compounds may be further enhanced in the future by combining chemical modification, bioconjugation, and nanocarrier-based drug delivery. Furthermore, by supporting regeneration and reducing neuroinflammation, an understanding of the mechanisms underlying mesenchymal stem cell-derived extracellular vesicles may enhance small-molecule therapies.

All things considered, indole is still a very special and adaptable pharmacophore for the development of ND drugs. To develop more effective and clinically relevant treatments targeted at slowing or stopping the progression of NDs, ongoing efforts to optimize structure, multi-target activity, and delivery strategies are crucial.

REFERENCES

ALNasser, M. N., Alboraiy, G. M., Alsowig, E. M. and Alqattan, F. M., Cholinesterase inhibitors from plants and their potential in Alzheimer's treatment: Systematic review, *Brain Sciences*, vol. **15**, no. 2, p. 215, February 19, 2025. DOI: 10.3390/brainsci15020215

Badawy, A. A.-B., Tryptophan metabolism: A versatile area providing multiple targets for pharmacological intervention, *Egyptian Journal of Basic and Clinical Pharmacology*, vol. **9**, 2019. DOI: 10.32527/2019/101415

Barresi, E., Robello, M., Baglini, E., Poggetti, V., Viviano, M., Salerno, S., Da Settimo, F. and Taliani, S., Indol-3-ylglyoxylamide as privileged scaffold in medicinal chemistry, *Pharmaceuticals*, vol. **16**, no. 7, p. 997, July 12, 2023. DOI: 10.3390/ph16070997

Benchekroun, M., Romero, A., Egea, J., León, R., Michalska, P., Buendía, I., Jimeno, M. L., et al., The antioxidant additive approach for Alzheimer's disease therapy: New ferulic (lipoic) acid plus melatonin modified tacrines as cholinesterases inhibitors, direct antioxidants, and nuclear factor (erythroid-derived 2)-like 2 activators, *Journal of Medicinal Chemistry*, vol. **59**, no. 21, pp. 9967–73, November 10, 2016. DOI: 10.1021/acs.jmedchem.6b01178

Ben-Shlomo, Y., Darweesh, S., Llibre-Guerra, J., Marras, C., Luciano, M. S. and Tanner, C., The epidemiology of Parkinson's disease, *The Lancet*, vol. **403**, no. 10423, pp. 283–92, January 20, 2024. DOI: 10.1016/S0140-6736(23)01419-8

Bernardo-Castro, S., Sousa, J. A., Brás, A., Cecília, C., Rodrigues, B., Almendra, L., Machado, C., et al., Pathophysiology of blood–brain barrier permeability throughout the different stages of ischemic stroke and its implication on hemorrhagic transformation and recovery, *Frontiers in Neurology*, vol. **11**, p. 594672, December 9, 2020. DOI: 10.3389/fneur.2020.594672

Bowroju, S. K., Mainali, N., Ayyadevara, S., Penthala, N. R., Krishnamachari, S., Kakraba, S., Reis, R. J. S. and Crooks, P. A., Design and synthesis of novel hybrid 8-hydroxy quinoline-indole derivatives as inhibitors of aβ self-aggregation and metal chelation-induced aβ aggregation, *Molecules*, vol. **25**, no. 16, p. 3610, January 2020. DOI: 10.3390/molecules25163610

Brew, B. J., Central And peripheral nervous system abnormalities, *Medical Clinics of North America*, vol. **76**, no. 1, pp. 63–81, January 1, 1992. DOI: 10.1016/S0025-7125(16)30371-6

Carroll, W. M., The global burden of neurological disorders, *The Lancet Neurology*, vol. **18**, no. 5, pp. 418–19, May 1, 2019. DOI: 10.1016/S1474-4422(19)30029-8

Chen, Z., Cao, Y., Huang, R., Yao, X., Xiao, F. and Deng, G.-J., Brønsted acid-promoted oxa-Pictet–Spengler reaction for the synthesis of indole-fused pyran derivatives, *The Journal of Organic Chemistry*, vol. **90**, no. 32, pp. 11529–37, August 15, 2025. DOI: 10.1021/acs.joc.5c01063

Cheng, S., Zheng, W., Gong, P., Zhou, Q., Xie, Q., Yu, L., Zhang, P., et al., (–)-Meptazinol–melatonin hybrids as novel dual inhibitors of cholinesterases and amyloid-β aggregation with high antioxidant potency for Alzheimer's therapy, *Bioorganic & Medicinal Chemistry*, vol. **23**, no. 13, pp. 3110–18, July 1, 2015. DOI: 10.1016/j.bmc.2015.04.084

Ciaglia, T., Miranda, M. R., Di Micco, S., Vietri, M., Smaldone, G., Musella, S., Di Sarno, V., et al., Neuroprotective potential of indole-based compounds: A biochemical study on antioxidant properties and amyloid disaggregation in neuroblastoma cells, *Antioxidants*, vol. **13**, no. 12, p. 1585, December 2024a. DOI: 10.3390/antiox13121585

Ciaglia, T., Miranda, M. R., Di Micco, S., Vietri, M., Smaldone, G., Musella, S., Di Sarno, V., et al., Neuroprotective potential of indole-based compounds: A biochemical study on antioxidant properties and amyloid disaggregation in neuroblastoma cells, *Antioxidants*, vol. **13**, no. 12, p. 1585, December 2024b. DOI: 10.3390/antiox13121585

Clark, G. L., A theory of local autonomy, *Annals of the Association of American Geographers*, vol. **74**, no. 2, pp. 195–208, March 1, 1984. DOI: 10.1111/j.1467-8306.1984.tb01448.x

Er-rajy, M., El Fadili, M., Alnajjar, R., Zarougui, S., Mujwar, S., Azzaoui, K., Abuelizz, H. A., Hammouti, B. and Elhallaoui, M., An in-depth study of indolone derivatives as potential lung cancer treatment, *Scientific Reports*, vol. **15**, no. 1, p. 2199, January 16, 2025. DOI: 10.1038/s41598-025-85707-7

Er-rajy, M., El Fadili, M., Hadni, H., Mrabti, N. N., Zarougui, S. and Elhallaoui, M., 2D-QSAR modeling, drug-likeness studies, ADMET prediction, and molecular docking for anti-lung cancer activity of 3-substituted-5-(phenylamino) indolone derivatives, *Structural Chemistry*, vol. **33**, no. 3, pp. 973–86, June 1, 2022. DOI: 10.1007/s11224-022-01913-3

Fadaeinasab, M., Basiri, A., Kia, Y., Karimian, H., Ali, H. M. and Murugaiyah, V., New indole alkaloids from the bark of *Rauvolfia reflexa* and their cholinesterase inhibitory activity, *Cellular Physiology and Biochemistry*, vol. **37**, no. 5, pp. 1997–2011, November 20, 2015. DOI: 10.1159/000438560

Feigin, V. L., Vos, T., Nichols, E., Owolabi, M. O., Carroll, W. M., Dichgans, M., Deuschl, G., Parmar, P., Brainin, M. and Murray, C., The global burden of neurological disorders: Translating evidence into policy, *The Lancet. Neurology*, vol. **19**, no. 3, pp. 255–65, March 2020. DOI: 10.1016/S1474-4422(19)30411-9

Gadhave, D. G., Sugandhi, V. V., Jha, S. K., Nangare, S. N., Gupta, G., Singh, S. K., Dua, K., Cho, H., Hansbro, P. M. and Paudel, K. R., Neurodegenerative disorders: Mechanisms of degeneration and therapeutic approaches with their clinical relevance, *Ageing Research Reviews*, vol. **99**, p. 102357, August 1, 2024. DOI: 10.1016/j.arr.2024.102357

Gandhi, S. and Abramov, A. Y., Mechanism of oxidative stress in neurodegeneration, *Oxidative Medicine and Cellular Longevity*, vol. **2012**, no. 1, p. 428010, 2012. DOI: 10.1155/2012/428010

Gribble, G. W., Recent developments in indole ring synthesis – Methodology and applications, *Journal of the Chemical Society, Perkin Transactions 1*, no. 7, pp. 1045–75, 2000. DOI: 10.1039/a909834h

Hagel, J. M., Chang, L., Li, J., Chen, X., Yu, L., Gallant, J. A. and Facchini, P. J., Bioproduction of a large-scale library of tryptamine derivatives for neuropsychiatric drug screening, *ACS Chemical Biology*, vol. **20**, no. 6, pp. 1212–31, June 20, 2025. DOI: 10.1021/acschembio.4c00857

Herraiz, T., González, D., Ancín-Azpilicueta, C., Arán, V. J. and Guillén, H., β-Carboline alkaloids in *Peganum harmala* and inhibition of human monoamine oxidase (MAO), *Food and Chemical Toxicology*, vol. **48**, no. 3, pp. 839–45, March 1, 2010. DOI: 10.1016/j.fct.2009.12.019

Hossain, M. S. and Hussain, M. H., Multi-target drug design in Alzheimer's disease treatment: Emerging technologies, advantages, challenges, and limitations, *Pharmacology Research & Perspectives*, vol. **13**, no. 4, p. e70131, June 18, 2025. DOI: 10.1002/prp2.70131

Inman, M. and Moody, C. J., Indole synthesis – Something old, something new, *Chemical Science*, vol. **4**, no. 1, pp. 29–41, November 29, 2012. DOI: 10.1039/C2SC21185H

Justino, A., Doring, T. H., Cruz, J. W. da, Braga, A. L. and Oliveira, A. S. de, Indole amide derivatives in Alzheimer's disease: Advancing from synthesis to drug discovery, *Discover Chemistry*, vol. **2**, no. 1, p. 175, August 10, 2025. DOI: 10.1007/s44371-025-00256-2

Kashyap, P., Kalaiselvan, V., Kumar, R. and Kumar, S., Ajmalicine and reserpine: Indole alkaloids as multi-target directed ligands towards factors implicated in Alzheimer's disease, *Molecules*, vol. **25**, no. 7, p. 1609, April 1, 2020. DOI: 10.3390/molecules25071609

Kaushik, N. K., Kaushik, N., Attri, P., Kumar, N., Kim, C. H., Verma, A. K. and Choi, E. H., Biomedical importance of indoles, *Molecules*, vol. **18**, no. 6, pp. 6620–62, June 6, 2013. DOI: 10.3390/molecules18066620

Kuriakose, D. and Xiao, Z., Pathophysiology and treatment of stroke: Present status and future perspectives, *International Journal of Molecular Sciences*, vol. **21**, no. 20, p. 7609, October 15, 2020. DOI: 10.3390/ijms21207609

Lamie, P. F., Abdel-Fattah, M. M. and Philoppes, J. N., Design and synthesis of new indole drug candidates to treat Alzheimer's disease and targeting neuro-inflammation using a multi-target-directed ligand (MTDL) strategy, *Journal of Enzyme Inhibition and Medicinal Chemistry*, vol. **37**, no. 1, pp. 2660–78, December 31, 2022. DOI: 10.1080/14756366.2022.2126464

Li, J. J., Gassman indole synthesis, in *Name reactions: A collection of detailed mechanisms and synthetic applications*, J. J. Li, Ed., Berlin, Heidelberg: Springer, accessed November 14, 2025a, from https://doi.org/10.1007/978-3-642-01053-8_108, pp. 251–52, 2009a.

Li, J. J., Reissert indole synthesis, in *Name reactions: A collection of detailed mechanisms and synthetic applications*, J. J. Li, Ed., Berlin, Heidelberg: Springer, accessed November 14, 2025b, from https://doi.org/10.1007/978-3-642-01053-8_216, pp. 463–64, 2009b.

Li, J. J., Bartoli indole synthesis, in *Name reactions: A collection of detailed mechanisms and synthetic applications fifth edition*, J. J. Li, Ed., Cham: Springer International Publishing, accessed November 14, 2025, from https://doi.org/10.1007/978-3-319-03979-4_12, pp. 24–25, 2014.

Liu, Z., Zhang, J., Jiang, F., Liu, C., Shao, Y. and Le, W., Biological effects of dietary restriction on Alzheimer's disease: Experimental and clinical investigations, *CNS Neuroscience & Therapeutics*, vol. **31**, no. 4, p. e70392, April 17, 2025. DOI: 10.1111/cns.70392

López-Iglesias, B., Pérez, C., Morales-García, J. A., Alonso-Gil, S., Pérez-Castillo, A., Romero, A., López, M. G., Villarroya, M., Conde, S. and Rodríguez-Franco, M. I., New melatonin–N,N-dibenzyl(N-methyl)amine hybrids: Potent neurogenic agents with antioxidant, cholinergic, and neuroprotective properties as innovative drugs for Alzheimer's disease, *Journal of Medicinal Chemistry*, vol. **57**, no. 9, pp. 3773–85, May 8, 2014. DOI: 10.1021/jm5000613

Losseff, N., Adams, M., Brown, M. M., Grieve, J. and Simister, R., Stroke and cerebrovascular diseases, in *Neurology*, John Wiley & Sons, Ltd, accessed November 16, 2025, from https://onlinelibrary.wiley.com/doi/abs/10.1002/9781118486160.ch5, pp. 133–85, 2016.

Ma, C., Liu, X., Li, X., Flippen-Anderson, J., Yu, S. and Cook, J. M., Efficient asymmetric synthesis of biologically important tryptophan analogues via a palladium-mediated heteroannulation reaction, *The Journal of Organic Chemistry*, vol. **66**, no. 13, pp. 4525–42, June 1, 2001. DOI: 10.1021/jo001679s

Marino, A., Battaglini, M., Moles, N. and Ciofani, G., Natural antioxidant compounds as potential pharmaceutical tools against neurodegenerative diseases, *ACS Omega*, vol. **7**, no. 30, pp. 25974–90, August 2, 2022. DOI: 10.1021/acsomega.2c03291

Mo, X., Rao, D. P., Kaur, K., Hassan, R., Abdel-Samea, A. S., Farhan, S. M., Bräse, S. and Hashem, H., Indole derivatives: A versatile scaffold in modern drug discovery—An updated review on their multifaceted therapeutic applications (2020–2024), *Molecules*, vol. **29**, no. 19, p. 4770, October 9, 2024. DOI: 10.3390/molecules29194770

Mohd Sairazi, N. S. and Sirajudeen, K. N. S., Natural products and their bioactive compounds: Neuroprotective potentials against neurodegenerative diseases, *Evidence-Based Complementary and Alternative Medicine*, vol. **2020**, no. 1, p. 6565396, 2020. DOI: 10.1155/2020/6565396

Nasri, S., Bayat, M., Miankooshki, F. R. and Samet, N. H., Recent developments in green approaches for sustainable synthesis of indole-derived scaffolds, *Molecular Diversity*, vol. **26**, no. 6, pp. 3411–45, December 1, 2022. DOI: 10.1007/s11030-021-10376-3

Pan, H., Qiu, H., Zhang, K., Zhang, P., Liang, W., Yang, M., Mou, C., et al., Fascaplysin derivatives are potent multitarget agents against Alzheimer's disease: In vitro and in vivo evidence, *ACS Chemical Neuroscience*, vol. **10**, no. 11, pp. 4741–56, November 20, 2019. DOI: 10.1021/acschemneuro.9b00503

Paredes, S. D., Barriga, C., Reiter, R. J. and Rodríguez, A. B., Assessment of the potential role of tryptophan as the precursor of serotonin and melatonin for the aged sleep-wake cycle and immune function: *Streptopelia risoria* as a model, *International Journal of Tryptophan Research : IJTR*, vol. **2**, pp. 23–36, January 14, 2009. DOI: 10.4137/ijtr.s1129

Parveen, S., Shah, M. S., Zaib, S., Gul, T., Khan, K. M., Iqbal, J. and Hassan, A., Modification of Bischler-Möhlau indole derivatives through palladium catalyzed Suzuki reaction as effective cholinesterase inhibitors, their kinetic and molecular docking studies, *Bioorganic Chemistry*, vol. **76**, pp. 166–76, February 1, 2018. DOI: 10.1016/j.bioorg.2017.11.003

Passos, C. S., Simões-Pires, C. A., Nurisso, A., Soldi, T. C., Kato, L., Oliveira, C. M. A. de, Faria, E. O. de, et al., Indole alkaloids of *Psychotria* as multifunctional cholinesterases and monoamine oxidases inhibitors, *Phytochemistry*, vol. **86**, pp. 8–20, February 1, 2013. DOI: 10.1016/j.phytochem.2012.11.015

Phukan, B. C., Roy, R., Gahatraj, I., Bhattacharya, P. and Borah, A., Therapeutic considerations of bioactive compounds in Alzheimer's disease and Parkinson's disease: Dissecting the molecular pathways, *Phytotherapy Research*, vol. **37**, no. 12, pp. 5657–99, 2023. DOI: 10.1002/ptr.8012

Pourparizi, A., Pourrajab, F., Nadri, H., Moradi, A., Rezaei, N. and Moghadamnia, M., Neuroprotection and NRF2 signaling cascade analysis of new synthetic cinnamic acid-eugenol hybrids, *Journal of Molecular Structure*, vol. **1292**, p. 136092, November 15, 2023. DOI: 10.1016/j.molstruc.2023.136092

Ribeiro, R. F. N., Santos, M. R., Aquino, M., Almeida, L. P. de, Cavadas, C. and Silva, M. M. C., The therapeutic potential of melatonin and its novel synthetic analogs in circadian rhythm sleep disorders, inflammation-associated pathologies, and neurodegenerative diseases, *Medicinal Research Reviews*, vol. **45**, no. 5, pp. 1515–39, 2025. DOI: 10.1002/med.22117

Robinson, B., The Fischer indole synthesis., *Chemical Reviews*, vol. **63**, no. 4, pp. 373–401, August 1, 1963. DOI: 10.1021/cr60224a003

Sánchez-Viesca, F., *On the Baeyer-Emmerling synthesis of indigo*, August 26, 2018. DOI: 10.12691/wjoc-6-1-2

Sharma, A., Rudrawar, S., Sharma, A., Bharate, S. B. and Jadhav, H. R., Unveiling the potential of novel indol-3-Yl-phenyl allylidene hydrazine carboximidamide derivatives as AChE/BACE 1 dual inhibitors: A combined in silico, synthesis and in vitro study, *RSC Advances*, vol. **14**, no. 33, pp. 23853–72, July 26, 2024a. DOI: 10.1039/D4RA04315D

Sharma, A., Rudrawar, S., Sharma, A., Bharate, S. B. and Jadhav, H. R., Unveiling the potential of novel indol-3-yl-phenyl allylidene hydrazine carboximidamide derivatives as AChE/BACE 1 dual inhibitors: A combined in silico, synthesis and in vitro study, *RSC Advances*, vol. **14**, no. 33, pp. 23853–72, July 26, 2024b. DOI: 10.1039/D4RA04315D

Singh, A. A., Yadav, D., Khan, F. and Song, M., Indole-3-carbinol and its derivatives as neuroprotective modulators, *Brain Sciences*, vol. **14**, no. 7, p. 674, July 2, 2024. DOI: 10.3390/brainsci14070674

Singh, M. and Jadhav, H. R., Melatonin: Functions and ligands, *Drug Discovery Today*, vol. **19**, no. 9, pp. 1410–18, September 1, 2014. DOI: 10.1016/j.drudis.2014.04.014

Sundberg, R., *The chemistry of indoles*, Elsevier, pp. 505, 2012.

Tang, S.-Q., Bricard, J., Schmitt, M. and Bihel, F., Fukuyama cross-coupling approach to isoprekinamycin: Discovery of the highly active and bench-stable palladium precatalyst POxAP, *Organic Letters*, vol. **21**, no. 3, pp. 844–48, February 1, 2019. DOI: 10.1021/acs.orglett.9b00031

Tois, J., Franzén, R. and Koskinen, A., Synthetic approaches towards indoles on solid phase recent advances and future directions, *Tetrahedron*, vol. **59**, no. 29, pp. 5395–5405, July 14, 2003. DOI: 10.1016/S0040-4020(03)00851-2

Yusufzai, S. K., Khan, M. S., Sulaiman, O., Osman, H. and Lamjin, D. N., Molecular docking studies of coumarin hybrids as potential acetylcholinesterase, butyrylcholinesterase, monoamine oxidase A/B and β-amyloid inhibitors for Alzheimer's disease, *Chemistry Central Journal*, vol. **12**, p. 128, December 4, 2018. DOI: 10.1186/s13065-018-0497-z

Zhan, Z.-J., Yu, Q., Wang, Z.-L. and Shan, W.-G., Indole alkaloids from *Ervatamia hainanensis* with potent acetylcholinesterase inhibition activities, *Bioorganic & Medicinal Chemistry Letters*, vol. **20**, no. 21, pp. 6185–87, November 1, 2010. DOI: 10.1016/j.bmcl.2010.08.123

Zhang, N., Chai, S. and Wang, J., Assessing and projecting the global impacts of Alzheimer's disease, *Frontiers in Public Health*, vol. **12**, accessed November 16, 2025a, from https://www.frontiersin.org/journals/public-health/articles/10.3389/fpubh.2024.1453489/full, January 15, 2025. DOI: 10.3389/fpubh.2024.1453489

Zhang, Y., Wang, N., Zhang, L., Zhuang, Y., Xin, Q., Gu, X., Jiang, C. and Wu, J., Serotonin (5-hydroxytryptamine): Metabolism, signaling, biological functions, diseases, and emerging therapeutic opportunities, *MedComm*, vol. **6**, no. 9, p. e70383, September 9, 2025b. DOI: 10.1002/mco2.70383

Zhao, B., Moochhala, S. M. and Tham, S., Biologically active components of *Physostigma venenosum*, *Journal of Chromatography B*, vol. **812**, no. 1, pp. 183–92, December 5, 2004. DOI: 10.1016/j.jchromb.2004.08.031

Zhao, Y., Ye, F., Xu, J., Liao, Q., Chen, L., Zhang, W., Sun, H., Liu, W., Feng, F. and Qu, W., Design, synthesis and evaluation of novel bivalent β-carboline derivatives as multifunctional agents for the treatment of Alzheimer's disease, *Bioorganic & Medicinal Chemistry*, vol. **26**, no. 13, pp. 3812–24, July 30, 2018. DOI: 10.1016/j.bmc.2018.06.018

9 Triazole Heterocycles as Multi-Target Ligands in Alzheimer's Disease

Hamid Aziz, Ismat Ullah Khan, Aamer Saeed, Akhtar Munir, and Saif Ur Rehman

9.1 INTRODUCTION

Azoles hold prevailing significance as propellants, pyrotechnics, pharmaco, and chemotherapeutics (Aziz et al., 2021). Triazoles termed as pyrrodiazoles are nitrogen-rich five-member aromatic heterocycles. Triazole ($C_2H_3N_3$), being the simplest member, is white to yellow crystalline solid, with boiling point of 260°C, and melting point of 120°C. Triazole is soluble in alcohol and possesses characteristic smell. Triazole display positional isomerism as 1,2,3-triazole, and 1,2,4-triazole, differ by nitrogen atoms within the ring. Each of the isomers has two tautomeric forms which differ by the relative position of hydrogen across nitrogen atoms (Figure 9.1) (Aziz et al., 2019, 2021).

9.1.1 Classification of Triazoles

Triazole heterocycles are classified based on the arrangements of nitrogen atoms within the ring.

9.1.1.1 1,2,3-Triazole Heterocycles

1,2,3-Triazoles contained adjacent nitrogen atoms and were recognized since the 19th century when Werner and Stiasny reported studies on 2-phenylbenzotriazole-1-oxide (Werner & Edm, 1899). 1,2,3-Triazoles are important in drug discovery, agrochemicals, dyes, and versatile molecular linkers. Their ability to form hydrogen bonds, combined with stability at physiological pH and resistance to oxidation, reduction, and enzymatic degradation, makes 1,2,3-triazoles valuable amide isosteres (Mishra, 2025). These heterocycles serve as drug carriers, agonists, anti-inflammatory, anticonvulsants, anticancer, antimicrobial, and β-lactamase inhibitors (da Silva et al., 2014). 1,2,3-Triazole-based materials show potential as super molecules, fluorescence probes, proteins, surface modification, and dendrimers (Huo et al., 2017).

9.1.1.2 1,2,4-Triazole Heterocycles

1,2,4-Triazole, first described by Bladin, features two adjacent nitrogen atoms with a third nitrogen atom masked by two carbon atoms. 1,2,4-Triazoles show promise across medicinal chemistry, agriculture, supramolecular chemistry, and materials science (Aggarwal & Sumran, 2020).

FIGURE 9.1 Tautomerism in different isomeric forms (1–4) of triazole.

DOI: 10.1201/9781003772019-9

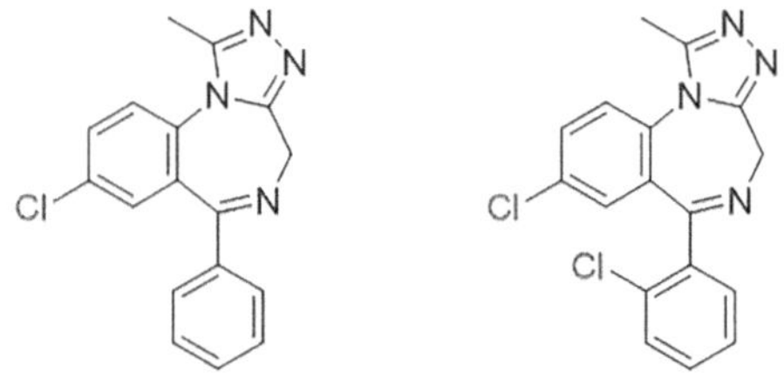

(5) (6)

FIGURE 9.2 Structures of Alprazolam (**5**) and Triazolam (**6**).

1,2,4-Triazoles are considered a wonder motif in heterocyclic chemistry owing to their ability to display a high degree of structural diversity around a rigid core, lesser toxicity, their ability to act as hydrogen bond donors or acceptors and as potent agonist or antagonist receptor ligands (Aggarwal & Sumran, 2020). The presence of three nitrogen atoms in the ring promotes effective intercalation and enhanced polarization to the chromophore moiety which assists in better DNA interaction (Matin et al., 2022). Therefore, 1,2,4-triazole moiety is incorporated in large number of therapeutically important heterocycles making them effective drug entrants. The 1,2,4-triazole-based antifungal drugs on the bases of sale values lie in the inventory of top 200 drugs (Aziz et al., 2021).

9.1.2 Characteristics of Triazole Heterocycles

Triazole heterocycles engage enzymes and receptors through covalent forces like hydrogen bonds, ion dipole interactions, van der Waals contacts, and hydrophobic effect, which turn triazole as broadly useful motifs. Triazole serves as effective isostere of thiazole, thiadiazole, pyrazole, imidazole, and oxazole. Thus, many triazoles are synthesized and screened as useful therapeutics. After clinical introduction of Alprazolam (**5**) and Triazolam (**6**) in 1980s, chemists have continued to design and optimize novel triazole scaffolds as potential anticonvulsant and CNS agents (Figure 9.2) (Song & Deng, 2018).

Triazole are pharmacologically relevant motifs as antiviral, antitumor, anti-oxidant, anti-ureases, histamine blockers, sedatives, and as cholinesterase inhibitors (Guo et al., 2021). The ambient nucleophilic center turns triazoles synthetic utility for bridged heterocycles. Triazoles transform therapeutically significant heterocycles into better drugs like Fluconazole (**7**), Voriconazole (**8**), and Posaconazole (**9**) as anxiolytic and CNS stimulants (Figure 9.3) (Aziz et al., 2019, 2021).

Aromatase enzyme holds significance in the biosynthesis of estrogens, responsible for breast cancer. Triazole-containing nonsteroidal aromatase drugs Vorozole (**10**), Letrozole (**11**) and Anastrozole (**12**) are used to prevent hormone-dependent breast cancer (Figure 9.4) (Janowska et al., 2024, Huda & Ihsan, 2022).

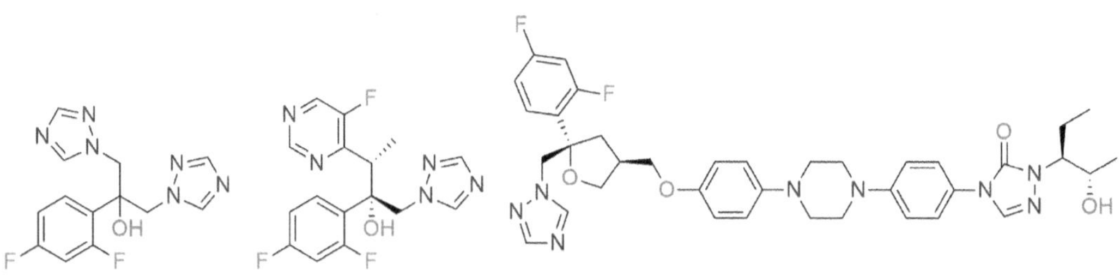

7 8 9

FIGURE 9.3 Structures of stimulants Fluconazole (**7**), Voriconazole (**8**), and Posaconazole (**9**).

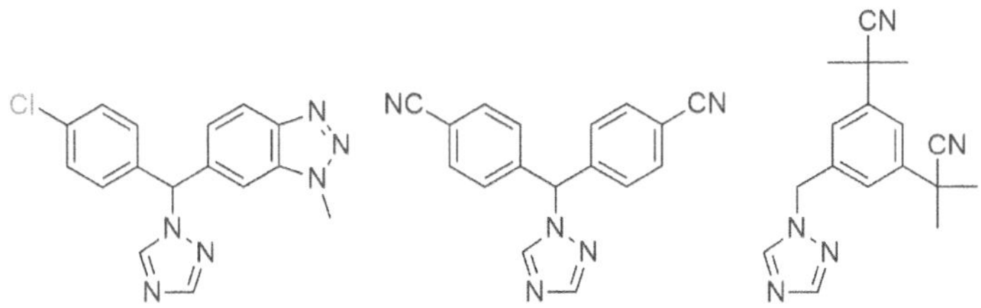

10 11 12

FIGURE 9.4 Structures of anticancer Vorozole (**10**), Letrozole (**11**), and Anastrozole (**12**).

9.2 ALZHEIMER'S DISEASE

Alzheimer's disease (AD), reported by Dr Alois Alzheimer in 1906, is a fatal neurodegenerative disorder characterized by progressive deterioration of memory and thinking (Dnyandev et al., 2024, Sang et al., 2022). AD is the most common cause of dementia (60–80%) and leads to gradual loss of independence from year to decade (Kamatham et al., 2024). Incidence and prevalence of AD rise sharply with age, especially after 60 years. Women are disproportionately affected because of longer average lifespan (Liu et al., 2021). Scientists consider AD as a complex and multifactorial disease caused by genetic, environmental, and endogenous factors. These factors encompass excessive protein misfolding and aggregation, loss of acetylcholine (ACh), biometals disorder, oxidative stress, impaired bioenergetics, mitochondrial abnormalities, and neuroinflammations (Ali et al., 2024). Therapies that directly target the molecular causes of AD are promising to prevent and slow its pathogenesis. However, the nebulous and complicated nature of the pathogenesis of AD makes it difficult to develop novel drugs. FDA-approved drugs focus on symptomatic treatment including selective acetylcholine esterase (AChE) inhibitors Tacrine (**13**), Galantamine (**14**), Donepezil (**15**), one hand AChE/BuChE inhibitors Rivastigmine (**16**), and one NMDA-receptor antagonist Memantine (**17**) (Figure 9.5). Choline esterase inhibitors treat mild-to-moderate AD, and NMDA-receptor antagonists treat moderate-to-severe AD. Notably, the inhibitors temporarily improve cognition but fail to prevent neuronal loss, worsening dementia, or modify AD progression, causing hepatotoxicity, CNS, gastrointestinal, and cardiac disorders (Sang et al., 2022).

The cholinergic hypothesis suggests neurotransmitter ACh effects nerve cells related to learning, memory, and attention. Cholinesterases (ChEs) are members of the serine hydrolase family and catalyze the hydrolysis of choline esters. Two distinct forms of ChEs are encoded by separate genes: acetylcholinesterase (AChE; EC 3.1.1.7) and butyrylcholinesterase (BChE; EC 3.1.1.8). AChE primarily hydrolyzes and inactivates acetylcholine (ACh), the principal neurotransmitter involved in CNS signaling. In contrast, BChE preferentially hydrolyzes carboxylic acid esters with bulky acyl groups, such as butyrylcholine and benzoylcholine, and plays a complementary role in cholinergic

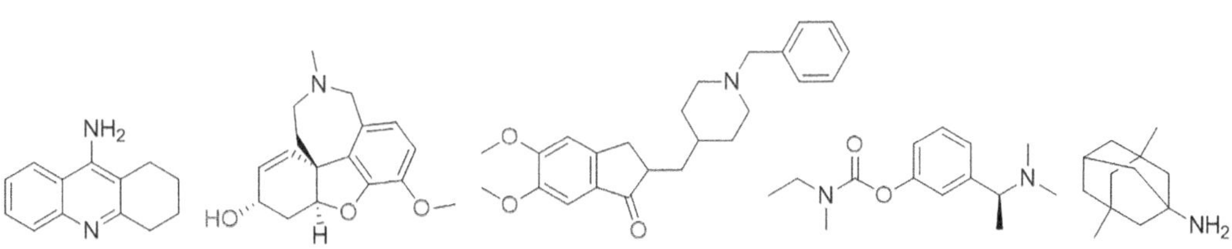

13 14 15 16 17

FIGURE 9.5 AChE inhibitors Tacrine (**13**), Galantamine (**14**), Donepezil (**15**), Rivastigmine (**16**), and Memantine (**17**).

regulation (Mishra et al., 2019). Deficiency of ACh declines overall cognitive function. Increased AChE accelerates hydrolysis of ACh to Choline and Acetate, contributing to memory deficits. Therefore, inhibition of AChE is an established therapeutic strategy for managing early-stages AD, by increasing ACh levels at synaptic junctions, thereby enhancing cholinergic neurotransmission and signal potentiation. Tacrine, Galantamine, Donepezil, and Rivastigmine are FDA-approved AChE inhibitors used in the symptomatic treatment of mild-to-moderate AD, aimed at enhancing cholinergic neurotransmission by increasing synaptic ACh levels. Despite their widespread clinical use, the inhibitors exhibit limited long-term efficacy, as they do not halt, prevent, or reverse the progression of AD, and are often associated with adverse effects such as nausea, vomiting, and diarrhea (Sang et al., 2022). Consequently, ongoing research has focused on developing more potent and safer AChE inhibitors against early-stage AD. BuChE acts as a co-regulator of cholinergic neurotransmission by hydrolyzing ACh. In moderate-to-severe stages of AD, BuChE becomes increasingly active and assumes a more prominent role in ACh hydrolysis, making it a promising therapeutic target for late-stage intervention. Moreover, selective inhibition of BuChE may help circumvent classical cholinergic toxicity, an adverse effect associated with AChE inhibition (Sang et al., 2022; Sobha et al., 2024). Triazoles and their hybrids are considered potential motifs owing to their facile synthesis and relevant profile. Triazole-based heterocycles represent promising motifs for AD drug discovery: serve as versatile linkers or pharmacophores in the synthesis of hybrid molecules, often improve metabolic stability and binding potency, and can be employed to overcome certain forms of drug resistance (Sobha et al., 2024).

9.3 TRIAZOLE HYBRIDS AGAINST AD

9.3.1 Coumarin–Triazole Hybrids

N.A. Dar et al. screened 16 coumarin–triazole hybrids against AChE **(18–33)** using Eserine as drug (IC_{50} = 8.05, 7.54 μM). Against AChE (IC_{50} = 2.18–67.89 μM), **26, 22,** and **20** emerged as most potent (IC_{50} = 2.18, 2.40, 3.27 μM). Inhibition decreased as: **26>22>20>33>18>S>19>32>25>31>30>29>23>27>21>24>28. Inhibitors 26, 22, 20, 33** and **18** have better AChE inhibition than drug. EWGs (Br, F) possess stronger interaction with receptor than EDGs (Figure 9.6) [Dar et al., 2025].

MD analysis of **26** and Eserine within the catalytic gorge of AChE revealed four binding poses for **26** (3.152–7.644 kcal/mol) and single pose for Eserine (7.949 kcal/mol). Interactions of **26** with key AChE residues, include hydrogen bonding with Tyr121 and His444, and π–π stacking with Tyr334 and Trp283. Binding free energy confirms favorable thermodynamic interactions between **26** and AChE (Dar et al., 2025).

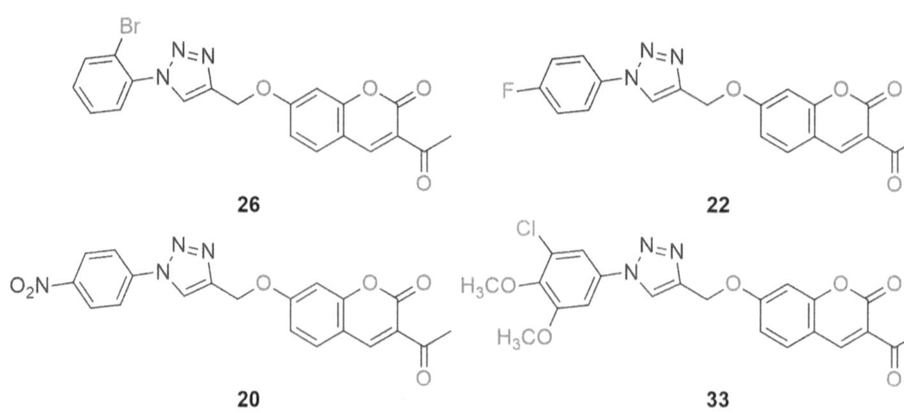

FIGURE 9.6 Structures of coumarin–triazole hybrids as potent AChE inhibitors.

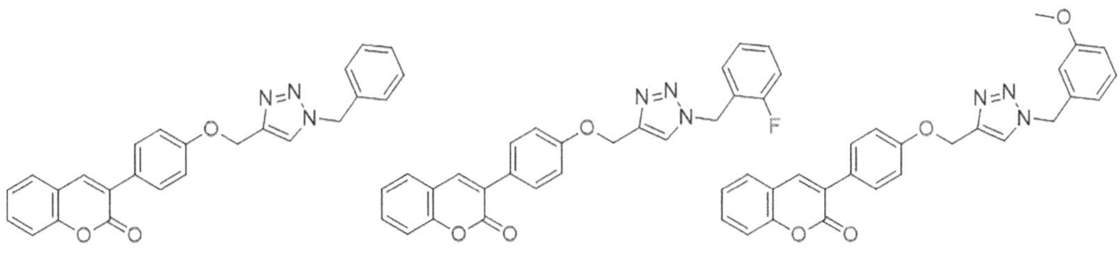

34a 34b 34l

FIGURE 9.7 Structures of coumarin–triazole hybrids as mild BuChE inhibitors.

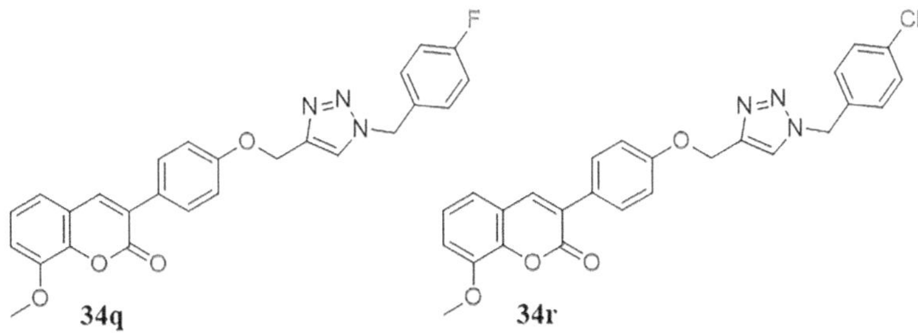

34q 34r

FIGURE 9.8 Structures of most potent coumarin–triazole hybrids against BuChE.

Pourabdi et al. screened 23 coumarin–triazole hybrids as AChE and BuChE inhibitors using Tacrine as drug. Inhibitors showed very weak potential against AChE and were found to possess moderate potential against BuChE. Inhibitors **34(a–b)** and **34l** showed mild BuChE inhibition (IC_{50} = 27.6, 36.5,19.5 μM) (Figure 9.7) (Pourabdi et al., 2022).

Among inhibitors with 8-methoxy group **34(n–s)** the highest inhibitory potential was exhibited by **34(q–r)** (IC_{50} = 18.9, 6.7 μM) (Figure 9.8).

Inhibitors **34(t–u)** with 6-bromocoumarin were inactive against BuChE. Inhibitor **34v** served as the best BuChE inhibitor (IC_{50} = 6.3 μM). Among unsubstituted aryl ring (**34a, 34n, 34v**), inhibition decreased as: **34v>34a>34n** against BuChE. To conclude, potent inhibition was observed against BuChE than AChE (Figure 9.9) (Pourabdi et al., 2022).

9.3.2 Thiazole–Triazole Hybrids

P. Rajesab et al. screened 12 thiazole–triazole hybrids **35(a–l)** against AChE and BuChE. Moderate inhibition was observed against AChE ((IC_{50} = 06.52 ± 0.20–10.52 ± 0.32 μM) compared to Donepezil (IC_{50} = 07.86 ± 0.18 μM). Inhibitors **35a, 35d, 35g,** and **35l** served as excellent inhibitors against AChE (IC_{50} = 07.25 ± 0.45, 07.45 ± 0.03, 05.23 ± 0.08, 06.52 ± 0.20 μM). Remaining inhibitors showed good to comparable inhibition compared to Donepezil (Rajesab et al., 2025) (Figure 9.10).

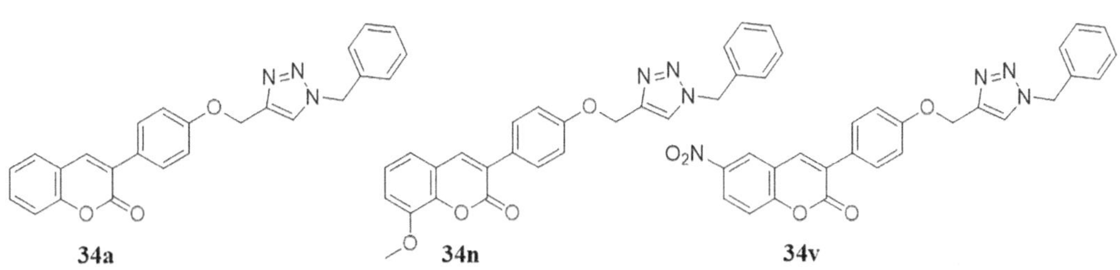

34a 34n 34v

FIGURE 9.9 Structures of the potent coumarin–triazoles against BuChE.

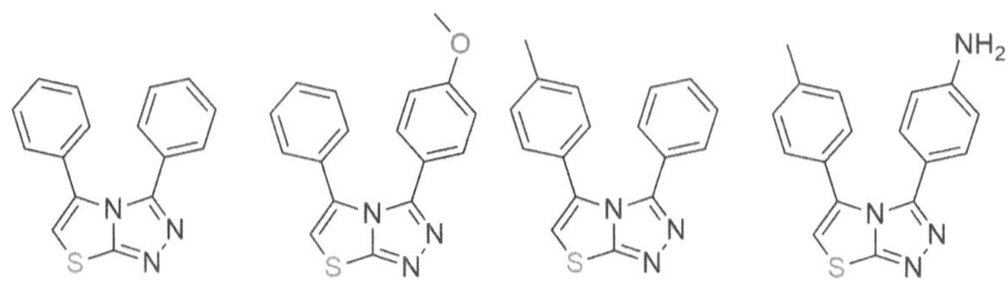

35a 35d 35g 35l

FIGURE 9.10 Structures of potent thiazole–triazole hybrids AChE inhibitors.

Against BChE, promising inhibitions were observed (IC_{50} = 10.24 ± 0.08–14.23 ± 0.21 μM) compared to Donepezil (IC_{50} = 12.52 ± 0.23 μM). Inhibitors **35g** and **35l** showed excellent inhibition (IC_{50} = 10.24 ± 0.08–11.02 ± 0.31 μM). Good to moderate inhibitions were observed for the remaining hybrids (Rajesab et al., 2025).

S. Khan et al. screened 19 thiazole-fused triazole/thiadiazoles **36(a–s)** for AChE and BuChE inhibition using Donepezil (IC_{50} = 4.60 ± 0.10 μM (AChE), 5.20 ± 0.71 μM (BuChE)) as drug. The inhibitors induced potent inhibitions (IC_{50} = 2.10 ± 0.20–14.10 ± 0.10 μM (AChE), 3.10 ± 0.50–15.30 ± 0.10 μM (BuChE)). Thus, **36o** (IC_{50} = 2.10 ± 0.20 μM (AChE), 3.10 ± 0.50 μM (BuChE)) showed highest inhibitions. Inhibitors **36b** (IC_{50} = 3.30 ± 0.20 μM (AChE), 3.70 ± 0.40 μM (BuChE)), **36g** (IC_{50} = 2.70 ± 0.10 μM (AChE), 3.24 ± 0.10 μM (BuChE)), **36l** (IC_{50} = 4.80 ± 0.30 μM (AChE), 5.40 ± 0.20 μM (BuChE)), and **36s** (IC_{50} = 4.40 ± 0.20 μM (AChE), 5.10 ± 0.10 μM (BuChE)) displayed better inhibitions than drug. SARs suggest **36o** as most potent inhibitor due to highly electronegative 2,5-difluoro substitutions, followed by **36g** (IC_{50} = 2.70 ± 0.10 μM (AChE), 3.240 ± 0.10 μM (BuChE)) (Figure 9.11) (Khan et al., 2024).

Among **36c, 36f, 36k, 36l, 36s,** inhibitor **36s** was the most potent (IC_{50} = 4.40 ± 0.20 μM (AChE), 5.10 ± 0.10 μM (BuChE)) followed by **36l** (IC_{50} = 4.80 ± 0.30 μM (AChE), 5.40 ± 0.20 μM (BuChE)). Inhibitor **36f** (IC_{50} = 6.30 ± 0.20 μM (AChE), 6.40 ± 0.30 μM (BuChE)) showed comparable inhibition. Inhibitor **36k** (IC_{50} = 7.10 ± 0.30 μM (AChE) and 7.70 ± 0.50 μM (BuChE)) possesses comparable potential to Donepezil. Mono-substituted **36c** (IC_{50} = 7.10 ± 0.40 μM (AChE), 8.80 ± 0.10 μM for (BuChE)) is less potent (Figure 9.12) (Khan et al., 2024).

Inhibitors **36b** (IC_{50} = 3.30 ± 0.20, 3.70 ± 0.40 μM) and **36i** (IC_{50} = 3.30 ± 0.30, 4.60 ± 0.20 μM) showed comparable AChE and BuChE inhibitions. Among methyl derivatives, **36q** (IC_{50} = 7.80 ± 0.10 μM (AChE), 8.40 ± 0.20 μM (BuChE)) exhibited higher potential than **36r** (IC_{50} = 8.50 ± 0.20 μM (AChE), 8.90 ± 0.40 μM (BuChE)). Inhibitors **36d** (IC_{50} = 9.20 ± 0.10 μM (AChE), 10.10 ± 0.20 μM for (BuChE)) and **36j** (IC_{50} = 9.20 ± 0.50 μM (AChE) and 10.60 ± 0.30 μM (BuChE)) showed comparable inhibition (Figure 9.13) (Khan et al., 2024).

Among **36a, 36e,** and **36p,** inhibitor **36a** possesses (IC_{50} = 5.70 ± 0.50 μM (AChE), 6.10 ± 0.50 μM (BuChE)) comparable potential to Donepezil. Inhibitor **36p** (IC_{50} = 11.20 ± 0.20 μM (AChE),

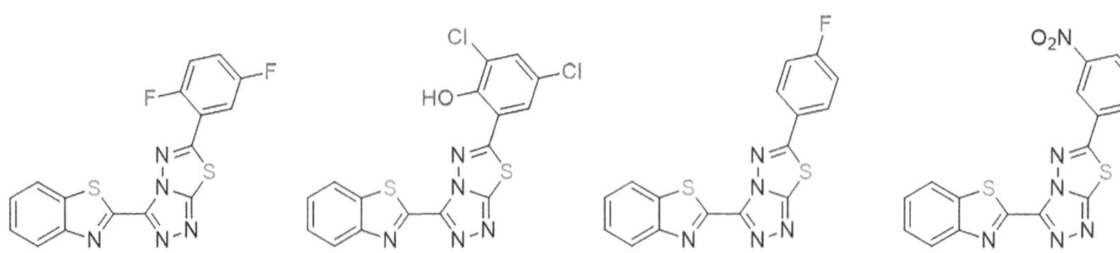

36o 36b 36g 36l

FIGURE 9.11 Structures of potent triazole–thiadiazoles hybrids than drug.

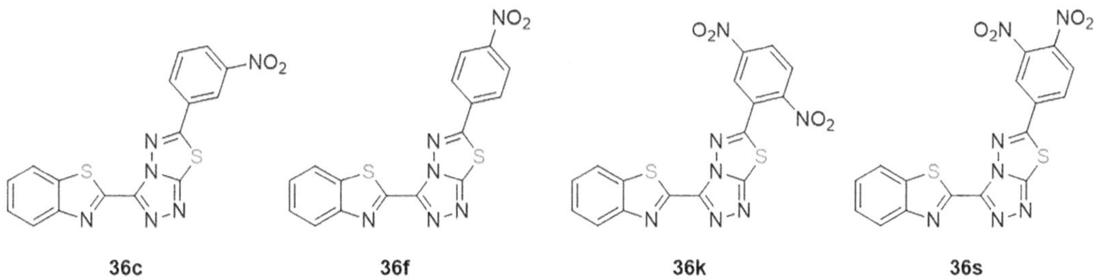

FIGURE 9.12 Structures of nitro-substituted potent triazole–thiadiazoles hybrids.

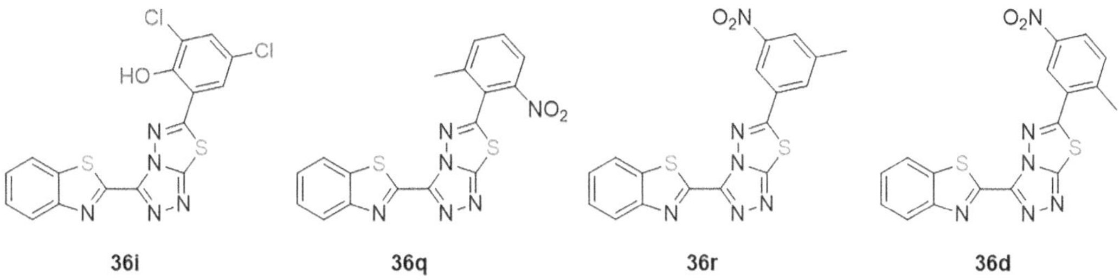

FIGURE 9.13 Structures of polysubstituted average triazole–thiadiazoles hybrids.

12.30 ± 0.10 μM (BuChE)) and **36e** (IC_{50} = 14.10 ± 0.10 μM (AChE), 15.30 ± 0.10 μM (BuChE)) showed lower inhibition than **36a** (Figure 9.14) (Khan et al., 2024).

9.3.3 Imidazole–Triazole Hybrids

Khan et al. screened 15 imidazole–triazole hybrids **37(a–o)** against AChE and BuChE. Thus, varied inhibitions were obtained (IC_{50} = 0.40 ± 0.05-19.60 ± 0.40 μM (AChE), IC_{50} = 1.50 ± 0.10-23.30 ± 0.50 μM (BuChE)) compared to Donepezil (IC_{50} = 0.016 ± 0.12 μM (AChE), IC_{50} = 4.5 ± 0.11 μM (BuChE)). SAR suggests inhibitory potential is highly dependent upon aryl ring substitution. EWGs increased AChE inhibition, strongly influenced by position. **Inhibitor 37g** (IC_{50} = 0.40 ± 0.05 μM) emerged as the most potent inhibitor followed by **37b** (IC_{50} = 0.60 ± 0.05 μM). Inhibitors **37g** inhibited AChE more than **37f** (IC_{50} = 0.80 ± 0.05 μM) and **37i** (IC_{50} = 1.10 ± 0.10 μM) (Figure 9.15) (Khan et al., 2022).

Similarly, among **37c, 37k,** and **37l** (IC_{50} = 1.30 ± 0.05, 2.40 ± 0.10, 19.60 ± 0.40 μM), **37c** showed higher potential than **37k** due to varied orientation of substituents (Figure 9.16) (Khan et al., 2022).

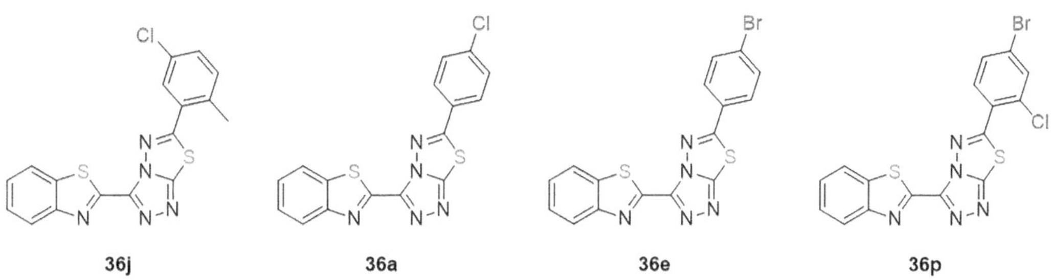

FIGURE 9.14 Structures of potent triazole–thiadiazoles to Donepezil.

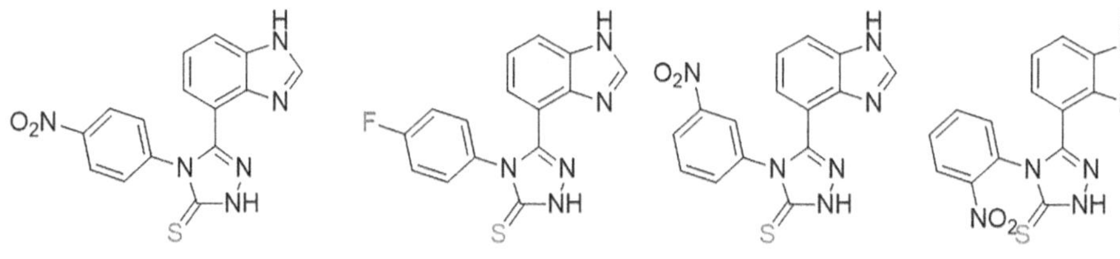

37g 37b 37f 37i

FIGURE 9.15 Structures of potent imidazole–triazole hybrids.

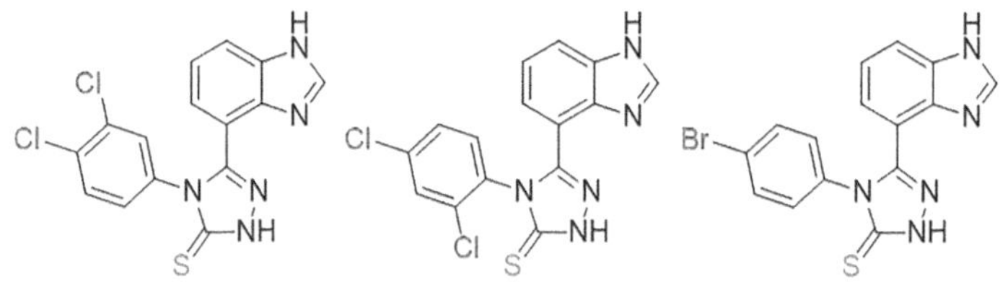

37c 37k 37l

FIGURE 9.16 Structures of halo-substituted imidazole–triazole hybrids.

Inhibitors **37n**, **37d**, **37e,** and **37a** (10.90 ± 0.20, 12.90 ± 0.20, 9.60 ± 0.10, 14.30 ± 0.30 μM) showed low AChE inhibition. The methyl group **37(d–e)** declined AChE inhibition, and hydrogen further decreased inhibition **37a** (Figure 9.17) (Khan et al., 2022).

In the case of BuChE, nitro, fluoro, and di-chloro groups at variable position of imidazole emerged as potent inhibitors. Inhibitor **37g** (IC_{50} = 1.50 ± 0.10 μM) showed superior BuChE inhibition, and decreased by changing the nitro group *meta/ortho*-positions in **37f** and **37i** (IC_{50} = 2.60 ± 0.10, 3.10 ± 0.10 μM). Inhibitor **37c** (IC_{50} = 3.20 ± 0.10 μM) displayed better BuChE inhibition, and declined by shifting chlorine to *ortho*-position **37k** (IC_{50} = 4.30 ± 0.10 μM). Inhibitor **37b** (IC_{50} = 2.10 ± 0.10 μM) showed superior BuChE potency than **37c** and **37k** (IC_{50} = 3.20 ± 0.10, 4.30 ± 0.10 μM). Changing the position of substituent induced a significant impact on inhibition. Inhibitor **37e** (IC_{50} = 17.30 ± 0.30 μM) showed higher inhibition than **37d**(IC_{50} = 22.20 ± 0.30 μM), which vanished in **37a** (IC_{50} = 27.60 ± 0.30 μM). Inhibitor **37n** (IC_{50} = 18.20 ± 0.30 μM) showed higher BuChE inhibition than **37e,** which greatly improved in **37b.** However, inhibition decreased highly in **37h** (Khan et al., 2022).

MD of **37b** showed different interactions for AChE such as Ser8, Ser200, His440, Asn85, Glu199 (hydrogen bond), Trp84 (π–π), Tyr121(π–π), and Phe331(π–S). Against BuChE, interactive residues were Leu286, Gly117, Ser198, Gly116, Phe329, Glu197 (hydrogen bond), Trp231(π–π), Trp82(π–S), Ala328 (π–R), His438(π–cation). Inhibitor **37g** being the most potent, showed significant interactions

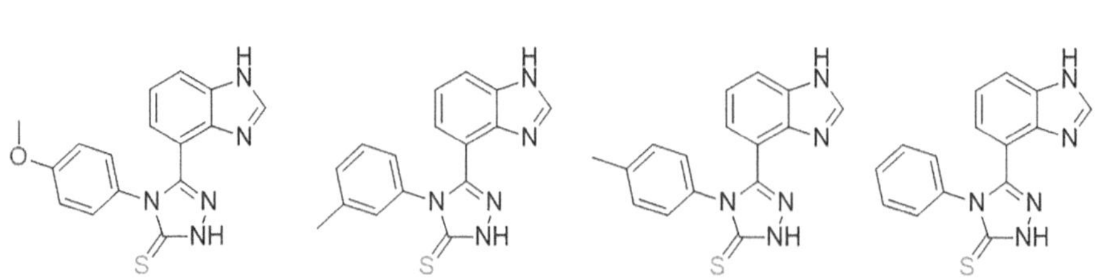

37n 37d 37e 37a

FIGURE 9.17 Structures of EWGs bearing imidazole–triazole hybrids.

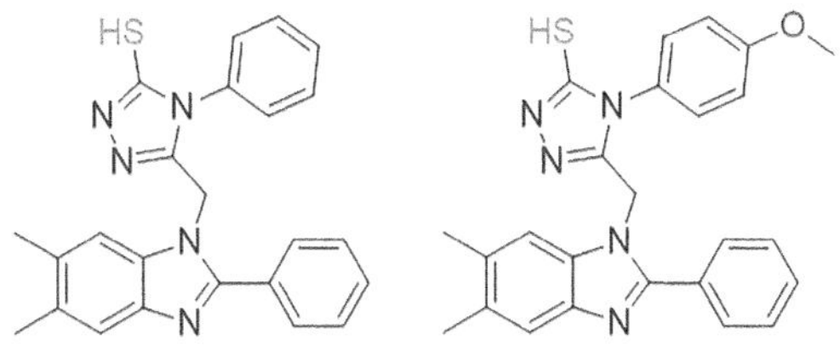

38a 38c

FIGURE 9.18 Structures of triazole-linked benzimidazole hybrids.

with AChE such as Tyr12, Trp84, Ser200, His440 (hydrogen bond), Trp334 (π–alkyl), Phe330 (π–alkyl), and Phe331 (π–R). Against BuChE, Ser198, Tyr332 (hydrogen bond), Phe329(π–alkyl) and His438 (π–cation). In terms of binding energies, inhibitors received higher energies than Donepezil. To conclude, **37b** and **37g** with EWGs showed better results against AChE and BuChE (Khan et al., 2022).

N.Ü. Karaali et al. screened three triazole-linked benzimidazoles **38(a–c)** as AChE inhibitors. Triazole hybrid **38a** served as the most potent inhibitor (IC_{50} = 8.0 ± 0.25 μg/mL) followed by **38c** (IC_{50} = 8.10±0.30 μg/mL) compared to Galantamine (IC_{50} = 20.75 ± 0.25 μg/mL) (Figure 9.18) (Karaali et al., 2025).

9.3.4 Schiff Base–Triazole Hybrids

A. Iraji et al. screened 16 quinazoline Schiff base–triazole hybrids **39(a–p)** against AChE and BuChE using Donepezil as drug. Thus, **39(a–q)** showed mild AChE inhibition, with **39(a–c)** as significant (IC_{50} = 30.85 ± 8.91, 11.81 ± 1.55, 18.42 ± 0.89 μM) and **39g** as moderate (IC_{50} = 47.04 ± 0.38 μM) inhibitor (Figure 9.19) (Iraji et al., 2025).

Against BChE, no significant inhibition was induced by **39(a–b)** (IC_{50}>100 μM). However, **39(c–d)** (IC_{50} = 21.25 ± 0.59, 28.03 ± 0.84 μM) served as good to moderate inhibitors. Among **39(e–h), 39(e–f)** displayed moderate inhibition (IC_{50} = 40.47, 32.95 μM), **39g** exhibited reduced inhibition than **39c** (IC_{50} = 5.28 ± 0.65 μM) and **39h** demonstrated moderate inhibition

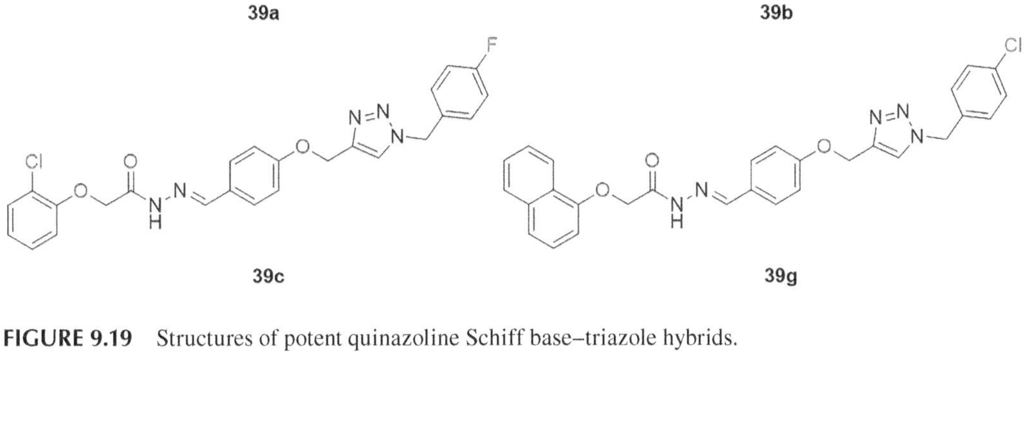

39a 39b 39c 39g

FIGURE 9.19 Structures of potent quinazoline Schiff base–triazole hybrids.

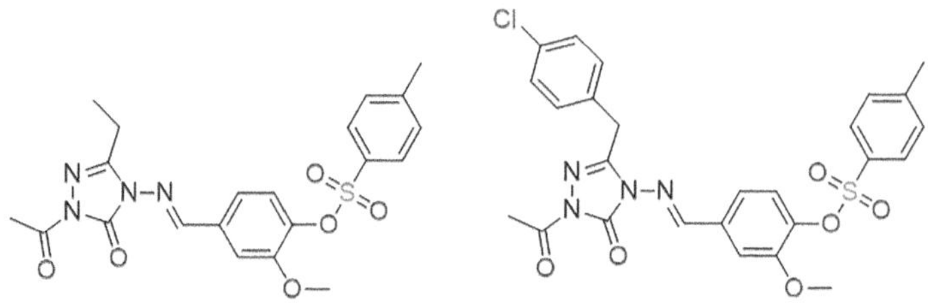

40b **40e**

FIGURE 9.20 Structures of Schiff bases bearing triazole and sulfonate ester.

(IC_{50} = 33.87 ± 0.76 μM). Among **39(i–l), 39i, 39j,** and **39l** were inactive (IC_{50}>100 μM), **39k** showed good inhibition (IC_{50} = 16.17 ± 0.69 μM). Among **39(m–p),** no inhibitions were observed (IC_{50}>100 μM) (Iraji et al., 2025).

H. Medetalibeyoğlu et al. screened six Schiff bases **40(a–f)** bearing triazole and sulfonate ester as AChE and BChE inhibitors. Tacrine, Donepezil, Rivastigmine, and Galantamine served as drugs. AChE inhibition (K_i = 0.67 ± 0.002–3.0 ± 0.008 μM) decreased as: **40b>40e>40c>40f>40d>40a.** Inhibitors **40b** and **40e** emerged as better inhibitors against AChE. Inhibitor **40b** showed higher inhibition (4.48-times) than **40a**. Inhibitors **40b** (IC_{50} = 3.05 μM, K_i = 0.67 μM) and **40e** (IC_{50} = 2.06 μM, K_i = 1.33 μM) were more potent than Tacrine (IC_{50} = 1.78 μM, K_i = 0.589 μM). Against BChE, inhibition (K_i = 2.14 ± 0.007–7.71 ± 0.14 μM) decreased as: **40b>40e>40d>40f>40c>40a.** Thus, **40b** showed 3.60 times (IC_{50} = 4.74 μM, K_i = 2.14 μM) and **40e** 3.01-fold (IC_{50} = 2.67 μM, K_i = 2.56 μM) higher inhibition than **40a** (IC_{50} = 9.12 μM, K_i = 7.71 μM). SAR suggests ethyl **40b** and chloro **40e** groups enhance inhibition against AChE and BChE. Tacrine, being more effective against BChE, **40b** and **40e** are promising inhibitors and need further optimization. MD suggests, **40b** exhibits maximum π–π interactions, and **40e** hydrophobic interactions. Inhibitor **40b** formed π–π interactions (Tyr86, Tyr337) and conventional hydrogen bonding (Phe295). Inhibitor **40e** developed hydrogen bond (Phe295), π–π stacking (Tyr124, Tyr341) to stabilize **40e** and hydrophobic interactions with nonpolar and aromatic residues (Tyr72, Trp286, Val294, Tyr337, Phe338, Trp439, and Tyr449). MD showed higher affinities for inhibitors **40(a–f)** within the active site of **BChE.** In **40b,** Trp82 and Phe329 forms π–π interaction. In **40e,** Trp231 and Phe329 show π–π stacking (Figure 9.20) (Medetalibeyoğlu, Tüzün et al., 2025).

I. Ozcan et al. screened seven Schiff base–triazole hybrids as AChE and BuChE inhibitors **41(a–h).** Thus, inhibitors were potent AChE inhibitors (K_i = 6.58 ± 0.98-20.33 ± 2.57 nM) compared to drug Tacrine (K_i = 26.38 ± 3.64 nM). Inhibitor **41c** being the most potent inhibitor (K_i = 6.58 ± 0.98 nM) can serve as effective medications against AD. BChE inhibition (54.34 ± 8.17–89.76 ± 8.20 nM) suggest **41c** (K_i = 54.34 ± 8.17 nM) as highly potent than Tacrine (83.07 ± 9.40). Inhibitor **41c** was highly selective for AChE over BChE (0.121). SAR suggest inhibitors bearing triazole group, azomethine moiety, and fluorine atoms at 4- or 2,4-positions showed varied inhibitions. Inhibitors **41g** and **41c** significantly inhibited AChE (IC_{50} = 8.32, 8.23 nM), while

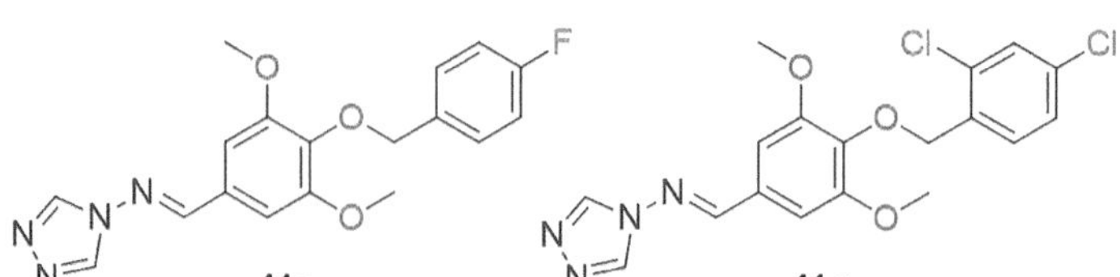

41c **41g**

FIGURE 9.21 Structures of Schiff base–triazole hybrids as AChE and BuChE inhibitors.

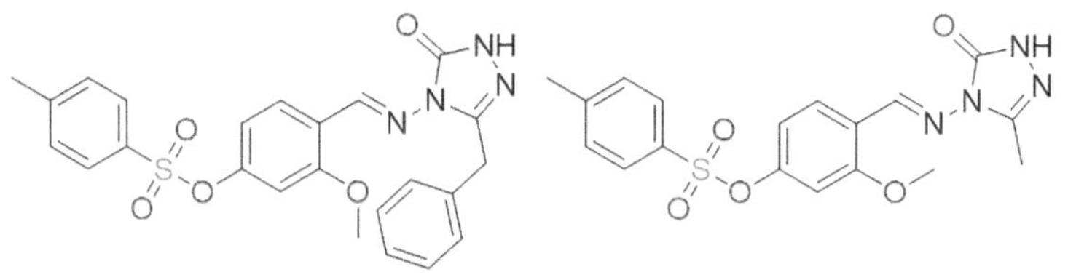

42d **42a**

FIGURE 9.22 Structures of the most potent Schiff base–triazole hybrids.

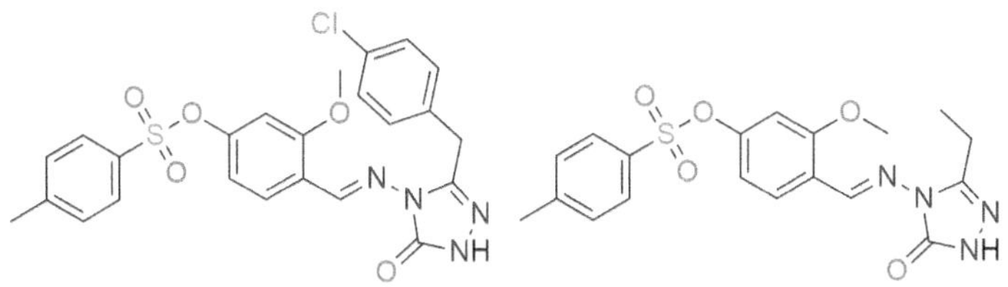

42g **42b**

FIGURE 9.23 Structures of the most potent Schiff base–triazole hybrids.

41b and **41d** were moderately active (IC_{50} = 12.84, 14.64 nM). Based on their lowest K_i values, **41c** and **41g** served as most potent AChE and BChE inhibitors (Figure 9.21) (Ozcan et al., 2025).

H. Medetalibeyoglu et al. screened eight Schiff base–triazole hybrids **42(a–h)** against AChE and BChE. Donepezil, Rivastigmine, and Galantamine served as drugs. Thus, significant inhibition was observed against AChE (K_i = 1.04 ± 0.003-4.03 ± 0.019 μM): **42d>42a>42f>42e–42g>42b>42h>42c.** Inhibitor **42d** showed 3.875 times greater inhibition than **42c** (Figure 9.22) (Medetalibeyoglu, Atalay et al., 2025).

Against BChE, inhibition (K_i = 0.607 ± 0.001–7.90 ± 0.25 μM) decreased as: **42g>42b>42e>42h>42d–42c>42a.** Inhibitors **42b** and **42g** emerged as better BChE inhibitors. Inhibitor **42g** was a 13-fold higher BChE inhibitor than **42a** (Figure 9.23) (Medetalibeyoglu, Atalay et al., 2025).

S. Khan et al. screened 16 triazole-derived Schiff bases bearing sulfonamides **43(a–p)** against AChE and BuChE. Inhibitors showed excellent to moderate AChE/BuChE potential (1.90 ± 0.40–14.50 ± 0.20 μM, 2.10 ± 0.10–14.80 ± 0.40 μM) compared to Donepezil (5.10 ± 0.20 (AChE), 5.70 ± 0.30 μM (BuChE)). Inhibitor **43d** served as most potent inhibitor (1.90 ± 0.40 (AChE), 2.10 ± 0.10 μM (BuChE)) followed by **43g** (IC_{50} = 3.10 ± 0.20, 3.30 ± 0.20 μM) (Figure 9.24) (Khan et al., 2024).

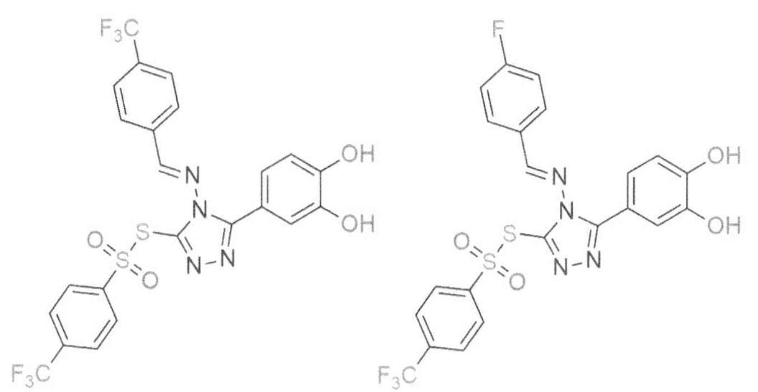

43d **43g**

FIGURE 9.24 Structures of the most potent Schiff base–triazole hybrids.

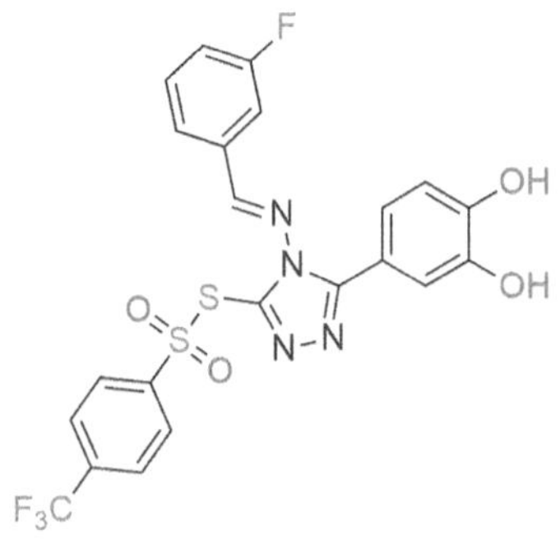

43m

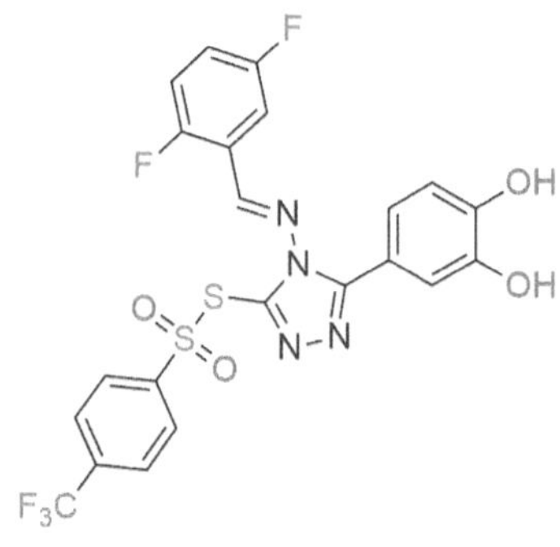

43o

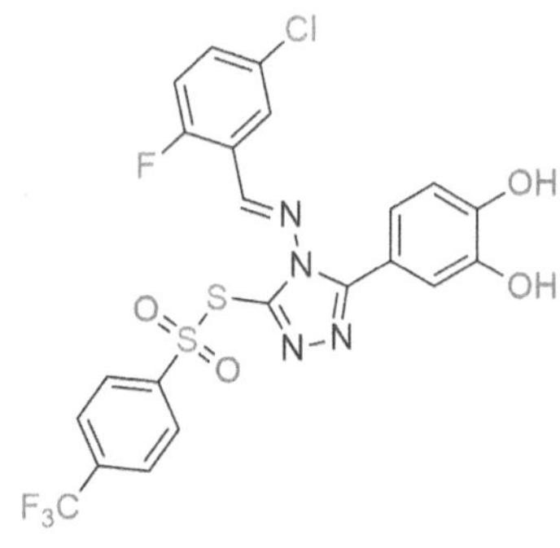

43j

FIGURE 9.25 Structures of the halo-substituted Schiff base–triazole hybrids.

Comparison of **43j** (IC_{50} = 6.20 ± 0.20, 7.20 ± 0.40 μM), **43m** (IC_{50} = 4.40 ± 0.10, 4.90 ± 0.60 μM), and **43o** (IC_{50} = 3.20 ± 0.80, 3.30 ± 0.90 μM) suggests **43o** showed highest inhibition. Inhibitor **43m** with 3-flouro moiety exhibits moderate, and **43j** bearing 3-chloro moiety decreased inhibition (Figure 9.25) (Khan et al., 2024).

Inhibitor **43i** bearing 3-chloro and 2,5-dihydroxyl moieties possess comparable inhibition (IC50=5.10±0.20, 5.70±0.10 μM) to **43n** (IC50=5.20±0.20, 5.50±0.30 μM) bearing 4-hydroxyl moiety (Figure 9.26) (Khan et al., 2024).

Inhibitors **43(a–b)** and **43k** were comparable inhibitors to Donepezil. Inhibitor **43a** (IC_{50} = 5.50 ± 0.80 (AChE), 6.80 ± 0.30 μM (BuChE)) bearing 3,5di-chloro group has greater inhibition than **43b** (IC50=6.30±0.10, 7.10±0.10 μM) and **43k** (IC50=7.70±0.10, 8.30±0.60 μM) (Figure 9.27) (Khan et al., 2024).

Nitro analogues **43c** (10.20 ± 0.20, 11.20 ± 0.20 μM), **43f** (7.60 ± 0.70, 8.90 ± 0.20 μM), and **43l** (6.90 ± 0.70, 7.10 ± 0.10 μM) were moderate AChE and BuChE inhibitors (Figure 9.28) (Khan et al., 2024).

Analogues **43e, 43h,** and **43p** were less potent due to the attached heavy groups (Figure 9.29) (Khan et al., 2024).

MD analysis of both enzymes was performed against respective codes of **1ACl** for AChE and **1P0P** for BuChE from **PDB (RSCB).** Effective protein ligand interaction of potent inhibitors **43d, 43g,** and **43o** was obtained. Docking of **43d** against AChE and BuChE suggest CF_3 moiety interact effectively via hydrogen bonding with both enzymes. Both hydroxyl groups interact through hydrogen bonding. MD investigation of **43g** against AChE and BuChE showed different interactions i.e. hydrogen bonding, π–lone pair, π–π T-shaped, π–π stacked, π–alkyl (Khan et al., 2024).

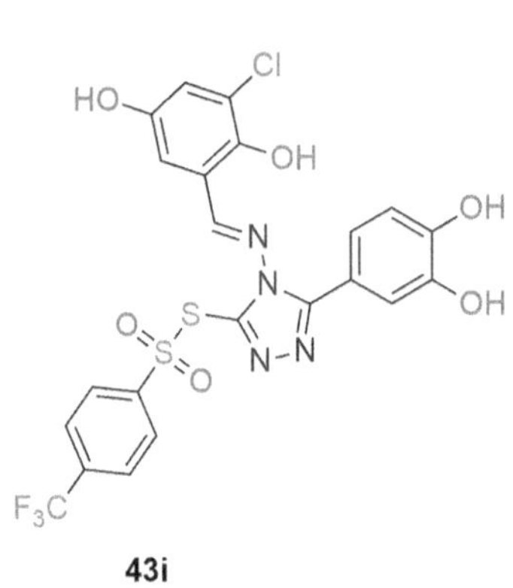

43i

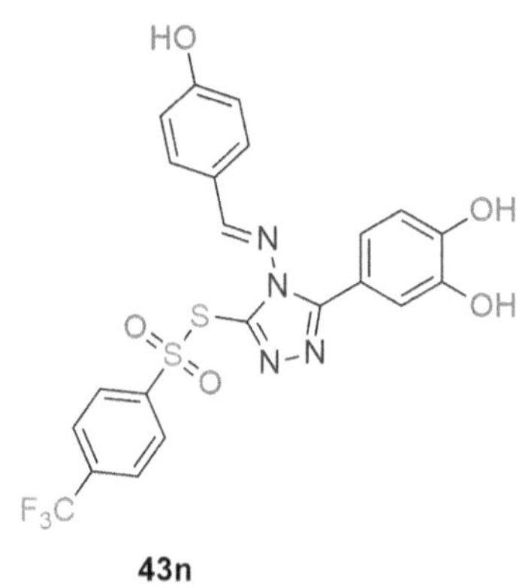

43n

FIGURE 9.26 Structures of the halo-substituted Schiff base–triazole hybrids.

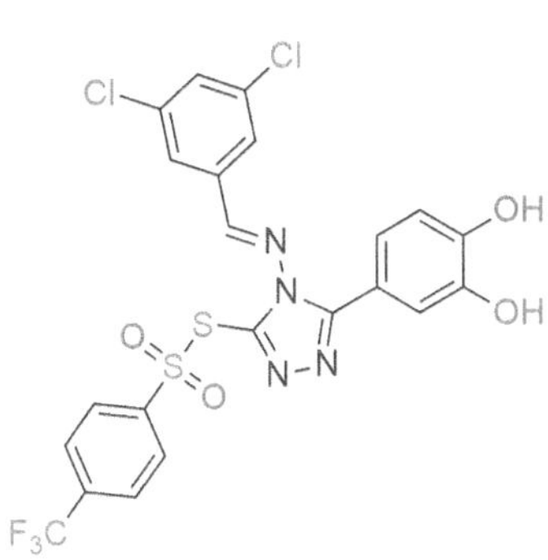
43a

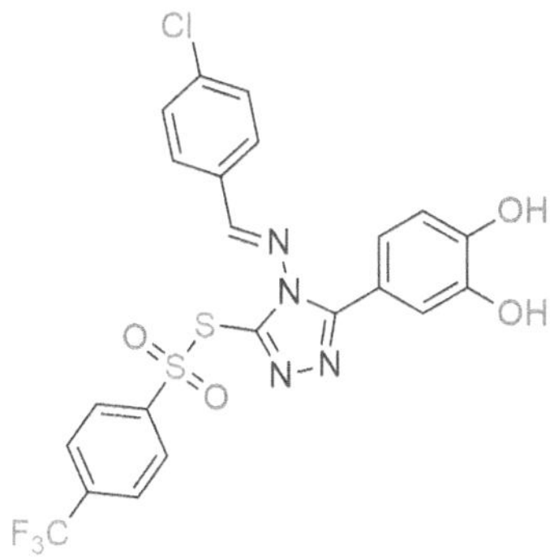
43b

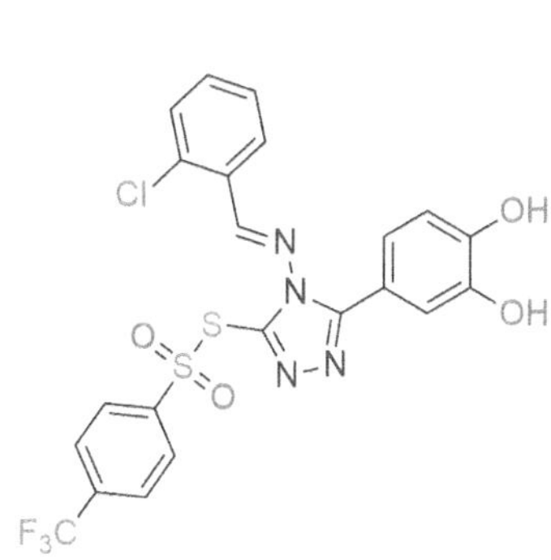
43k

FIGURE 9.27 Structures of the chloro-substituted Schiff base–triazole hybrids.

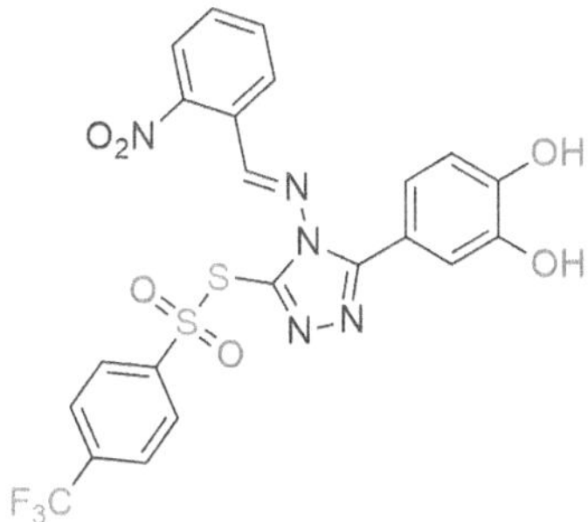
43c

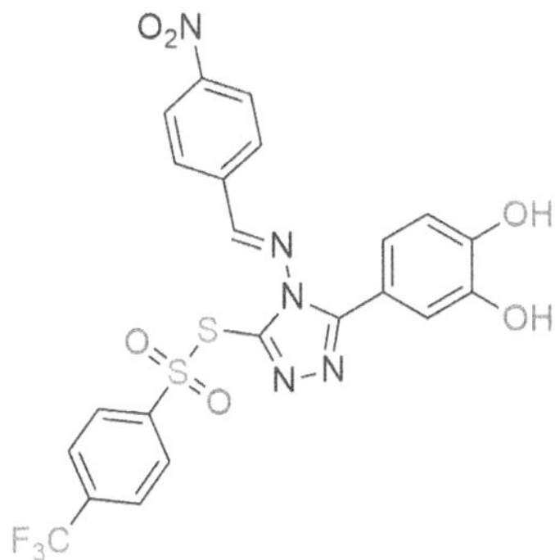
43f

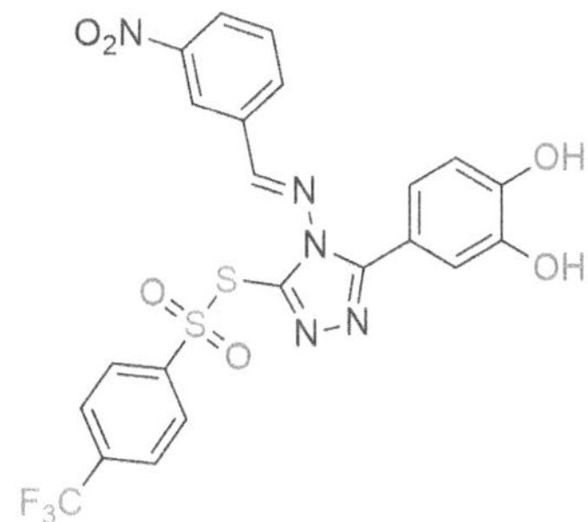
43l

FIGURE 9.28 Structures of the nitro-substituted Schiff base–triazole hybrids.

9.3.5 Amide–Triazole Hybrids

C. Efeoglu et al. screened eight novel bis-triazoles **44(a–h)** against AChE. Inhibitor **44a** served as moderate inhibitor (K_I = 405.10 ± 43.39 nM) than Tacrine (K_I = 255.44 ± 59.13 nM). Inhibitor **44b** (K_I = 509.40 ± 64.05 nM) turns two-fold less potent than drug. Inhibitor **44c** (K_I = 375.50 ± 41.96 nM) is slightly better inhibitor though less effective than drug. Inhibitor **44d** increases

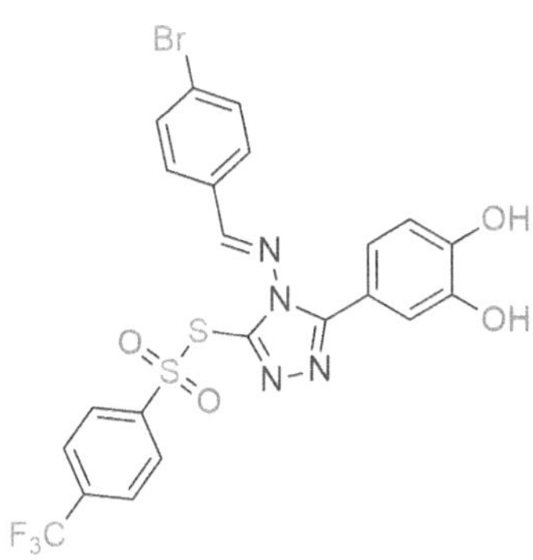
43e

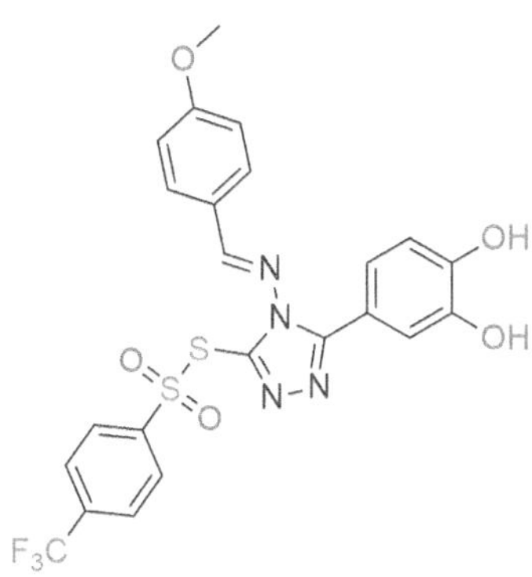
43h

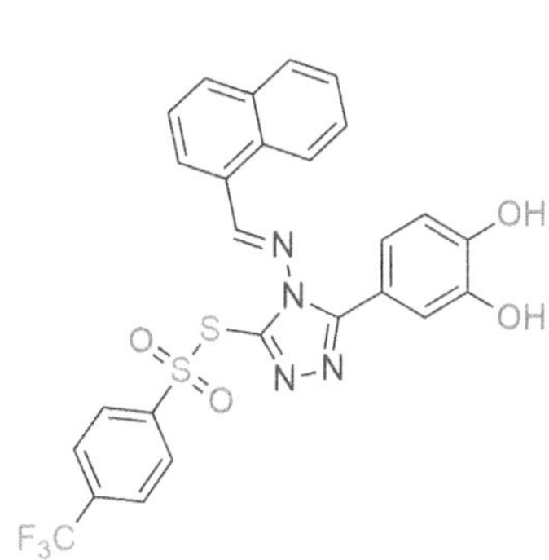
43p

FIGURE 9.29 Structures of the least potent Schiff base–triazole hybrids.

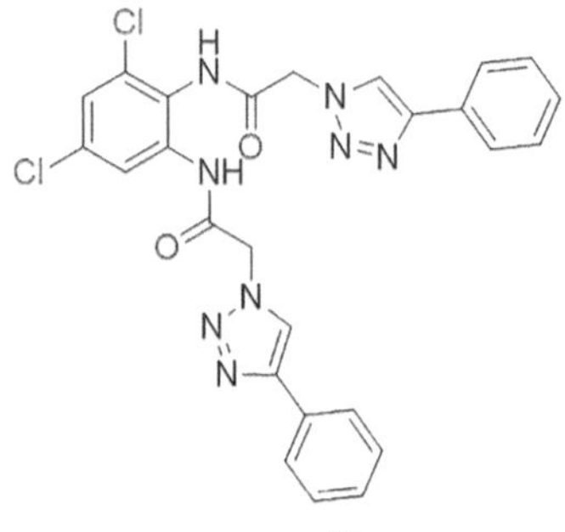

44g

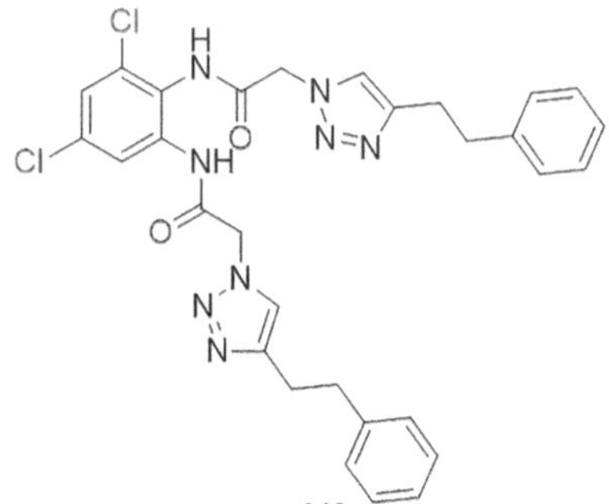

44f

FIGURE 9.30 Structures of the most potent bis-triazoles.

potency (K_I = 247.00 ± 35.28 nM) than drug. Inhibitor **44e** decreases potency slightly (K_I = 297.50 ± 35.02 nM) than **44d** still 1.16-fold more potent than drug. Inhibitors **44(f–g)** (K_I = 205.10 ± 24.84, 125.90 ± 13.49 nM) showed higher inhibition than drug. Inhibitor **44h** revealed weakest inhibition (K_I = 539.90 ± 59.13 nM) (Figure 9.30). MD analysis of **4g** against AChE **(PDB:7XN1)** revealed higher binding (DS = −8.93 kcal/mol), indicating stable complex with cholinergic enzyme. Complex network of interactions anchored ligand within aromatic gorge of active site. Amide NH forms hydrogen bond with Tyr124, followed by hydrogen bond with Tyr33 and Ser293. Aryl ring of **44g** interact Trp86 (Efeoglu et al., 2025).

S.Z. Siddiqui et al. screened 11 amide–triazole hybrids **45(a–k)** as AChE and BChE inhibitors. Eserine and DMSO served as drugs. Inhibitor **45c** showed potent AChE inhibition AChE (IC_{50} = 9.52 ± 0.25 μM), and **45g** as robust inhibitor against BChE (IC_{50} = 31.51 ± 0.38 μM), compared to Eserine (IC_{50} = 0.19 ± 0.05, 0.62 ± 0.08) (Figure 9.31) (Siddiqui et al., 2023).

SAR based on substituted phenyl ring of *N*-acetamide suggests **45(a–d)** (IC_{50} = 27.75 ± 0.28, 62.87 ± 0.42, 9.52 ± 0.25, 21.25 ± 0.37 μM) showed excellent AChE inhibition. Thus, EDGs-induced potent inhibition observed in **45c** as most potent AChE inhibitor. Inhibitors **45(a–d)** (IC_{50} = 57.94 ± 0.43, 136.83 ± 0.41, 61.18 ± 0.34, 136.83 ± 0.41 μM) are moderate BChE inhibitors compared to Eserine (IC_{50} = 0.62 ± 0.08 μM). Thus, EDGs (methyl) at *meta*-position of as critical for AChE and BChE inhibition. Inhibitors **45(e–h)** bearing di-substituted ring are moderate to good inhibitors. Inhibitor **45h** bearing 3,4- dimethyl substituents is the most potent against AChE (IC_{50} = 18.46 ± 0.14 μM), in **45g** with 3,5-dimethyl group, inhibition decreased (IC_{50} = 22.53 ± 0.32 μM), while inhibition decreased highly, with di-*ortho* methyl group in **45e** (IC_{50} = 408.23 ± 0.45 μM). Contrary, **45f** (IC_{50} = 408.23 ± 0.45, 406.58 ± 0.53 μM) displayed low inhibition due to *ortho*-substitutions. Most inhibitors were less potent against BChE, but **45(g and h)** (IC_{50} = 31.51 ± 0.38, 42.26 ± 0.22 μM) were the most potent against BChE (Figure 9.32) (Siddiqui et al., 2023).

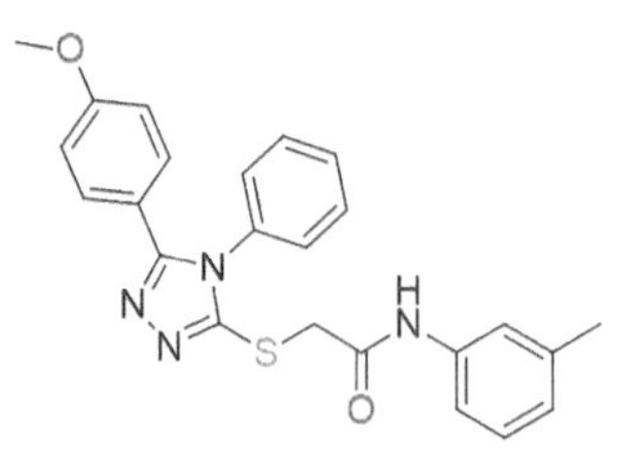

45c

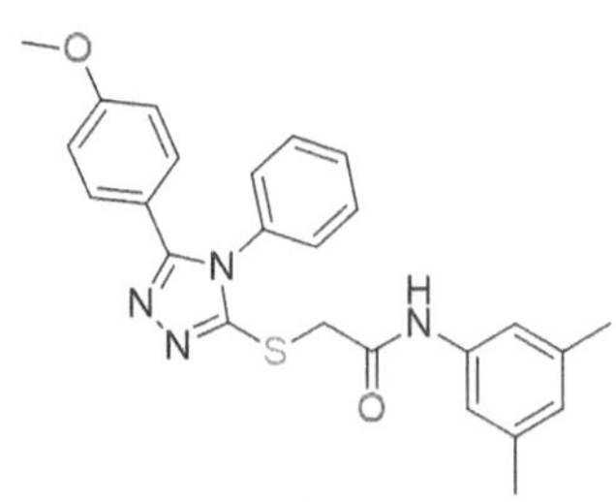

45g

FIGURE 9.31 Structures of the most potent amide–triazole hybrids.

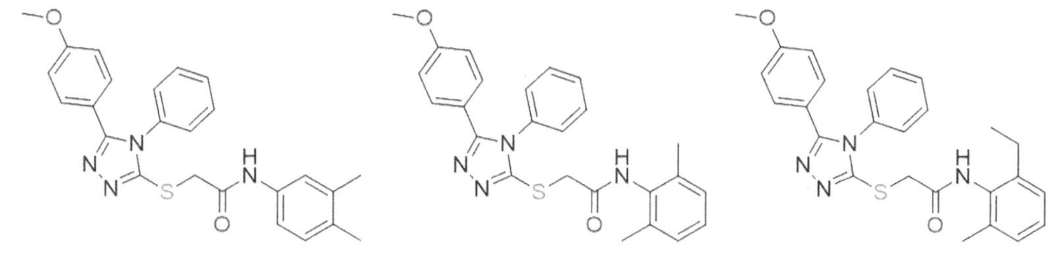

45h 45e 45f

FIGURE 9.32 Structures of amide–triazole hybrids with EDGs.

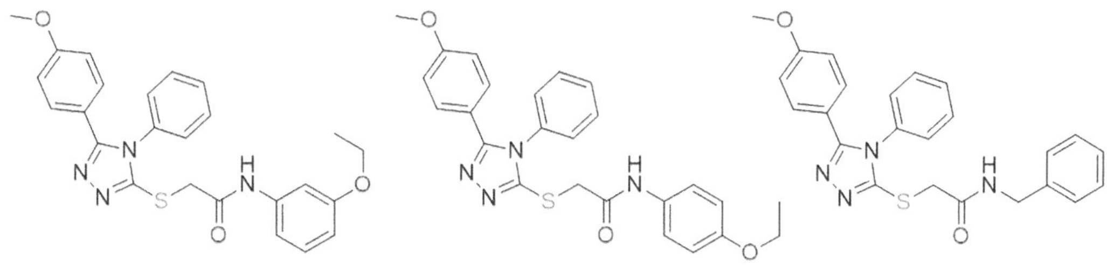

45i 45j 45k

FIGURE 9.33 Structures of moderately potent amide–triazole hybrids.

Inhibitors with ethoxy and benzyl groups were moderate AChE and BChE inhibitors. Inhibitor **45i** (IC_{50} = 56.93 ± 0.25 μM) was potent and inhibition decreased significantly in **45j** (IC_{50} = 125.43 ± 0.42 μM). Inhibitor **45k** (IC_{50} = 62.42 ± 0.38 μM), with benzyl group, displayed moderate AChE inhibition. Inhibitors **45(i–j)** (IC_{50} = 121.35 ± 0.36, 142.13 ± 0.36 μM) showed 3/4-ethoxy groups improved BChE inhibition, and **45k** was inactive (Figure 9.33) (Siddiqui et al., 2023).

SAR suggests *meta*-substituent as good inhibitors against both enzymes. However, inhibition deteriorates by removing or changing the *meta*-group (Siddiqui et al., 2023).

S. Hosseini, S.A. Pourmousavi, M. Mahdavi et al. screened 13 naphthoquinone conjugated to aryl triazole–acetamide hybrids as AChE and BuChE inhibitors using Tacrine as drug. Generally, hybrids showed promising (K_i = 10.16–65.31 nM (AChE), K_i = 8.04–60.58 nM (BuChE)) inhibitions. Inhibitor **46a** showed K_i = 34.88 ± 4.82 nM against AChE. Evaluations of halogen-based inhibitors **46(b–e)** (K_i = 10.16, 31.74, 12.84, 40.53 nM) suggest 2-Cl and 3-Br enhanced inhibition than **46a** (Figure 9.34) (Hosseini et al., 2022).

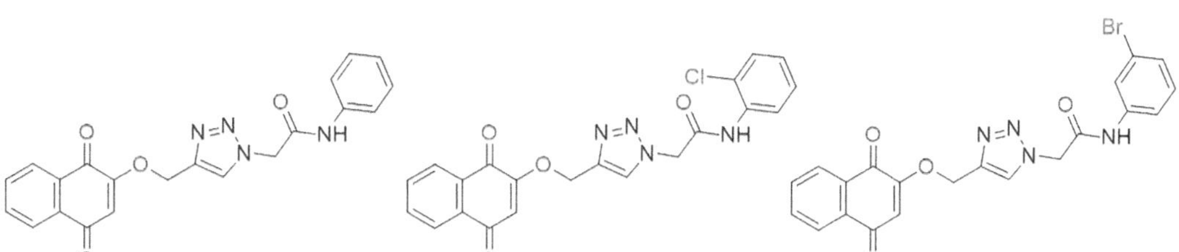

46a 46b 46d

FIGURE 9.34 Structures of potent naphthoquinone conjugated amide–triazole hybrids.

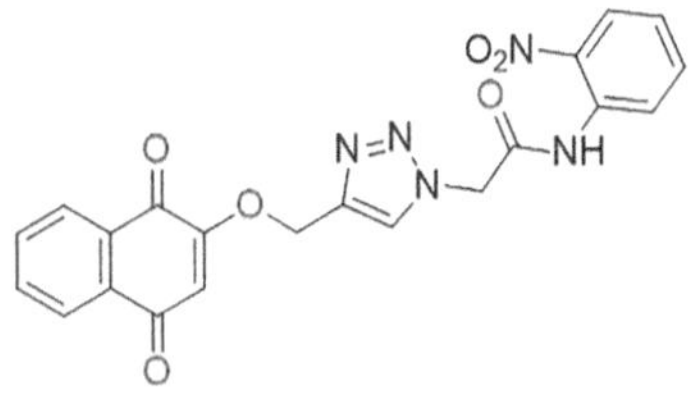

46f

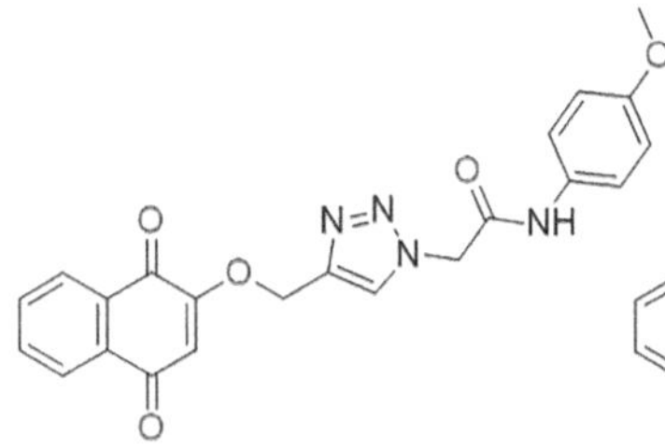

46g

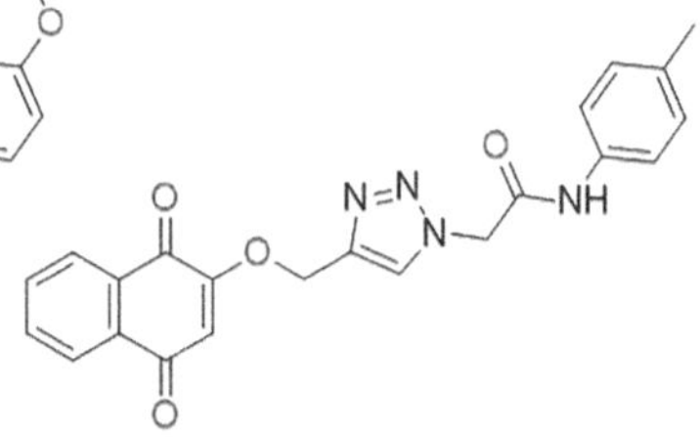

46h

FIGURE 9.35 Structures of least potent naphthoquinone conjugated amide–triazole hybrids.

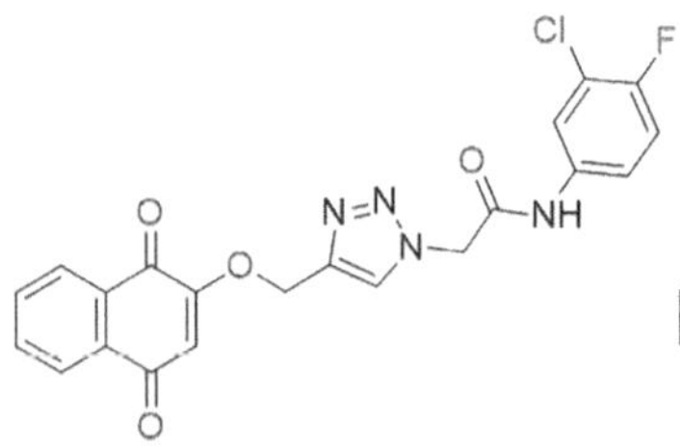

46l

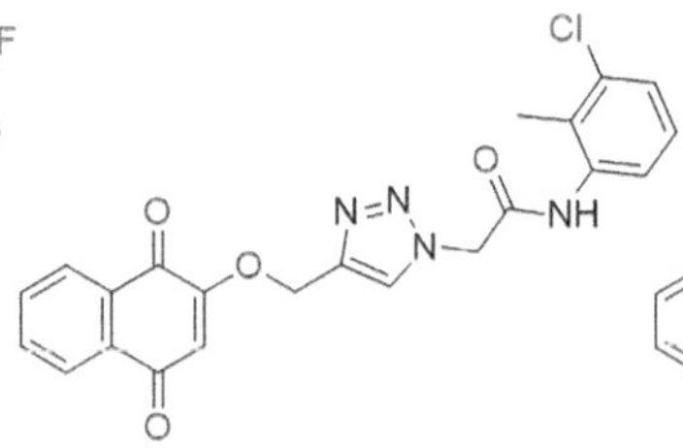

46j

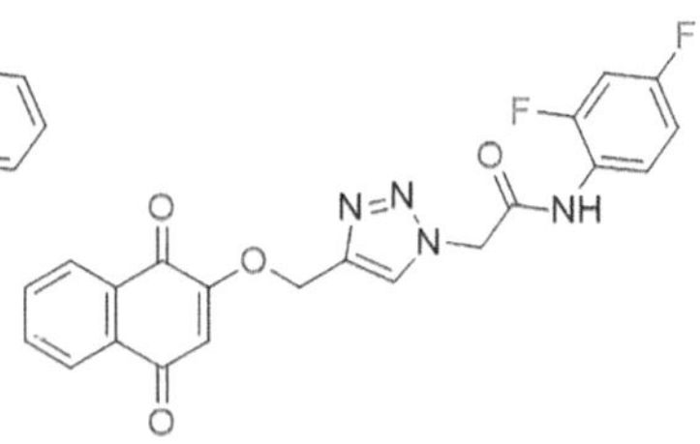

46k

FIGURE 9.36 Structures of di-substituted naphthoquinone conjugated amide–triazole hybrids.

In **46f** (K_i = 41.72 ± 9.42 nM), and EDGs **46(g–h)** (K_i = 48.62 ± 8.44, 40.31 ± 7.43 nM), inhibitions were not improved in reference to **46a** (Figure 9.35) (Hosseini et al., 2022).

In di-substituted inhibitors **46(i–l),** inhibition decreased as: **46l** (K_i = 15.95 nM)>**46j**(K_i = 20.51 nM)>**46k** (K_i = 25.93 nM)≫**46i** (K_i = 65.31 ± 10.63 nM) (Figure 9.36) (Hosseini et al., 2022).

With increase in space between acetamide and phenyl ring **46m,** increased inhibition (K_i = 28.63 nM) (Figure 9.37) (Hosseini et al., 2022).

To conclude, **46b** and **46d** showed higher inhibition suggesting halogens as crucial for AChE inhibition. BuChE inhibition followed similar trends as AChE inhibition. Inhibitor **46a** served as the most potent BuChE inhibitor (K_i = 28.37 μM). Inhibitors **46(b–e)** with halogen group, improved inhibition in **46b** (K_i = 8.04 nM) eight-fold than Tacrine followed by **46d** (K_i = 11.77 nM). Inhibitor **46f** showed lower inhibition (K_i = 34.94 nM) than **46a**. EDGs **(46g, 46h)** showed no increase in inhibitory potential compared to **46a.** In di-substituted inhibitors **46l** and **46j,** inhibition increased than **46a**, while decreased in **46i** (K_i = 60.58 nM). Length of the linker suggests adding one carbon linker results in **46m** (K_i = 22.95 μM) with improved inhibition. In a similar fashion, 2-Cl or 3-Br groups improved BuChE inhibition. MD analysis showed **46b** span, the length of the enzyme cavity reaching the

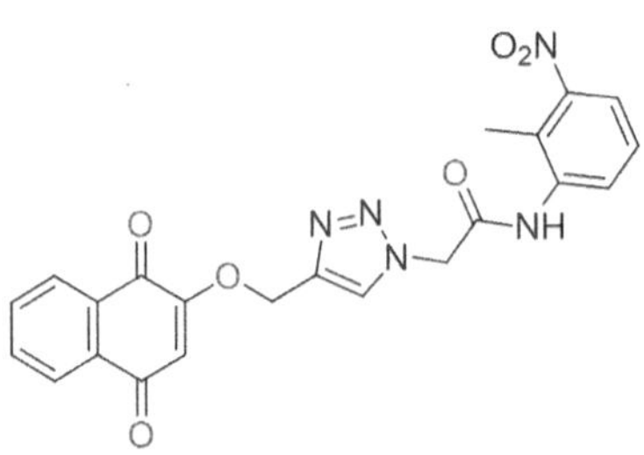

46i

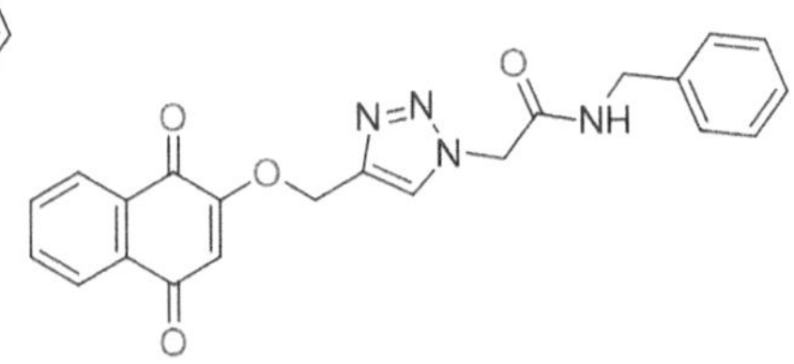

46m

FIGURE 9.37 Structures of potent naphthoquinone conjugated amide–triazole hybrids.

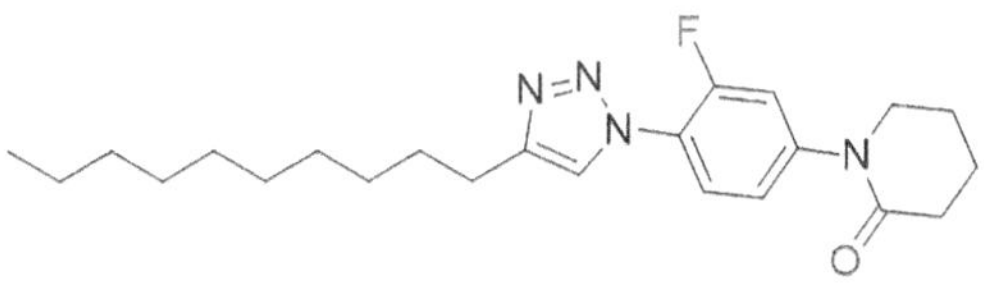

47a

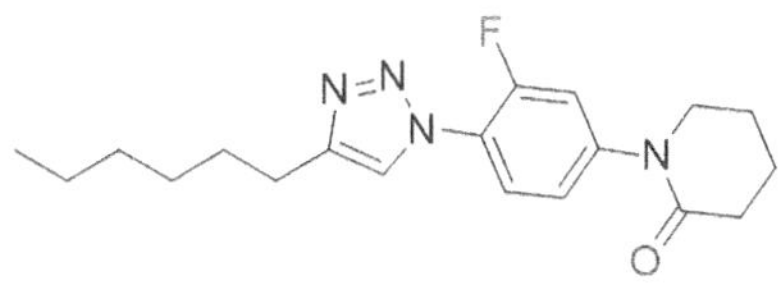

47b

FIGURE 9.38 Structures of weak lactam–triazole hybrids.

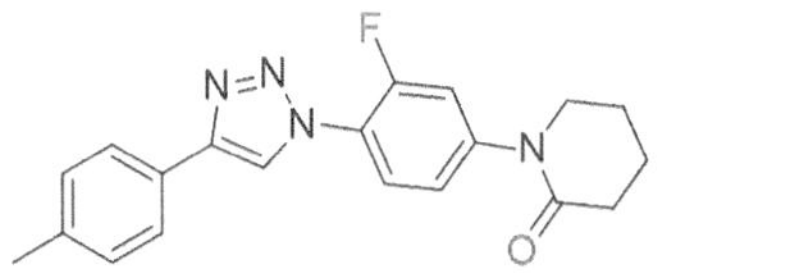

47d

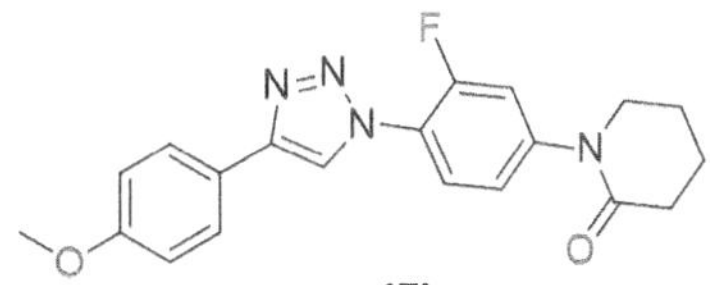

47h

FIGURE 9.39 Structures of EDGs attached lactam–triazole hybrids.

deeper catalytic site. Naphthalene-1,4-dione moiety establishes hydrogen bonding (Tyr124) and π–π stacking (Trp286, Tyr341). Triazole–acetamide form π–π stacking interactions (Phe338, Tyr341), π–lone pair, and van der Waals interaction (Tyr337). Chlorobenzene ring stabilized through three π–alkyl, van der Waals, and π–π stacking (Trp86). Against BuChE, **46b** showed dual binding by U shape conformation. Naphthalene-1,4-dione moiety interacted with aromatic interactions (His438, Ala328). Triazole–acetamide group showed three hydrogen bounds (Trp82, Thr120, Ser198), and van der Walls interactions (Gly115, Gly116, Glu197). Chlorobenzene interacted (Pro84) by two π–alkyl and Thr120 through van der Waals interaction (Hosseini et al., 2022).

V. Cavallaro et al. screened 15 lactam–triazole hybrids **47(a–o)** against AChE and BChE, using Tacrine and Donepezil as drugs. Thus, **47a** exhibited weak (IC_{50} = 100 μM) and **47b** moderate inhibition (IC_{50} = 21.4 μM) (Figure 9.38) (Cavallaro et al., 2025).

The substituted aryl ring increased inhibition significantly such as **47d** and **47h** showed strong AChE inhibition (IC_{50} = 8.1, 0.9 μM) (Figure 9.39) (Cavallaro et al., 2025).

EDGs improve inhibition by optimizing electronic properties and binding efficiency particularly at *para*-position in **47(d–e)** and **47(f–h).** *Meta*-substitutions like **47e** and **47g** are less potent (Figure 9.40) (Cavallaro et al., 2025).

Inhibitors **47i** was less potent, **47j** showed good inhibition, and **47k** improved inhibition than **47c**. Inhibitor **47m** strongly inhibited AChE and BChE. Inhibitor **47m** showed superior inhibition due to larger size and moderate electronegativity of Chlorine to enhance hydrophobic interactions within active site. Inhibitor **47n** with bromine's larger atomic radius and higher polarizability contribute to solubility challenges and no inhibition (Figure 9.41) (Cavallaro et al., 2025).

BChE followed similar trends like AChE. Inhibitors **47(a–b)** showed weak and moderate BChE inhibition (IC_{50} = 100, 42.9 μM). Aromatic groups enhanced BChE inhibition. Inhibitor **47h** displayed significant dual inhibition (IC_{50} = 8.4 μM), **47d**, **47j**, and **47l** exhibited strong inhibitions (IC_{50} = 8-9 μM). Inhibitor **47i** exhibited strong inhibitions (IC_{50} = 8.0 μM). Inhibitor **47k** contributes

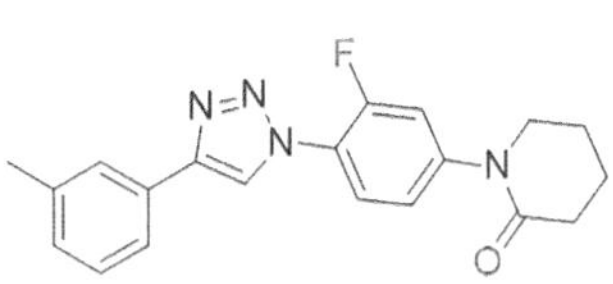

47e

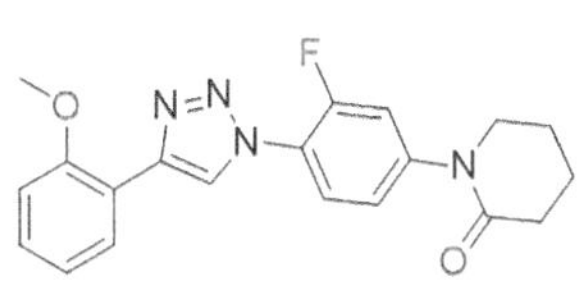

47f

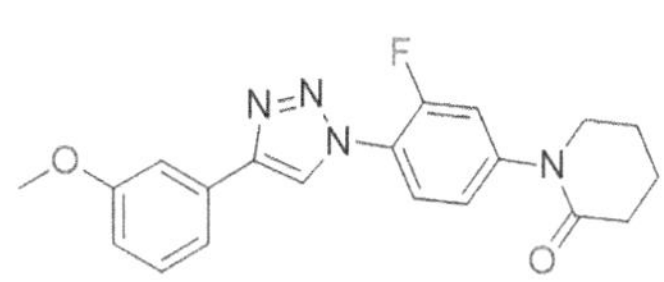

47g

FIGURE 9.40 Structures of *meta/para* substituted lactam–triazole hybrids.

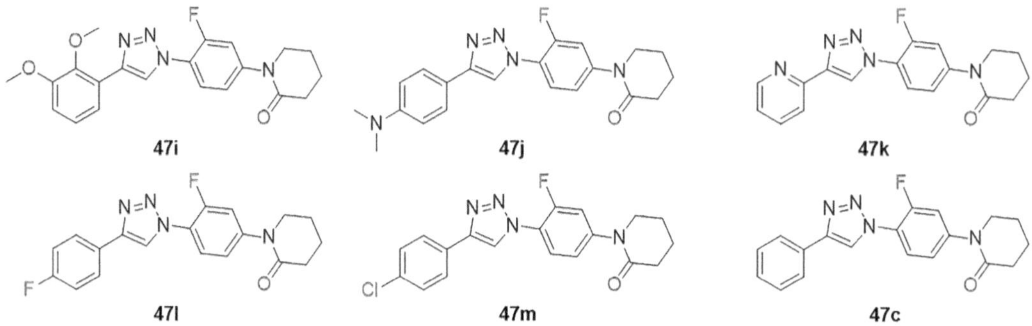

FIGURE 9.41 Structures of substituted lactam–triazole hybrids.

moderate and significant dual inhibition. Inhibitor **47m** emerged as most potent (IC_{50} = 0.7 μM (AChE), 0.2 μM (BChE). Inhibitor **47o** exhibited remarkable dual inhibition (IC_{50} = 0.8 μM) for both enzymes like **47h**. Inhibitor **47m** and **47o** served as dual inhibitors, potent than Tacrine and Donepezil, indicate that the structural modifications introduced in this study enhance enzyme affinity and selectivity while potentially reducing toxicity (Cavallaro et al., 2025).

9.3.6 Indole–Triazole Hybrids

M. Ashram et al. evaluated novel indole–triazole hybrids as AChE and BuChE inhibitors. SARs suggest inhibitors **48–77** as potent AChE (IC_{50} = 0.589-3.870 μM) and BuChE (IC_{50} = 0.050–4.598 μM) inhibitors and were selective toward BuChE due to polar substituents. Inhibitors **49, 51, 53,** and **54** (IC_{50} = 0.624, 0.371, 0.504, 0.642 μM) inhibited BuChE significantly than **52** (IC_{50} = 1.900 μM). Inhibitors **54** and **75** showed appreciable AChE inhibition (IC_{50} = 1.002, 0.897 μM). Benzyl group **(56–63, 73, 76)** reduced AChE and increased BuChE, except **62** and **59** (IC_{50} = 0.771, 0.796 μM), as notable AChE inhibitors. Inhibitors **56–63** strongly inhibited AChE than **48–55.** Inhibitors with triazole groups **(64–70)** exhibited increased potency against BuChE. In triazole hybrids **(64–71, 74, 77),** significant inhibitions were obtained especially against BuChE, for fluoro, chloro, bromo, iodo, and nitro (IC_{50} = 0.05, 0.097, 0.444, 0.359, 0.54 μM). Ester, sulfamide, and methoxy groups reduced inhibitions. Inhibitors **48–55, 56–63,** and **64–70** showed respective selectivity (AChE/BuChE) of 1.56–4.02, 1.1–8.32, 0.29–21.21. Introducing methyl-acetate group **(72–73),** decreased AChE and increased BuChE (SI≈ 4.6) inhibition. Contrary, **74** increased AChE inhibition, incorporating isatin moiety **(75–77)** improved AChE and BuChE inhibitions (Figure 9.42) (Ashram et al., 2025).

Electronic properties of substituted phenyl triazole or indole–triazole-hybrid modulate inhibitions. Thus, *para*-chlorine highly effects BuChE inhibition than AChE **(48–71).** Strong EWG (NO_2) instead of halogens, and methoxy group **(48–71),** enhance inhibition of AChE and BuChE. Notably, nitro group at *para*-position enhanced AChE inhibition more than two-fold **(49, 57)** versus **54** and **70.** Inhibitors with EWGs like halogens **(48–51, 56–59, 64–67),** showed reduced AChE and higher BuChE inhibition. Inhibitors with nitro groups **(54, 62, 70)** and isatin moieties **(75–77)** displayed improved AChE inhibition. Inhibitors **64–65** were highly selective (18.39, 21.21) BuChE inhibitors than AChE and underscore significance of EWGs, fluorine **(64),** and chlorine **(65).** Remaining inhibitors **(48–63, 66–77)** showed varied inhibitions and selectivity against AChE and BuChE, which reflects sensitivity of enzyme-ligand interactions to their structural modifications (Ashram et al., 2025).

9.3.7 Cinnamamide/Ester–Triazole Hybrids

L.-J. Tan et al. screened 37 cinnamamide/ester–triazole hybrids **78(a–s)/79(a–r)** against AChE and BuChE. Inhibitors **78(a–s)** and **79(a–r)** were evaluated against AChE using Donepezil and

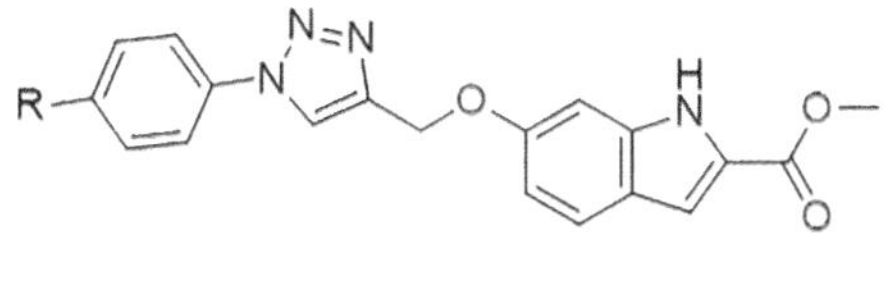

(48-55)

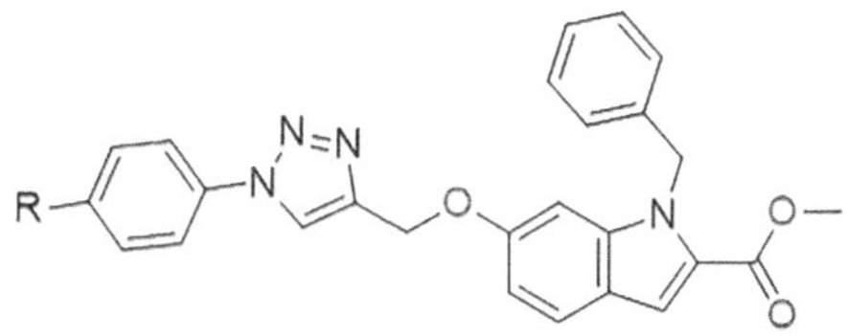

(56-63)

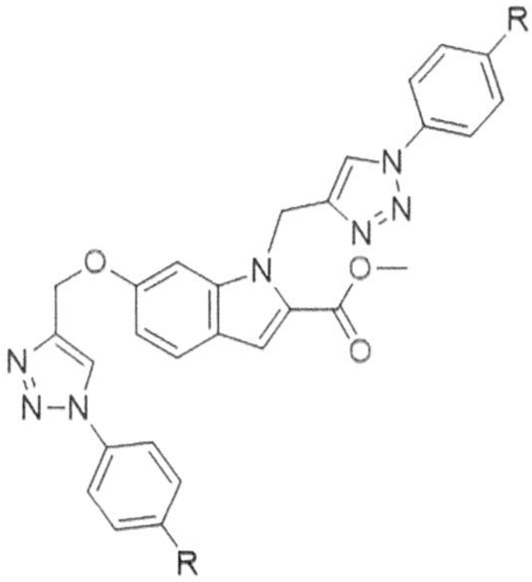

(64-71)

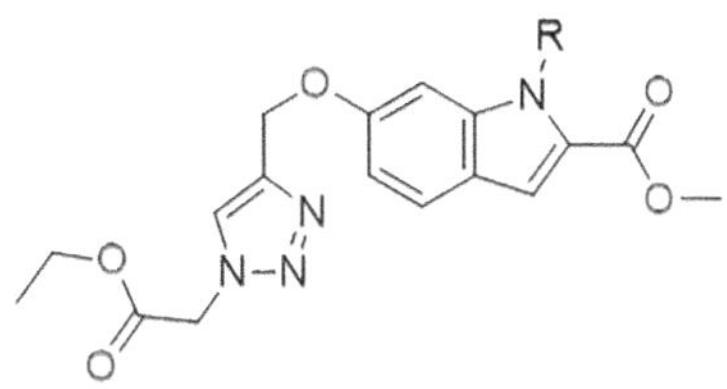

(72-74)

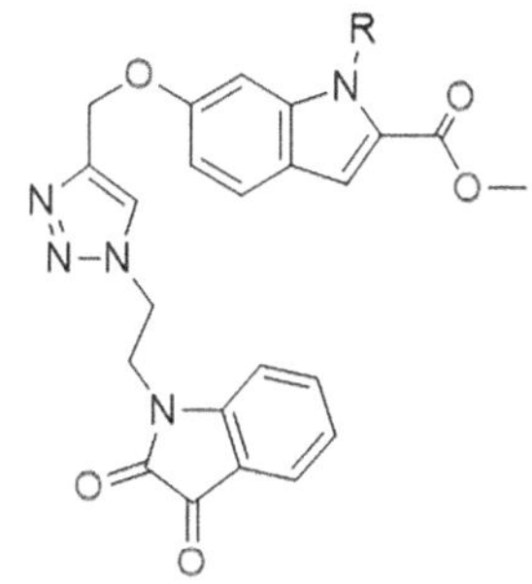

(75-77)

FIGURE 9.42 General structures of indole–triazole hybrids.

Memantine as drugs. Thus, Cinnamamide triazole hybrid **78o** served as the most (IC_{50} = 0.15 ± 0.01 µM). Cinnamamide-ester–triazole hybrid **79k,** showed highest inhibition (IC_{50} = 0.71 ± 0.40 µM) (Figure 9.43) (Tan et al., 2024).

Inhibitors **78f, 78(h–j), 78p, 79b, 79d,** and **79m** inhibited AChE (IC_{50} = 2.90–56.60 µM) (Figure 9.44) (Tan et al., 2024).

Against BuChE, **78i, 78q, 78h, 78i, 78n, 78o,** and **78p** induced highest inhibition (40%) (Figure 9.45) (Tan et al., 2024).

9.3.8 Piperidine–Triazole Hybrids

H. M. A. Asif et al. screened eleven piperidine-based triazole hybrids **80(a–k)** as potent AChE inhibitors. In terms of IC_{50} values (µM), excellent inhibition was shown by **80j, 80h,** and **80k** (IC_{50} = 0.031 ± 0.85, 0.115 ± 1.14, 0.169 ± 0.97 µM) compared to Esrine (IC_{50} = 0.04 ± 0.001 µM) (Figure 9.46) (Asif et al., 2023).

Methyl derivatives **80b** and **80d** (IC_{50} = 0.528 ± 0.92, 0.158 ± 1.22 µM) served as the most potent inhibitors. Chloro derivative **80e** (IC_{50} = 0.250 ± 1.15 µM), bromo derivative **80h** (IC_{50} = 0.115 ± 1.14 µM), and fluoro derivative **80j** emerged as the most potent inhibitors. MD with AChE co-crystal structure **(PDB ID:1NEN)** predicted **80j** showed conventional hydrogen bonding with

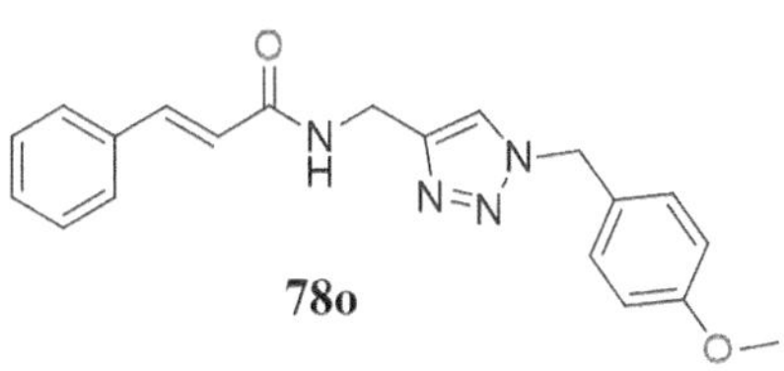

78o

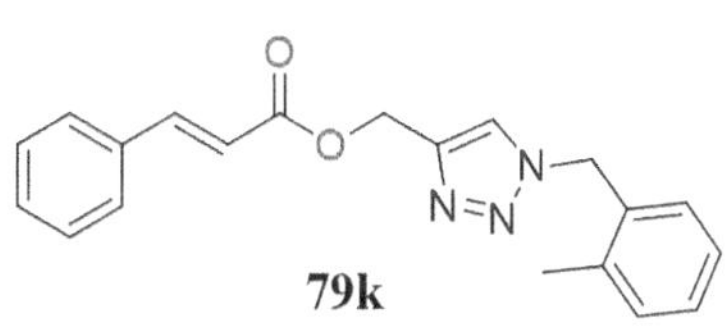

79k

FIGURE 9.43 Structures of potent cinnamamide/ester–triazole hybrids.

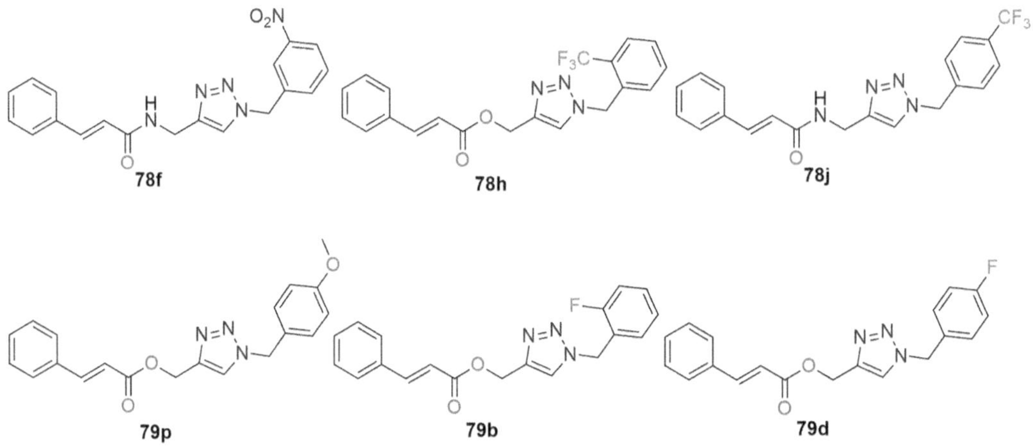

FIGURE 9.44 Structures of cinnamamide/ester–triazole hybrids.

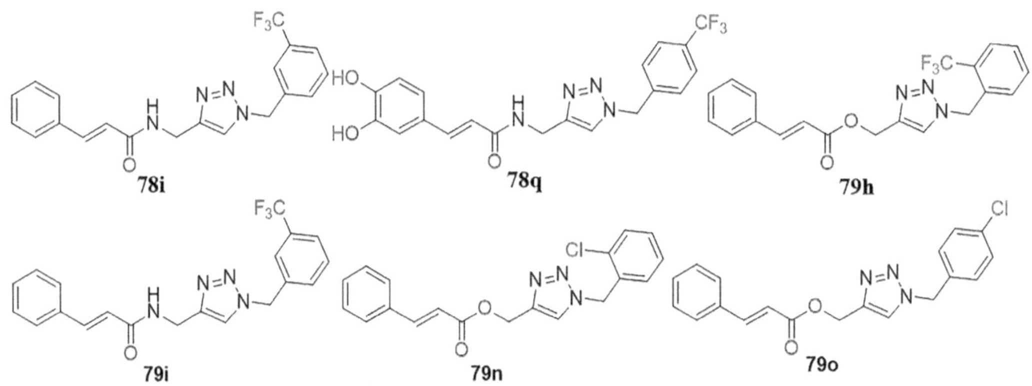

FIGURE 9.45 Structures of potent BuChE cinnamamide/ester–triazole hybrids.

amino acid HIS 45, THR 46, ASN 218, ALA 201, electrostatic interaction with HIS 354, ARG 399, GLU 388, and van der Waals interactions with HIS 354, ALA 49, LEU 252, LYS 38, HIS 354, ALA49, LYS 38. Moreover, π–π T-shaped, alkyl and π–alkyl interactions anchored ligand within active site (Figure 9.47) (Asif et al., 2023).

9.3.9 Thiosemicarbazone–Triazole Hybrids

F. Rahim et al. screened 21 thiosemicarbazone–triazole hybrids **81(a–u)** as AChE and BuChE inhibitors. Donepezil served ($IC_{50} = 2.16 \pm 0.12$ μM (AChE), $IC_{50} = 4.5 \pm 0.11$ μM (BuChE)) as drug.

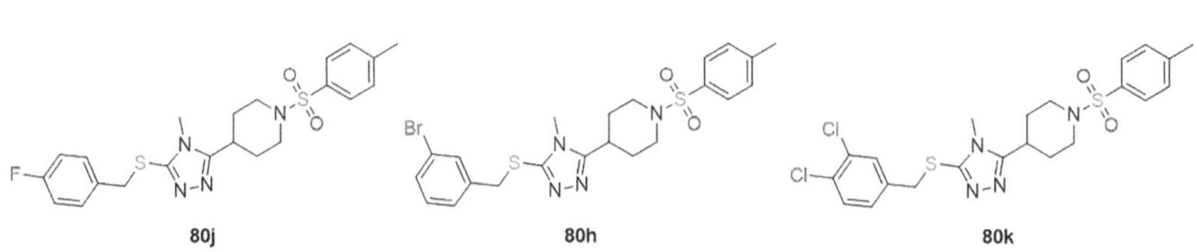

FIGURE 9.46 Structures of potent piperidine–triazole hybrids.

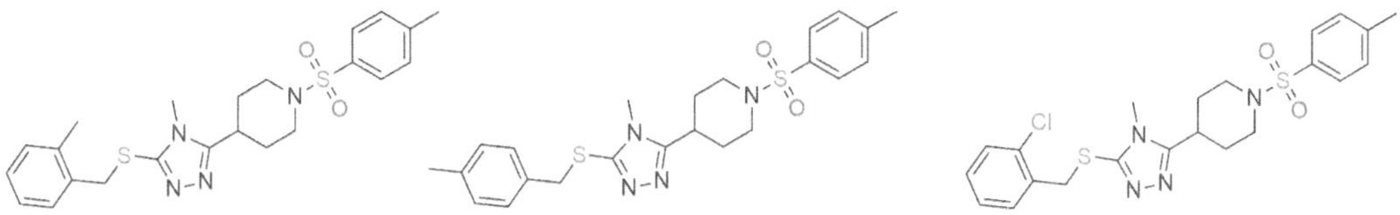

80b 80d 80e

FIGURE 9.47 Structures of potent methylated piperidine–triazole hybrids.

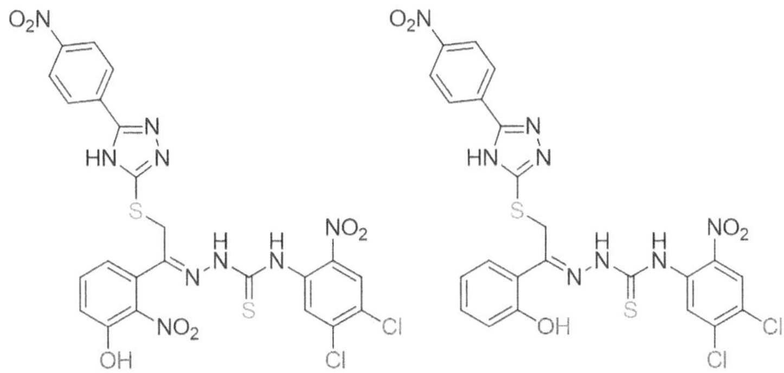

81i 81b

FIGURE 9.48 Structures of potent thiosemicarbazone–triazole hybrids.

Inhibitors showed very weak potential against AChE and were found to possess moderate potential against BuChE. Inhibitors showed good to moderate potential (IC_{50} = 0.10 ± 0.050–12.20 ± 0.30 μM (AChE), IC_{50} = 0.20 ± 0.050–14.10 ± 0.40 μM (BuChE)). Inhibitor **81i** (IC_{50} = 0.10 ± 0.050 (AChE), IC_{50} = 0.20 ± 0.050 μM (BuChE)) served as the most potent inhibitor of AChE and BuChE, followed by **81b** (IC_{50} = 0.20 ± 0.10 μM (AChE), IC_{50} = 0.30 ± 0.10 μM (BuChE)) (Figure 9.48) (Rahim et al., 2023).

Inhibitor **81k** was found to be least potent (IC_{50} = 12.20 ± 0.30, 14.10 ± 0.40 μM) inhibitor (Figure 9.49) (Rahim et al., 2023).

Inhibition enhanced in **81(a–b)** and **81l** against AChE and BuChE. Inhibitor **81b** (IC_{50} = 0.20 ± 0.10, 0.30 ± 0.10 μM) showed greater inhibition than **81a** (IC_{50} = 5.10 ± 0.20, 6.40 ± 0.20 μM) and **82l** (IC_{50} = 4.60 ± 0.010, 5.90 ± 0.10 μM) (Figure 9.50) (Rahim et al., 2023).

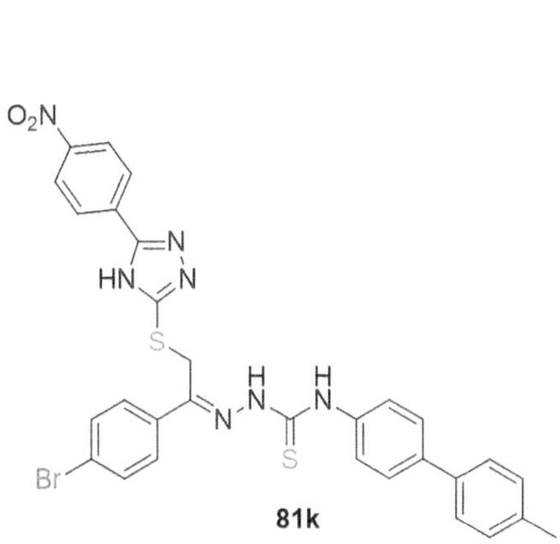

81k

FIGURE 9.49 Structures of least potent thiosemicarbazone–triazole hybrid.

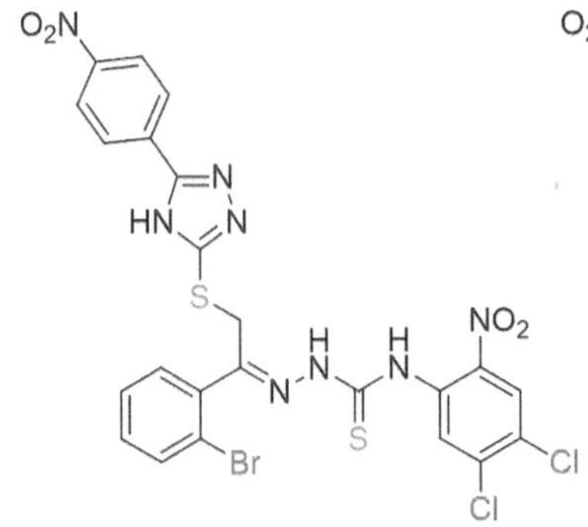

81a

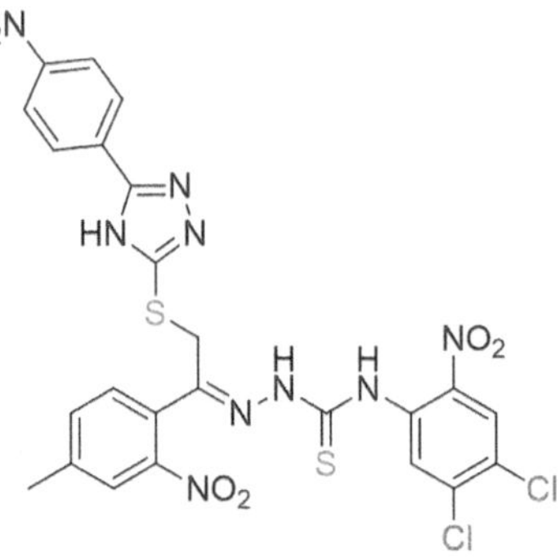

81l

FIGURE 9.50 Structures of potent thiosemicarbazone–triazole hybrids.

Inhibitor **81g** (IC_{50} = 2.10 ± 0.10, 4.30 ± 0.10 μM) has better inhibition than **81f** (IC_{50} = 2.40 ± 0.10, 4.70 ± 0.10 μM). Minor changes in inhibition of **81h** (IC_{50} = 9.10 ± 0.20, 11.20 ± 0.30 μM), **81m** (IC_{50} = 2.70 ± 0.10, 3.80 ± 0.10 μM), **81r** (IC_{50} = 1.30 ± 0.050, 2.20 ± 0.10 μM), **81t** (IC_{50} = 1.90 ± 0.10, 2.50 ± 0.10 μM) reflects changes in nature and position of the groups across aryl ring B (Figure 9.51) (Rahim et al., 2023).

Similarly, differences in inhibition of **81n** (IC_{50} = 0.70 ± 0.05, 1.70 ± 0.050 μM), **81s** (IC_{50} = 1.40 ± 0.050, 2.30 ± 0.10 μM), and **81u** (IC_{50} = 2.90 ± 0.10, 3.70 ± 0.10 μM) are due to different nature and position of substituents on ring B (Figure 9.52) (Rahim et al., 2023).

MD analyzed binding mechanisms of inhibitors against both enzymes. Inhibitors actively oriented within the active sites. Inhibitor **81i** being most potent, interacted catalytic residues of AChE via Tyr332 (π–sigma, π–π stacked), Trp82 (π–anion), Glu197 (hydrogen bond), Phe329 (π–sulfur),

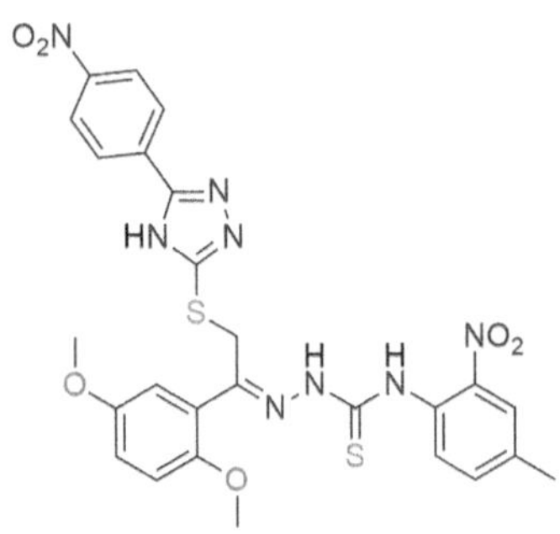

81f

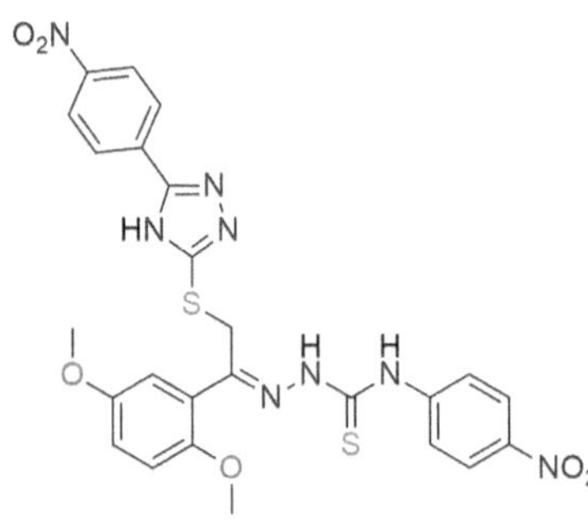

81g

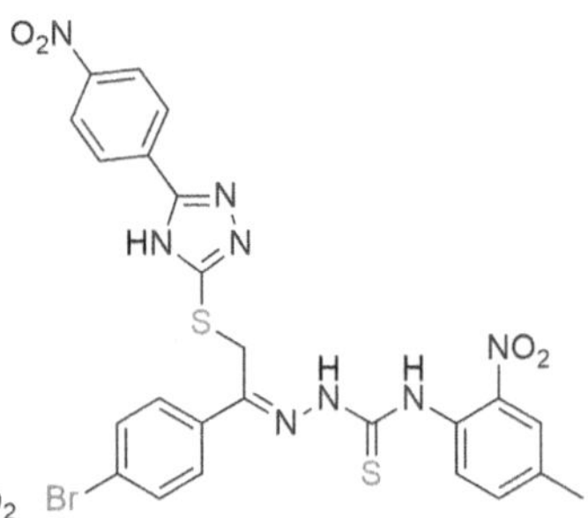

81h

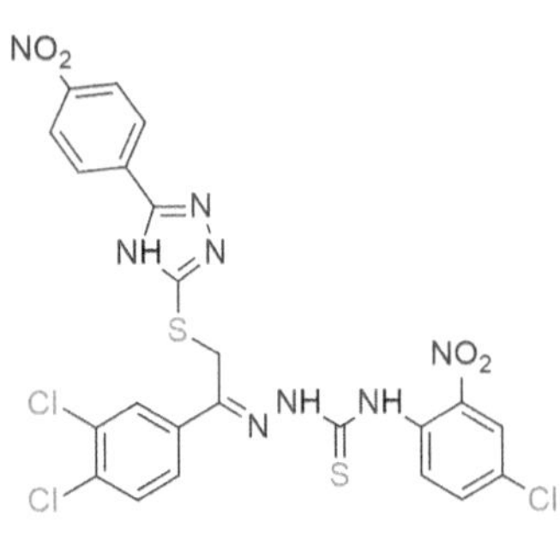

81m

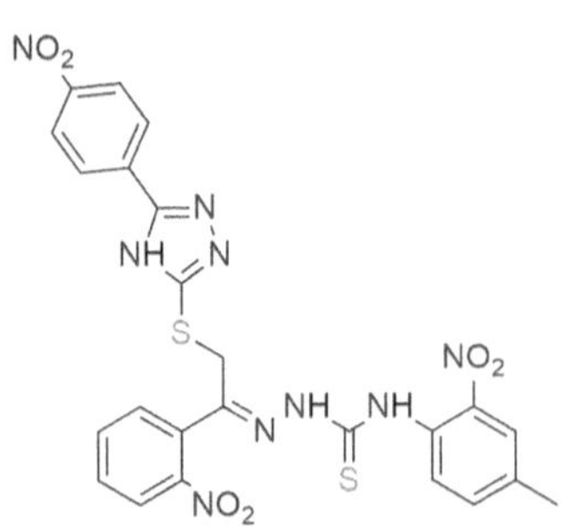

81r

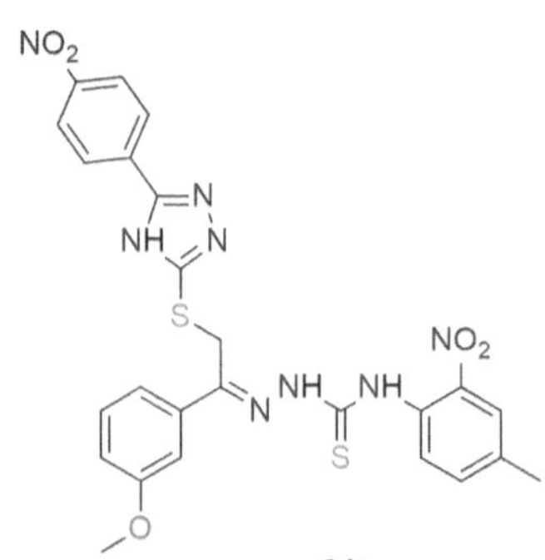

81t

FIGURE 9.51 Structures of thiosemicarbazone–triazole hybrids with comparable inhibitions.

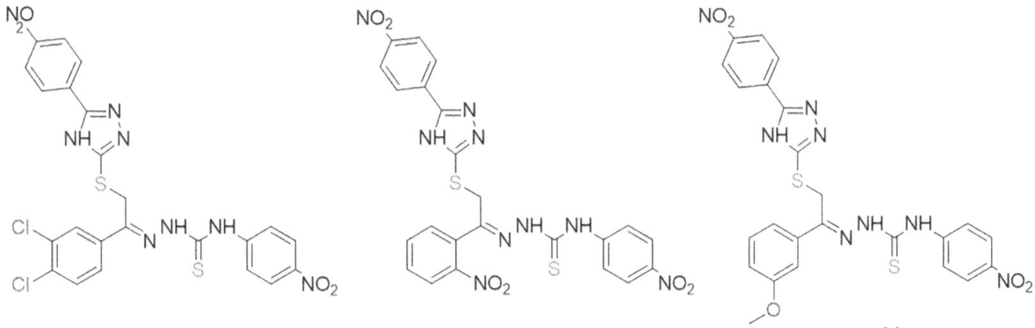

FIGURE 9.52 Structures of EWGs bearing thiosemicarbazone–triazole hybrids.

Ser198, His438, Asn289, Asn68 (hydrogen bond), and Asp70 (π–cation). Against BuChE, interactions include Tyr116 (π–alkyl), Tyr130 (π–alkyl), His440 (C–H bond), Phe330 (π–anion), Asp72 (π–cation), Asn85 (hydrogen bond), Tyr121 (π–sulfur), Trp279 (π–π shaped and amide–π stacked), Tyr70 (hydrogen bond), Tyr334 (π–π stacked), Trp84 (π–π T shaped stacked), Gly118 (hydrogen bond), and Gly117 (π–π stacked) (Rahim et al., 2023).

9.3.10 1-Deoxynojirimycin–Triazole Hybrids

T.C.S. Evangelista et al. screened four 1-deoxynojirimycin and triazole hybrids as AChE and BuChE inhibitors using Galantamine as drug. Thus, heterodimers **82a, 82c,** and **82d** showed negligible inhibition against AChE. Inhibitor **82b** showed inhibition AChE in micromolar range. Against BuChE, **82 (a–d)** showed better inhibition in micromolar range (IC_{50} = 7–50 μM) highly selective for BuChE (SI index = 9.7). Inhibitor **82b** served as 1.5-fold more BuChE inhibition, while **82a, 82c,** and **182d** were 1.7–5 times less potent BuChE inhibitors than 1-deoxynojirimycin. MD analysis of **82b** suggest benzotriazole and 1-deoxynojirimycin moiety is accommodated in respective peripheral anionic site and catalytic anionic site of AChE. However, in binding analysis with BuChE, benzotriazole moiety was accommodated in catalytic anionic site (Figure 9.53) (Evangelista et al., 2022).

9.3.11 β-Carboline–Triazole Hybrids

W. Liu et al. screened 17 β-carboline-triazole hybrids as AChE and BuChE inhibitors using Tacrine as drug. Thus, inhibitors **83(a–j)** showed weak to good potential (3.40–94%) against AChE. Substitution at C-1 **83(a–e)** possess better inhibition than C-3 **83 (f–j)** compared to Donepezil (98.79%). Inhibitors with cyclopropyl carboxamide group at C-1 **(83a),** and C-3 **(83f)** showed higher AChE inhibitions (94.0 and 84.70%) (Figure 9.54) (Liu et al., 2022).

With increase in ring size, AChE inhibition decreased **(94–21.2% 83(a–e), (84.70–3.4%) 83(f–j))** (Figure 9.55) (Liu et al., 2022).

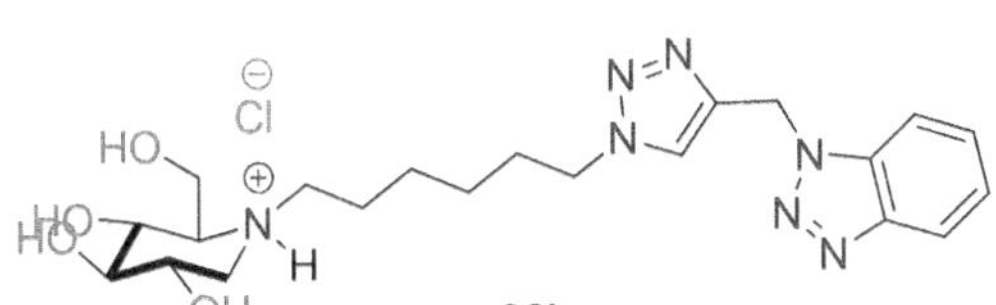

FIGURE 9.53 Structures of potent 1-deoxynojirimycin–triazole hybrid.

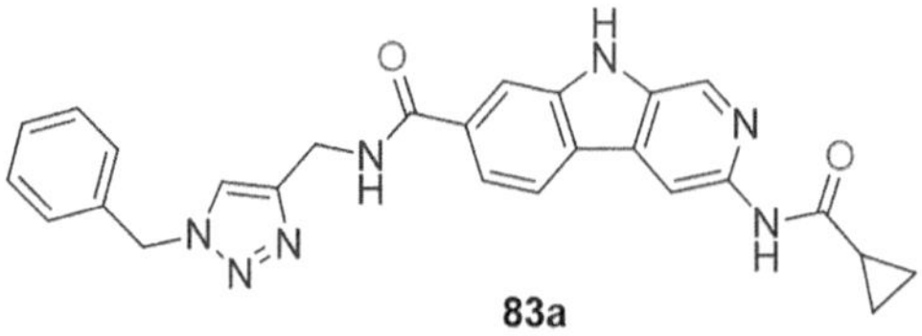

83a

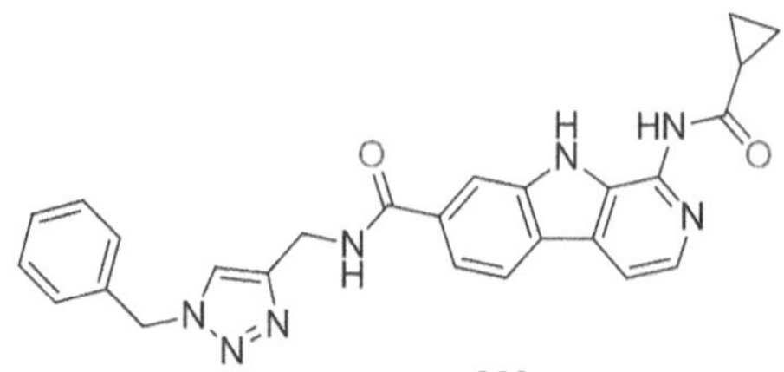

83f

FIGURE 9.54 Structures of potent β-carboline–triazole hybrids.

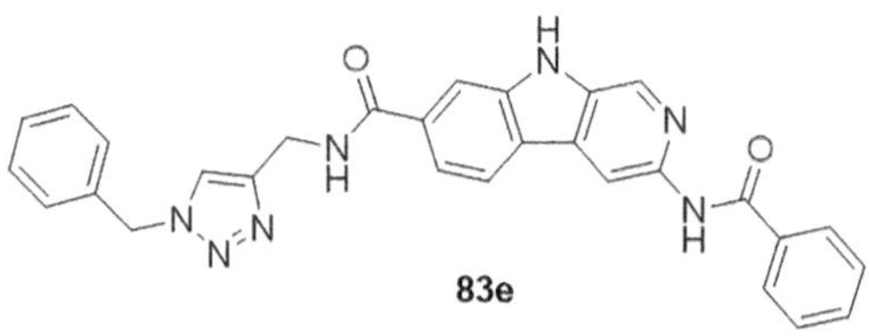

83e

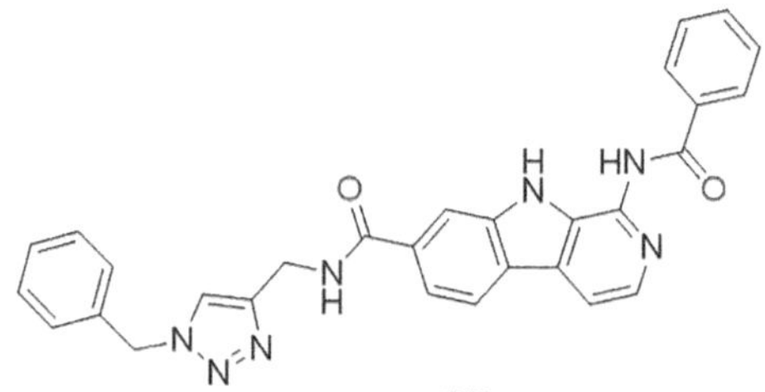

83j

FIGURE 9.55 Structures of weak β-carboline–triazole hybrids.

Inhibitors **83(k–q)** were highly potent AChE (IC_{50} = 0.20–0.61 μM) than Harmine (IC_{50} = 10.10 ± 1.14 μM), with **83f** and **83k** (IC_{50} = 0.21 ± 0.03, 0.20 ± 0.02 μM) superior to ZLWH-23 (IC_{50} = 0.27 ± 0.03 μM). Inhibitors **83f** and **83(k–q)** showed halogen **(83(k–m) vs 83(k–q),** and 1,2,3-triazole as linker improved AChE inhibition (**83(k–q)** vs ZLWH-23). Thus, inhibition increased as: **H>2-F>3-F>4-F>4-CH3>4-CF3>3,5-di-2F>4-Cl.** Generally, inhibitors showed selectivity for AChE and were less potent BuChE inhibitors (IC_{50}>10 μM). Evaluation of inhibitor **83k** against hAChE (IC_{50} = 0.34 μM) suggests **83k** as novel, effective, and selective AChE inhibitor (Figure 9.56) (Liu et al., 2022).

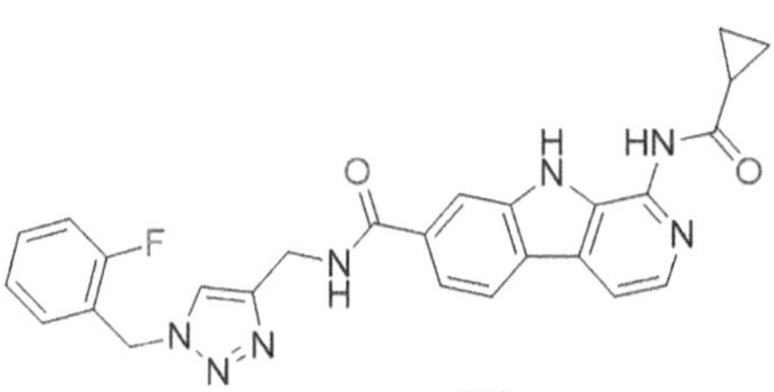

83k

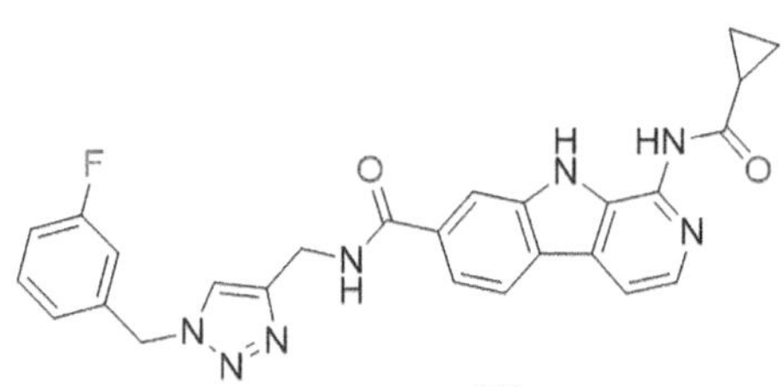

83l

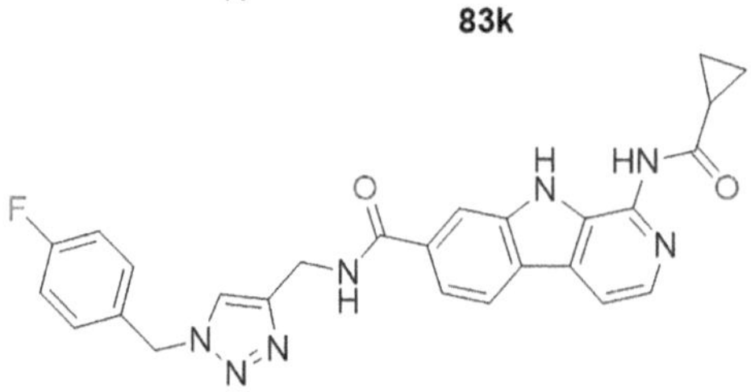

83n

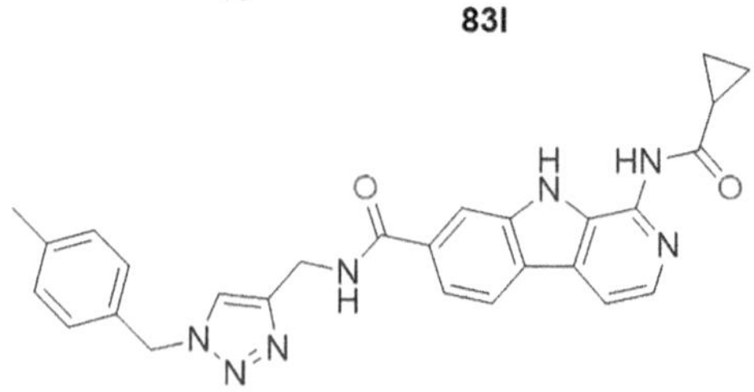

83q

FIGURE 9.56 Structures of β-carboline-based triazoles as selective AChE inhibitors.

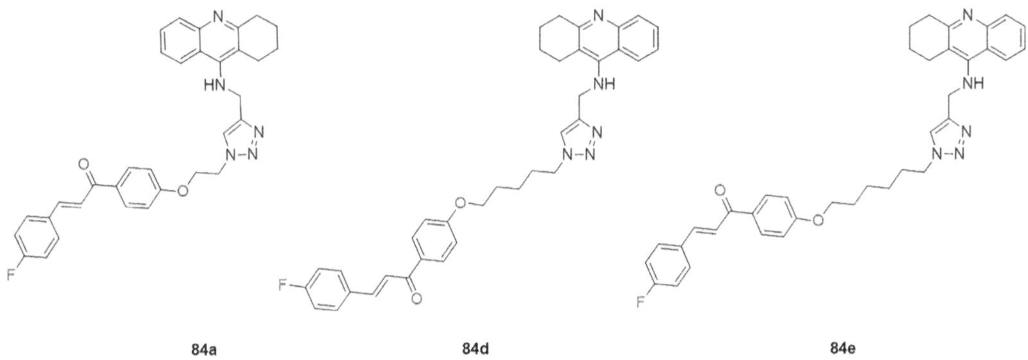

FIGURE 9.57 Structures of triazole tacrine–chalcone conjugates as weak AChE inhibitors.

MD analysis of **83f** and **83k** with AChE **(PDB: 4EY7)** suggest, benzyl group bind to CAS site of AChE, β-carboline is anchored to the PAS site through π–π stackings, where aryl ring of N-benzyl interacts indole of Trp86 in CAS, β-carboline interacts Trp286 of PAS, and 1,2,3-triazole interacts Tyr337 and indole of Tyr341. π–π stacking is present between terminal benzyl in **83f** and **83k** combines with imidazole of His447 (Liu et al., 2022).

9.3.12 Chalcone–Triazole Hybrids

Rani et al. screened 28 triazole tacrine–chalcone conjugates **84(a–x)** and **85(a–f)** against AChE using Tacrine as drug. Inhibition is dependent upon substitution across chalcone and spacer length of the pharmacophores. Thus, inhibition of hybrids **84(a–f)** (IC_{50} = 0.531–0.987 μM) decreased with increase in chain length i.e. **84a, 84d,** and **84e** (IC_{50} = 0.694, 0.940, 0.987 μM) (Figure 9.57) (Rani et al., 2021).

Chlorine on chalcone ring increased AChE inhibition **84(h and i)** (IC_{50} = 0.259, 0.372 μM) (Figure 9.58) (Rani et al., 2021).

In **84(g–l),** inhibition decreased with increase in alkyl chain lengths except **84k** (IC_{50} = 0.376 μM). In **84(m–r),** EDGs (methoxy) reduced inhibition (IC_{50} = 0.911–1.015 μM). Trimethoxy group in **84(s–x)** did not improve inhibition except **84t** (IC_{50} = 0.376 μM). Ferrocenyl conjugates **85(a–e)** did not improve inhibition except **85f** (IC_{50} = 0.327 ± 0.012). Against BuChE, **84i** showed low inhibition (IC_{50} = 5.328 μM) in comparision to Tacrine (IC50 = 0.208 μM) suggesting the hybrids as selective AChE inhibitors (Figure 9.59) (Rani et al., 2021).

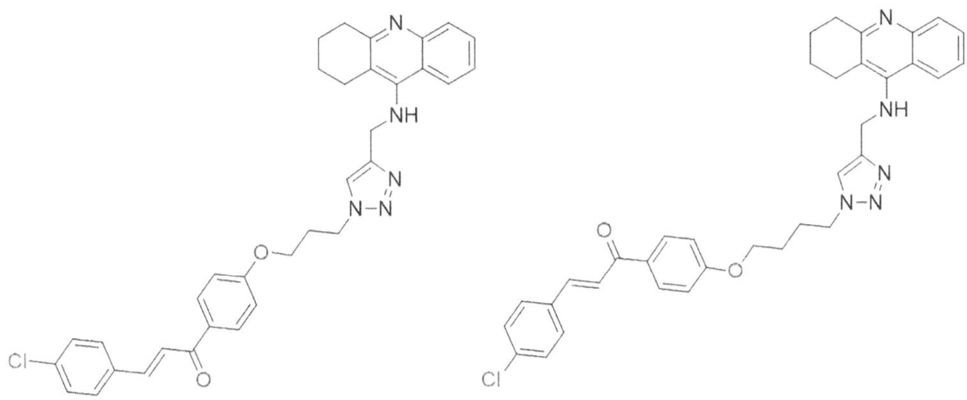

FIGURE 9.58 Structures of triazole tacrine–chalcone conjugates bearing chlorine.

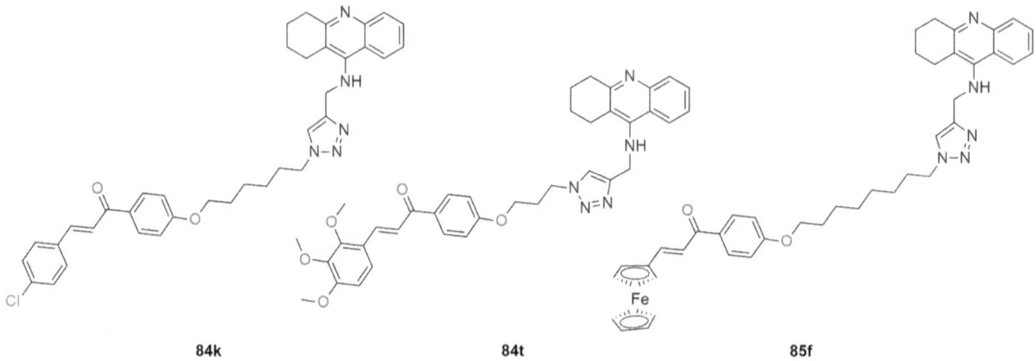

FIGURE 9.59 Structures of triazole tacrine–chalcone conjugates as selective AChE inhibitors.

9.3.13 Quinazoline–Triazole Hybrids

G. Le-Nhat-Thuy et al. screened 15 quinazoline–triazole hybrids against AChE using Donepezil as drug. Inhibitors are divided into 3-nitrophenylamine **(86a, 87(a–d))**, benzylamine **(86b, 88(a–d))**, and *N*-methylpiperazyl **(86c, 89(a–d))** groups (Figure 9.60) (Le-Nhat-Thuy et al., 2020).

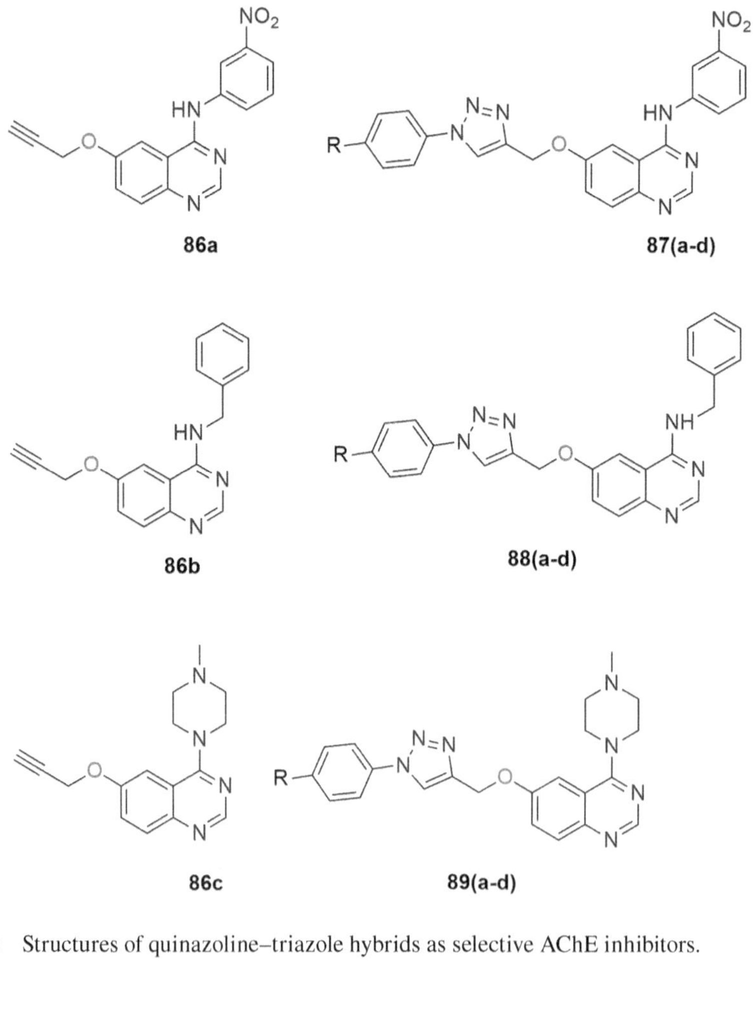

FIGURE 9.60 Structures of quinazoline–triazole hybrids as selective AChE inhibitors.

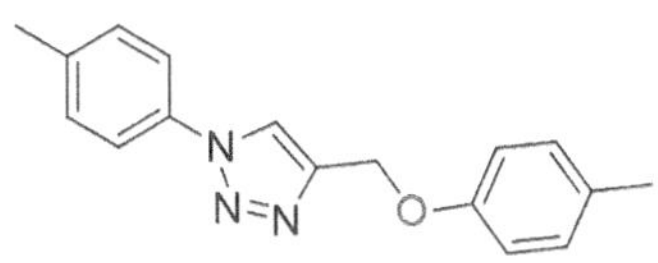

90a

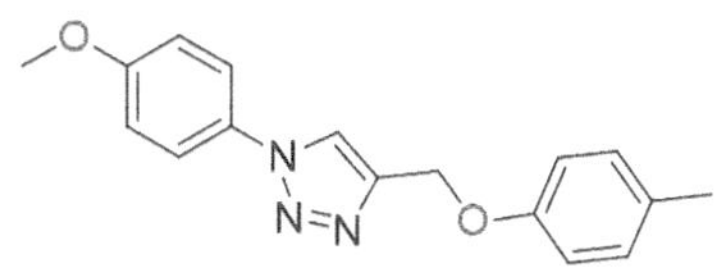

90b

FIGURE 9.61 Structures of potent *p*-cresol–triazole hybrids.

Among inhibitors **86(a–c), 86b** served as moderate (IC_{50} = 47.73 μM), **86c** as lower (IC_{50} = 83.90 μM), and **86a** remained inactive (IC_{50}>200 μM). Inhibitors **86(a–c)** improved AChE inhibition in triazole quinazoline hybrids **87a–89d**. Thus, **88a** displayed potent (IC_{50} = 0.23 μM), **88b** reduced (IC_{50} = 1.10 μM), and **88c** lost AChE inhibition (IC_{50}>200 μM) compared to Donepezil (IC_{50} = 0.12 μM). Introducing *N*-methylpiperazyl, **89(a–b)** were less potent than **88(a–b)** (IC_{50} = 37.38, 15.79 μM). Inhibitors **89(c–d)** were inactive as **88c**. In **87(a–d),** only **87a** was found to be active (IC_{50} = 2.06 μM). Thus, the nature of amines at C-4 of quinazoline effected AChE inhibitions. Regarding electronic properties of substituents at triazole, nitro group induced better inhibition than 3-trifluoromethyl-4-nitrile group which reduced inhibition in **87d**, **88d,** and **89d** (Le-Nhat-Thuy et al., 2020).

9.3.14 Cresol–Triazole Hybrids

S. Onur et al. screened eight *p*-cresol–triazole hybrids as AChE and BuChE inhibitors **90(a–h).** Galantamine and Donepezil served as drugs (IC_{50} = 8.05, 7.54 μM). Significant AChE inhibitions, higher than drugs were obtained. EDGs (methoxy, methyl) enhanced AChE inhibition. Inhibitor **90b** showed highest AChE inhibition (IC_{50} = 1.66 μM), highly selective (SI = 27.38). Inhibitor **90a** improved inhibition compared to drug (IC_{50} = 5.29 μM) (Figure 9.61) (Onur et al., 2025).

Inhibitors with EWGs showed potent AChE inhibition and selectivity. Inhibitor **90e** possesses very low inhibition (IC_{50} = 1.64 μM) and highest selectivity (SI = 33.63) followed by **90f** (IC_{50} = 4.90 μM, SI = 9.23) for AChE. Inhibitors with halogens, **90e** showed superior (IC_{50} = 1.64 μM), followed by **90d** as moderate (IC_{50} = 5.56 μM) AChE inhibitor (Figure 9.62) (Onur et al., 2025).

Inhibitor **90g** induced highest potency (IC_{50} = 1.57 μM), followed by **90c** (IC_{50} = 1.96 μM). Conversely, **90h** showed weakest AChE inhibition (IC_{50} = 8.73 μM) (Figure 9.63) (Onur et al., 2025).

To conclude, **90g**, **90e**, **90b**, and **90c** served as the most potent AChE inhibitors than drugs. Inhibitors **90e** and **90b** exhibited the highest selectivity for AChE over BuChE (Onur et al., 2025).

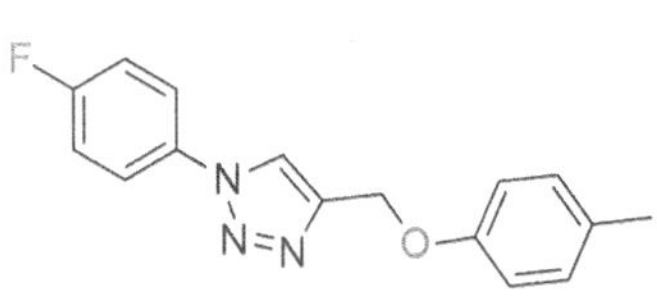

90d

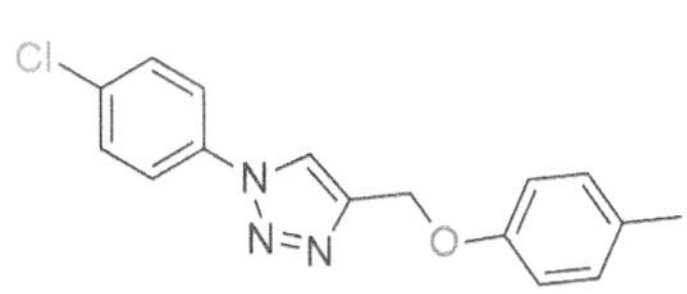

90e

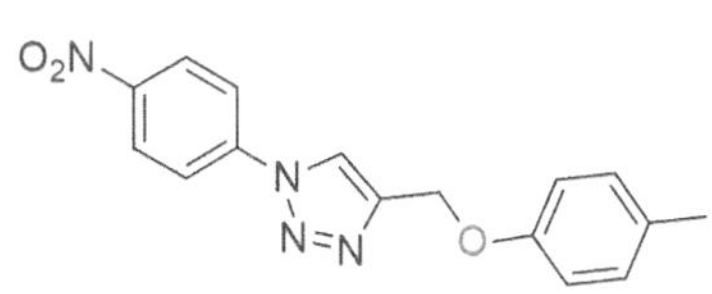

90f

FIGURE 9.62 Structures of potent *p*-cresol–triazole hybrids bearing EWGs.

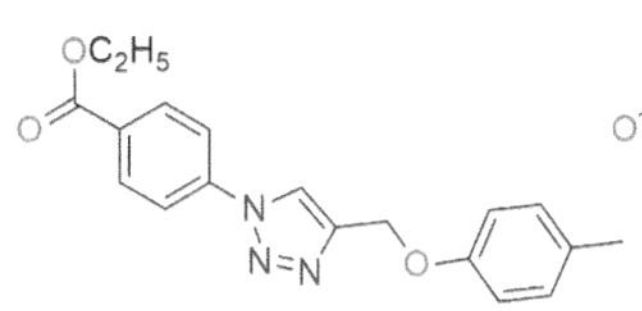

90g

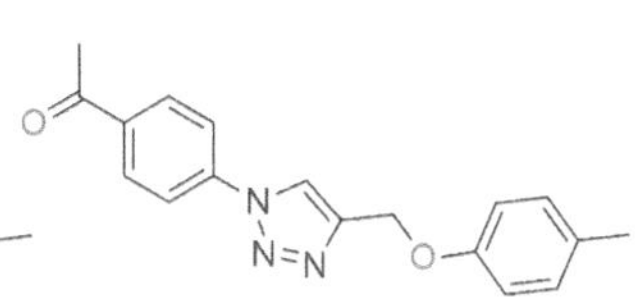

90h

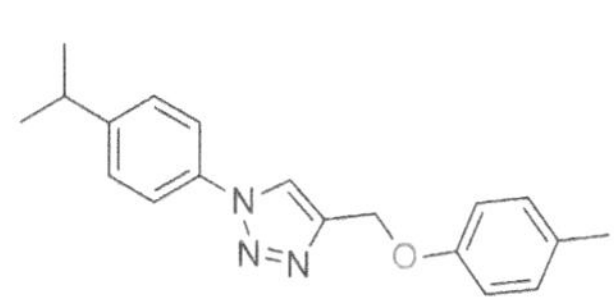

90c

FIGURE 9.63 Structures of *p*-cresol–triazole hybrids.

9.4 CONCLUSION AND FUTURE REMARKS

Triazoles and their hybrids constitute nitrogen-rich pharmacophores with broad bioisosteric and therapeutic role in drug design. Numerous 1,2,3- and 1,2,4-triazole hybrids are reported as potent acetylcholinesterase and butyrylcholinesterase inhibitors than drugs. Through enzyme inhibition, inhibitors can enhance synaptic acetylcholine level and improve cognitive function augmenting their development as promising candidates for Alzheimer's disease and multi-target therapeutic strategies. Structure–activity relationship suggests nature and position of substituents across triazole core, linker length, and the electronic properties of appended aryl or heteroaryl rings critically influence potency and selectivity for inhibition of esterase enzymes. Triazole scaffolds offer convenient synthetic accessibility and are readily amenable to systematic structural modification, enabling rapid exploration of structure–activity relationships, and optimization of potency, selectivity, and overall drug-like properties. Their modular synthesis allows triazoles to serve as versatile linkers between distinct pharmacodynamic motifs within a single chemical entity, thereby facilitating the design of multi-target-directed ligands. Thus, the present study contributes to the rational design and synthesis of novel triazole hybrids that are both effective and safe inhibitors of AD-related pathways. Continued research into triazole hybrids is expected to broaden their utility and generate innovative strategies for the treatment of neurodegenerative diseases.

REFERENCES

Aggarwal, R. & Sumran, G (2020). An insight on medicinal attributes of 1,2,4-triazoles. *European Journal of Medicinal Chemistry, 27*(205), 112652. **https://doi.org/10.1016/j.ejmech.2020.112652**

Ali, J., Choe, K., Park, J. S., Park, H. Y., Kang, H., Park, T. J., & Kim, M. O. (2024). The interplay of protein aggregation, genetics, and oxidative stress in Alzheimer's disease: Role for natural anti-oxidants and immunotherapeutics. *Antioxidants, 13*(7), 862. **https://doi.org/10.3390/antiox13070862**

Ashram, M., Mustafa, A. A., Habashneh, A. Y., Al-Alawneh, M. H., Sha'er, M. A. A., Awwadi, F. F., & Ashram, I. (2025). Novel 1,2,3-triazole-indole hybrid: Synthesis, anti-cholinesterases, antioxidant activities, kinetic and molecular docking analysis. *Journal of Molecular Structure, 1350*, 144035. **https://doi.org/10.1016/j.molstruc.2025.144035**

Asif, H. M. A., Kamal, S., Rehman, A., Bibi, I., Masoud, N.A., Alomar, T. S., & Iqbal, M. (2023). Synthesis characterization and evaluation of novel triazole based analogs as a acetylcholinesterase and α-glucosidase inhibitors. *Arabian Journal of Chemistry, 16*, 104626. **https://doi.org/10.1016/j.arabjc.2023.104626**

Aziz, H., Saeed, A., Jabeen, F., Ullah, N., & Rehman, A.U. (2021). Synthesis, characterization, in vitro biological and molecular docking evaluation of N,N'-(ethane-1,2-diyl)bis(benzamides). *Journal of the Iranian Chemical Society, 18,* 2425–2436. https://doi.org/10.1007/s13738-021-02199-8

Aziz, H., Saeed, A., Jabeen, F., Wadood, A., Rehman, A. U., Majid, M., & Haq, I. U. (2019). Synthesis, characterization and in vitro antioxidant assay of 4-(benzylideneamino)-5-pentadecyl-2H-1,2,4-triazol-3(4H)-ones. *Journal of the Iranian Chemical Society, 16*, 2143–2157. **https://doi.org/10.1007/s13738-019-01686-3**

Aziz, H., Saeed, A., Rehman, A. U., Jabeen, F., Nasir, B., Khan, A. U., & Khan, I. U. (2021). Synthesis, characterization, *in vitro* biological and computational evaluation of 5-benzyl-4-(benzylideneamino)-2H-1,2,4-triazole-3(4H)-thiones. *Journal of the Iranian Chemical Society, 18*(8), 1965–1977. **https://doi.org/10.1007/s13738-021-02156-5**

Cavallaro, V., Bjerg, E. E., Rojas, M., Romo, F. S., Murray, A. P., Radivoy, G., Duarte, Y., Zacconi, F. C., & Moglie, Y. (2025). Dual cholinesterase inhibition by lactam-1,2,3-triazole hybrids: A click chemistry approach for drug discovery. *Bioorganic Chemistry, 163*, 108643. **https://doi.org/10.1016/j.bioorg.2025.108643**

Dar, N. A., Wani, O. W., Wang, Y., Khan, F. I., Ganie, B. A., Shah, S. W. A., Dar, T. A., & Ismail, T. (2025). Novel triazole-based coumarin compounds as acetylcholinesterase inhibitors: Evidence and mechanism of 3-acetyl coumarin tethered (2-bromophenyl)-1,2,3 triazole as a potential mixed type inhibitor. *European Journal of Medicinal Chemistry Reports, 15*, 100289. **https://doi.org/10.1016/j.ejmcr.2025.100289**

da Silva, F. C., do Carmo Cardoso, M. F., Ferreira, P. G., & Ferreira, V. F. (2014). *Biological properties of 1H-1,2,3- and 2H-1,2,3-triazoles, topics in heterocyclic chemistry*. Springer International Publishing, 40:117–165. **https://doi.org/10.1007/7081_2014_124**

Efeoglu, C., Demir, Y., Türkes, C., Yabalak, E., Seferoglu, Z. & Nural, Y. (2025). New 1,2,3-triazole derivatives as acetylcholinesterase and carbonic anhydrase inhibitors: Synthesis, molecular docking, and solubility. *Archives of Biochemistry and Biophysics, 771*, 110515. **https://doi.org/10.1016/j.abb.2025.110515**

Evangelista, T. C. S., López, O., Puerta, A., Fernandes, M. X., Ferreira, S. B., Padrón, J. M., Bolaños, J. G. F., Sydnes, M. O., & Lindbäck, E. (2022). A hybrid of 1-deoxynojirimycin and benzotriazole induces preferential inhibition of butyrylcholinesterase (BuChE) over acetylcholinesterase (AChE). *Journal of Enzyme Inhibition and Medicinal Chemistry, 37*, 2395–2402. **https://doi.org/10.1080/14756366.2022.2117912**

Gadhave, D. G., Sugandhi, V. V., Jha, S. K., Nangare, S. N., Gupta, G., Singh, S. K., Dua, K., Cho, H., Hansbro, P. M., & Paudel, K. R. (2024). *Ageing Research Reviews*, 99, 102357. **https://doi.org/10.1016/j.arr.2024.102357**

Guo, H. Y., Chen, Z.A., Shen, Q. K., & Quan, Z.S. (2021). Application of triazoles in the structural modification of natural products. *Journal of Enzyme Inhibition and Medicinal Chemistry,* 36, 1115-1144. **https://doi.org/10.1016/j.arr.2024.102357**

Hosseini, S., Pourmousavi, S. A., Mahdavi, M., & Taslimi, P. (2022). Synthesis, and *in vitro* biological evaluations of novel naphthoquinone conjugated to aryl triazole acetamide derivatives as potential anti-Alzheimer agents. *Journal of Molecular Structure, 1255*, 132229. **https://doi.org/10.1016/j.molstruc.2021.132229**

Huo, J., Hu, H., Zhang, M., Hu, X., Chen, M., Chen, D., Liu, J., Xiao, G., Wanga, Y., & Wen, Z. (2017). A mini review of the synthesis of poly-1,2,3-triazole-based functional materials. *RSC Advances,7*, 2281–2287. **https://doi.org/10.1039/c6ra27012c**

Iraji, A., Sharghei-Boroujeni, D., Aliabadi, A., Dara, M., Hashempur, M. H., Hariri, R., Khassaki, D. A. M., Saeedi, M., & Akbarzadeh, T. (2025). Design and synthesis of new substituted benzyl-1,2,3-triazole-methoxybenzylidene coupled to arylacetohydrazides against Alzheimer's disease. *Journal of Molecular Structure, 1348*, 143337. **https://doi.org/10.1016/j.molstruc.2025.143337**

Janowska, S., Holota, S., Lesyk, R., & Wujec, M. (2024). Aromatase inhibitors as a promising direction for the search for new anticancer drugs. *Molecules, 29*(2), 346. **https://doi.org/10.3390/molecules29020346**

Kamatham, P. T., Shukla, R., Khatri, D. K., & Vora, L. K. K. (2024). Pathogenesis, diagnostics, and therapeutics for Alzheimer's disease: Breaking the memory barrier. *Ageing Research Reviews. 101*, 102481. https://doi.org/10.1016/j.arr.2024.102481

Karaali, N. U., Akyüz, G., Uluçay, B., & Emirik, M. (2025). Aromatic carbothioamide-, triazole-, thiadiazole-, and thiazolidinone-derived benzimidazoles: Synthesis, characterization, urease and AChE inhibition properties, and docking study. *Journal of Heterocyclic Chemistry*, 1–11. **https://doi.org/10.1002/jhet.70117**

Khan, S., Hussain, R., Khan, Y., Iqbal, T., Khan, M. B., Ahmary, K. M. A., & Mhyawi, S. R. A. Insight into role of triazole derived Schiff base bearing sulfonamide derivatives in targeting Alzheimer's disease: Synthesis, characterization, in vitro and in silico assessment. *Journal of Molecular Structure*, 1315, 138845. **https://doi.org/10.1016/j.molstruc.2024.138845**

Khan, S., Hussain, R., Khan, Y., Iqbal, T., Chinnam, S., Akif, M., & Ayed, A. D. (2024). Hybrid benzothiazole derived fused triazole/thiazole derivatives as versatile anti-Alzheimer agents: Synthesis, characterization, biological evaluation and molecular docking studies. *Journal of Molecular Structure, 1318*, 139200. **https://doi.org/10.1016/j.molstruc.2024.139200**

Khan, Y., Rehman, W., Hussain, R., Khan, S., Malik, A., Khan, M., Liaqat, A., Rasheed, L., Begum, F., Fazil, S., Khan, I., & Latif, M. H. A. (2022). New biologically potent benzimidazole-based-triazole derivatives as acetylcholinesterase and butyrylcholinesterase inhibitors along with molecular docking study. *Journal of Heterocyclic Chemistry, 59*, 2225–2239. **https://doi.org/10.1002/jhet.4553**

Kharb, R., Sharma, P.C., & Yar, M.S. (2010). Pharmacological significance of triazole scaffold. *Journal of Enzyme Inhibition and Medicinal Chemistry, 26*, 1–21. **https://doi.org/10.3109/14756360903524304**

Liu, W., Deng, W., Gong, X., Ou, J., Yu, S., & Chen, S. (2021). Global burden of Alzheimer's disease and other dementias in adults aged 65 years and over, and health inequality related to SDI, 1990–2021: Analysis of data from GBD 2021. *BMC Public Health, 25*, 1256. **https://doi.org/10.1186/s12889-025-22378-z**

Liu, W., Tian, L., Wu, L., Chen, H., Wang, N., Liu, X., Zhao, C., Wu, Z., Jiang, X., Wu, Q., Xu, Z., Liu, W., & Zhao, Q. (2022). Discovery of novel β-carboline-1,2,3-triazole hybrids as AChE/GSK-3β dual inhibitors for Alzheimer's disease treatment. *Bioorganic Chemistry, 129*, 106168. **https://doi.org/10.1016/j.bioorg.2022.106168**

Matin, M. M., Matin, P., Rahman, M. R., Hadda, T. B., Almalki, F. A., Mahmud, S., Ghoneim, M. M., Alruwaily, M., & Alshehri, S. (2022). Triazoles and their derivatives: Chemistry, synthesis, and therapeutic applications. *Frontiers in Molecular Biosciences*, 9, 864286. **https://doi.org/10.3389/fmolb.2022.864286**

Medetalibeyoglu, H., Atalay, A. U., Saglamtas, R., Manap, S., Ortaakarsu, A. B., Ekinci, E., Yüksek, H., & Tüzün, B. (2025). Synthesis, design, and cholinesterase inhibitory activity of novel 1,2,4-triazole Schiff bases: A combined experimental and computational approach. *International Journal of Biological Macromolecules, 306*, 141350. **https://doi.org/10.1016/j.ijbiomac.2025.141350**

Medetalibeyoğlu, H., Tüzün, B., Atalay, A., Ortaakarsu, A.B., Sağlamtaş, R., Manap, S., Akyıldırım, O., & Yüksek, H. (2025). Novel triazole-based N-acetyl Schiff bases: Synthesis, computational analysis, and acetylcholinesterase and butyrylcholinesterase inhibitory potential for Alzheimer's disease. *Journal of Molecular Structure*, 143176. **https://doi.org/10.1016/j.molstruc.2025.143176**

Mishra, K. B. (2025). 1,5-Disubstituted 1,2,3-triazoles: Molecular scaffolds for medicinal chemistry and biomolecular mimetics. *European Journal of Medicinal Chemistry, 291*, 117614. **https://doi.org/10.1016/j.ejmech.2025.117614**

Mishra, P., Kumar, A., & Panda, G. (2019). Anti-cholinesterase hybrids as multi-target-directed ligands against Alzheimer's disease (1998-2018). *Bioorganic and Medicinal Chemistry, 27*(6), 895–930. **https://doi.org/10.1016/j.bmc.2019.01.025**

Onur, S., Çesme, M., & Tümer, F. (2025). Exploring the multi-faceted potential of novel p-cresol-triazole hybrids: Anticancer, anticholinesterase, and antioxidant properties with in silico insights. *Bioorganic Chemistry, 164*, 108916. **https://doi.org/10.1016/j.bioorg.2025.108916**

Ozcan, I., Alici, H., Taslimi, P., & Tahtaci, H. (2025). Novel 1,2,4-triazole-derived Schiff base derivatives: Design, synthesis, and multi-enzyme targeting potential for therapeutic applications. *Bioorganic Chemistry, 157*, 108246. **https://doi.org/10.1016/j.bioorg.2025.108246A**

Pourabdi, L., Küçükkılınç, T. T., Khoshtale, F., Ayazgök, B., Nadri, H., Alashti, F. F., Forootanfar, H., Akbari, T., Shafiei, M., Foroumadi, A., Sharifzadeh, M., Ardestani, M. S., Abaee, M. S., Firoozpour, L., Khoobi, M., & Mojtahedi, M. M. (2022). Synthesis of new 3-arylcoumarins bearing N-benzyl triazole moiety: Dual lipoxygenase and butyrylcholinesterase inhibitors with anti-amyloid aggregation and neuroprotective properties against Alzheimer's disease. *Frontiers in Chemistry, 9*, 810233. **https://doi.org/10.3389/fchem.2021.810233**

Rahim, F., Ullah, H., Taha, M., Hussain, R., Sarfraz, M., Iqbal, R., Iqbal, N., Khan, S., Shah, S. A. A., Albalawi, M. A., Abdelaziz, M. A., Alatawi, S. F., Alasmari, A., Sakran, M. I., Zidan, N., Jafri, I. & Khan, K. M. (2023). Synthesis of new triazole-based thiosemicarbazone derivatives as anti-Alzheimer's disease candidates: Evidence-based in vitro study. *Molecules, 28*, 21. **https://doi.org/10.3390/molecules28010021**

Rajesab, P., Mathada, B. S., Niranjan, V., Sinha, L., Setlur, A. S., Maurya, A., Chandrashekar, K., & Prasad, O. (2025). Unveiling a novel thiazolo[2,3-c][1,2,4]triazole scaffolds as a dual cyclooxygenase and cholinesterase inhibitors: Spectral analysis, DFT calculations, in vitro and in silico studies. *Journal of Molecular Structure, 1345*, 143166. **https://doi.org/10.1016/j.molstruc.2025.143166**

Rani, A., Singh, A., Kaur, J., Singh, G., Bhatti, R., Gumede, N., Kisten, P., Singh, P., & Kumar, S. V. (2021). 1*H*-1,2,3-triazole grafted tacrine-chalcone conjugates as potential cholinesterase inhibitors with the evaluation of their behavioral tests and oxidative stress in mice brain cells. *Bioorganic Chemistry, 114*, 105053. https://doi.org/10.1016/j.bioorg.2021.105053

Rashdan, H.R.M., & Shehadi, I.A. (2022). Triazoles Synthesis & Applications as Nonsteroidal Aromatase Inhibitors for Hormone-Dependent Breast Cancer Treatment. *Heteroatom Chemistry,* 1-16. https://doi.org/10.1155/2022/5349279

Sang, Z., Wang, K., Dong, J., & Tang, L. (2022). Alzheimer's disease: Updated multi-targets therapeutics are in clinical and in progress. *European Journal of Medicinal Chemistry, 238*, 114464. **https://doi.org/10.1016/j.ejmech.2022.114464**

Siddiqui, S. Z., Arfan, M., Abbasi, M. A., Rehman, A., Shah, S. A. A., Parveen, R., Ashraf, M., Solangi, M., Hussain, S., & Khan, K. M. (2023). Design, synthesis of triazole-based scaffolds, N-(substitutedphenyl)-2-(5-(4-methoxyphenyl)-4-phenyl-4H-1,2,4-triazol-3-ylthiol)acetamides: As potential anti-cholinesterase agents for neurodegenerative diseases. *Journal of Molecular Structure, 1289*, 135885. **https://doi.org/10.1016/j.molstruc.2023.135885**

Sobha, A., Ganapathy, A., Mohan, S., Madhusoodanan, N., Babysulochana, A. D., Alaganandan, K., & Somappa, S. B. (2024). Novel small molecule-based acetylcholinesterase (AChE) inhibitors: From biological perspective to recent developments. *European Journal of Medicinal Chemistry Reports, 12*, 100237. **https://doi.org/10.1016/j.ejmcr.2024.100237**

Song, M. X., & Deng, X. Q. (2018). Recent developments on triazole nucleus in anticonvulsant compounds: A review. *Journal of Enzyme Inhibition and Medicinal Chemistry, 33*, 453–478. **https://doi.org/10.1080/14756366.2017.1423068**

Tan, L. J., Lei, W. J., Liu, M. M., Cai, Z. D., Jiang, H. L., Liu, R., & Li, Z. R. (2024). Discovery of cinnamamide/ester triazole hybrids as potential treatment for Alzheimer's disease. *Bioorganic Chemistry*, 150, 107584. **https://doi.org/10.1016/j.bioorg.2024.107584**.

Thuy, G. L. N., Thi, N. N., The, H. P., Thi, T. A. D., Thi, H. N., Thi, T. H. N., Hoang, S. N., Nguyen, T. V. (2020). Synthesis and biological evaluation of novel quinazoline-triazole hybrid compounds with potential use in Alzheimer's disease. *Bioorganic & Medicinal Chemistry Letters, 30*, 127404. **https://doi.org/10.1016/j.bmcl.2020.127404**

Werner, A., & Edm, S. (1899). On nitro derivatives of azo-, azoxy- and hydrazo-benzene. *Reports of the German Chemical Society, 32*(3), 3256–3282.

10 Thiazoles as Emerging Therapeutic Agents Against Neurodegenerative Disorders

Ismat Ullah Khan, Naveed Ullah, Aamer Saeed, and Hamid Aziz

10.1 INTRODUCTION

Neurodegenerative disorders (NDs) are one of the most significant categories of chronic, progressive diseases that can be defined as the selective degeneration of the neuronal structure and functioning of the central nervous system (CNS). The disorders involve Alzheimer's disease (AD) (Shiradkar et al., 2007), Parkinson's disease (PD), Huntington's disease (Ghotbi et al., 2020) and amyotrophic lateral sclerosis (Bensimon et al., 1994) and are linked to debilitating cognitive, motor and behavioral disabilities that can significantly lower the quality of life and escalate the socio-economic burden of the world (Selkoe & Hardy, 2016; Querfurth & LaFerla, 2010). Neurodegenerative diseases have been on the increase, especially because of the longer life expectancy and the growing aging populations, and in the context of this, there is AD that has affected more than 50 million people around the world and is expected to triple this number by 2050 (Butterfield & Halliwell, 2019). Although there has been a lot of research, the therapeutic interventions that are presently used are mostly symptomatic and do not provide much of an effect in stopping or reversing the neuronal degeneration. This shows the fact that new treatment approaches to address the pathological processes of neurodegeneration are urgently needed.

The pathophysiology of NDs is deemed to be complex and multifactorial. Some of the major mechanisms involved in the pathogenesis of the disease are oxidative stress, excitotoxicity, protein misfolding, aggregation, neuroinflammation, poor neurotransmitter signaling, and malfunctioning of mitochondria (Poewe et al., 2017). As an example, amyloid-β plaques and hyperphosphorylated tau tangles in AD, or α-synuclein aggregates in PD, are examples of the role of abnormal protein homeostasis in neuronal death (Schapira & Jenner, 2011). Small-molecule therapies, e.g. cholinesterase inhibitors in AD or dopaminergic agents in PD, only treat or alleviate symptoms and do not alter the course of the disease. Therefore, more and more attention is paid to the discovery and development of disease-modifying agents that can act on several pathogenic pathways at the same time.

Over the past decades, heterocyclic compounds have become an efficient family of bioactive molecules with high therapeutic potential in a diverse variety of diseases. The heterocycles (HCs), where at least one atom in the cycle is not a carbon atom, exhibit a variety of pharmacological and chemical characteristics, which render them a useful construct to base the design of drug (Ross & Tabrizi, 2011). Of them, the thiazoles, which are a five-membered ring system, including both nitrogen and sulfur atoms, have been a particular object of interest due to their large array of bio-potential activities, such as antimicrobial, anticancer, anti-inflammatory and neuroprotective effects (Hardiman et al., 2017). The selective approach of interaction with various biological targets, such as enzymes, receptors and nucleic acids, is possible because thiazoles have an electronic configuration and a structural rigidity, allowing them to selectively react with biological targets. Similarly, thiazole has good potential in the management of neurodegenerative diseases.

DOI: 10.1201/9781003772019-10

A number of research have been indicating the possibility of thiazole derivatives in regulating major pathological pathways involved in NDs. These compounds have been observed to have antioxidant activity which inhibits the attack of reactive oxygen species and oxidative destruction of neuronal cells. Additionally, thiazoles have also been shown to be protein aggregation inhibitor, including amyloid-β, α-synuclein which is central to the pathogenesis of Alzheimer's and PDs, respectively. Additionally, other thiazole analogs possess anti-inflammatory properties because of the suppression of the release of pro-inflammatory cytokines, and this lowers neuroinflammatory processes which may lead to the death of neurons (Heneka et al., 2015). These complicated mechanisms concentrate on the curative potential of thiazoles as a neuroprotectant agent as an alternative to traditional monotherapeutic forms of treatment.

The purpose of this chapter is to discuss novel therapeutic properties of thiazoles against NDs. It will cover the chemistry of thiazoles, structure/activity relation of the compounds, mechanistic insights to their neuroprotective properties and preclinical studies that demonstrate the effectiveness of the compounds. The chapter will also discuss the existing challenges in the transfer of such findings to clinical practice and analyze the future perspective of drug development based on thiazole. This chapter aims to demonstrate the possibility of using thiazoles as an effective form of disease-modifying therapy in neurodegenerative diseases by considering chemical, biological and pharmacological knowledge.

10.2 CHEMISTRY OF THIAZOLES

Thiazoles refer to five-membered aromatic HCs that are nitrogen- and sulfur-based which have two isomers, i.e., 1,2-thiazole (isothiazole) and 1,3-thiazole (thiazole). The thiazoles have nitrogen and sulfur at the 1,3-position of the ring as mentioned in Figure 10.1 (**1, 2**). On the same note, the thiadiazole ligand has four forms of isomers (Ayati et al., 2015; Maradiya & Patel, 2003; Mishra et al., 2015).

HCs based on thiazole are available in abundance in both marine and terrestrial micro-organisms, natural products and certain pharmaceutically useful skeletons. They are present in several natural products and biologically important alkaloids such as bacillamide B (**3**) and nostocyclamide (**4**) (Figure 10.2) (Mishra et al., 2015).

Thiazoles are ideal constituent of thiamine (vitamin B1, **5**, Figure 10.3), which is needed during the production of acetylcholine to allow CNS to work normally (Hutchinson et al., 2002).

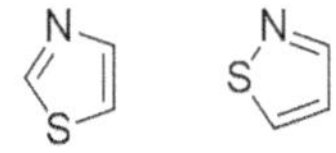

1 **2**

FIGURE 10.1 Structures of 1,3-thiazole (**1**), 1,2-thiazole (**2**).

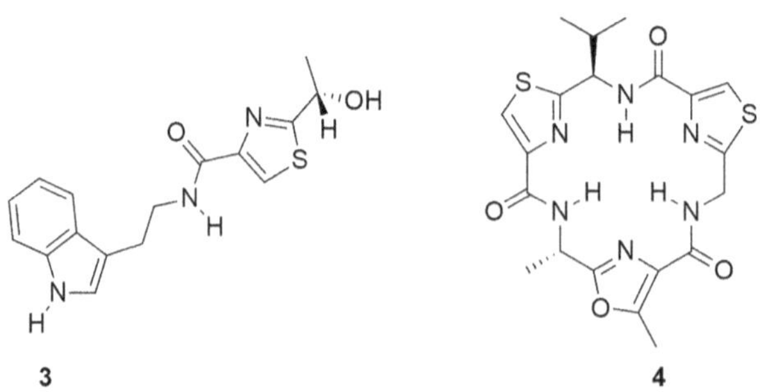

3 **4**

FIGURE 10.2 Structures of bacillamide B (**3**) and nostocyclamide (**4**).

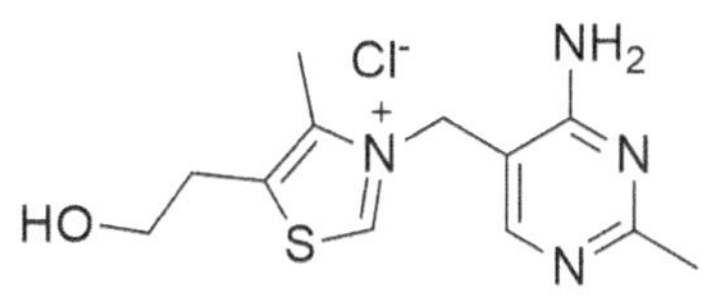

5

FIGURE 10.3 Structure of vitamin B1 (**5**).

10.2.1 Synthesis of 1,3-thiazoles

Review of literature indicates that there are many synthetic pathways to synthesize the thiazole ring. Among all of these, the most typical, effective and simple strategies of synthesis are given below.

10.2.1.1 Hantzsch Synthesis

The efficient synthesis of thiazole was first reported in 1887 by Hantzsch who reacted 3-haloketone or an aldehyde with thioamides (Ayati et al., 2015; Gupta et al., 2013). This simple synthetic methodology can also be carried out with thousands of ammonium dithiocarbonate in place of thioamides, as illustrated in **Schemes 10.1–10.2**.

The most common thiazole synthesis strategy developed by Hantzsch is the one that takes place through condensation and cyclization of ketones with α-halo substituents and thioamides.

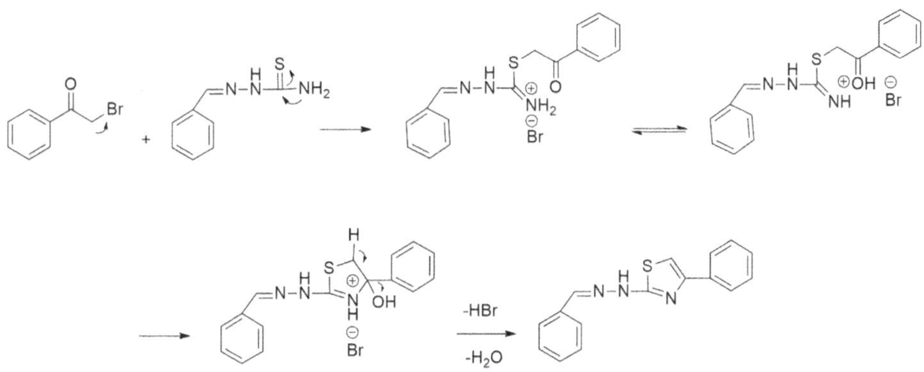

SCHEME 10.1 Thiazole synthesis by Hantzsch's method using thioamides and α-haloketones.

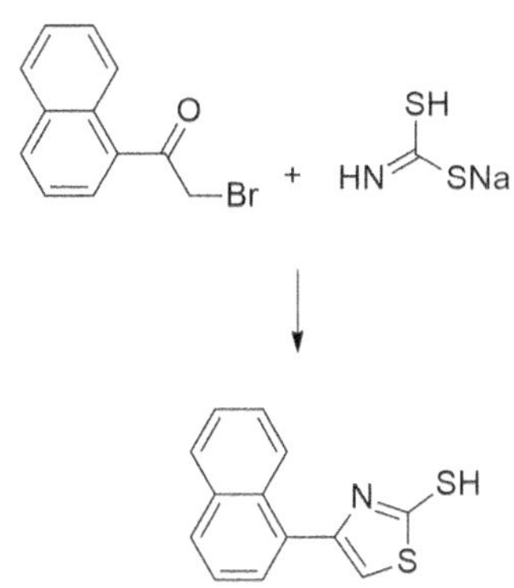

SCHEME 10.2 The preparation of thiazole by Hantzsch's method using α-haloketone and ammonium dithiocarbonate.

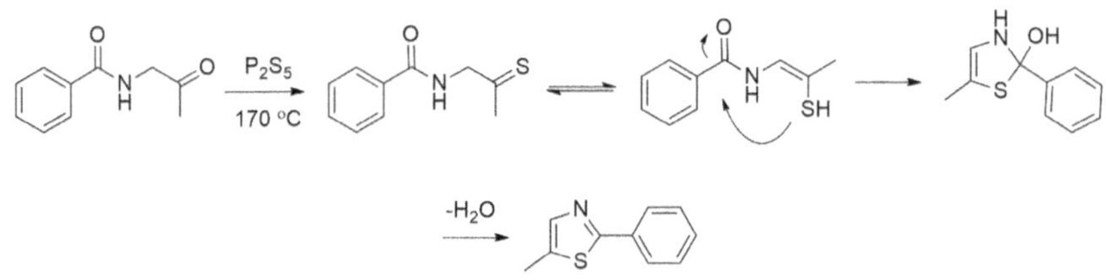

SCHEME 10.3 Synthesis of thiazole by the Gabriel method with the help of α-acylamino ketones and P_2S_5.

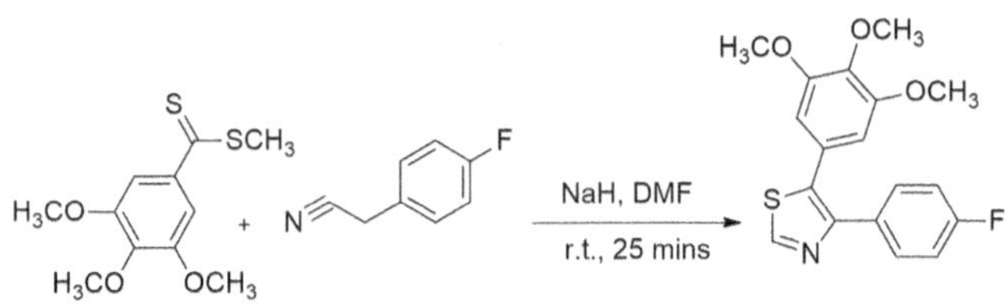

SCHEME 10.4 Formation of thiazole by use of isocyanides, ketones and arenecarbodithioates.

This reaction starts off with the attack of a sulfur atom of thioamide upon the α-carbon atom of α-haloketone that leads to the production of α-thioketone intermediate. Subsequently, the process is followed by dehydration that form the respective thiazole ring (H. Aziz, 2026).

10.2.1.2 Gabriel Synthesis

The synthesis of 2,5-disubstituted thiazoles is done through Gabriel synthesis. This is done by reacting amino acetals or α-acylamino ketones and P_2S_5 (phosphorus penta-sulfide) in stoichiometric proportions. The process is illustrated in **Scheme 10.3** (Gupta et al., 2013; Ayati et al., 2015).

10.2.1.3 Click Synthesis

Cyclization of isocyanides containing arenecarbodithioates and active methylene leads to an effective and simplistic synthesis of substituted thiazoles, as shown in **Scheme 10.4**. In this process the base to be employed is sodium hydride so as to produce 4,5-disubstituted thiazoles. This low-cost, high-efficient and fast synthesis is carried out under mild conditions of reaction and is more yielding (Lingaraju et al., 2012).

10.2.1.4 Domino Synthesis

In the presence of microwave when thioureas underwent a domino alkylation and cyclization reaction, the result was 2-aminothiazoles, as illustrated in **Scheme 10.5**. This is synthesized in the presence of K_2CO_3 as a base to carry out the reaction within a short duration and to produce good yields (Castagnolo et al., 2009).

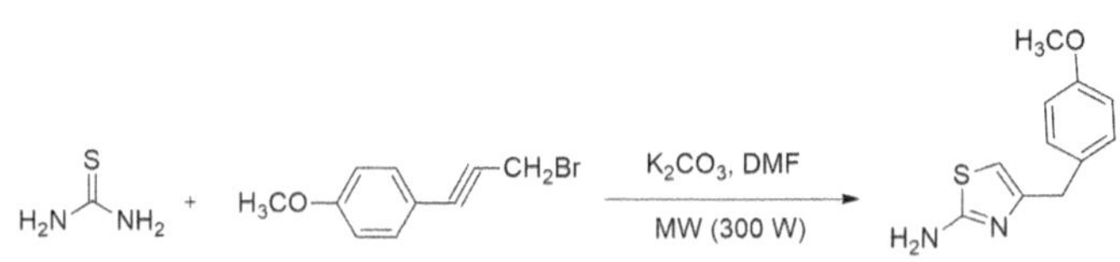

SCHEME 10.5 Synthesis of thiazoles by the domino method with the help of microwave radiations.

10.3 THIAZOLE DERIVATIVES AGAINST DEGENERATIVE DISEASES

10.3.1 1,3-Thiazole Base Compounds in Alzheimer's Disease

AD is a life-threatening neurodegenerative illness in which a sufferer experiences severely impaired cognitive functions, memory as well as incapability to operate in the chronic phases. This disease is largely caused by cell death of neurons which is caused by a large number of factors. It has been demonstrated that the brain of patients with AD contains low concentrations of acetylcholine which can be caused by the buildup of protein, called beta-amyloid (Aβ) fragments. AD is also associated with chemical modification in tau proteins that aid in the maintenance of the microtubules and cause cell death. In patients with AD, oxidative stress is also known to have a precarious role in the death of brain cells. Chemical, toxin, pesticide and trauma damage might also result in inflammation which may also be another causative agent of AD (Wollen, 2010; Zhang et al., 2012).

AChE (acetylcholinesterase) inhibitors, JNK-3 (c-Jun N-terminal kinase-3) inhibitors, NMDA (N-methyl-D-aspartate) antagonists, anti-inflammatory drugs, antihypertensive drugs and secretase inhibitors are various drugs that are currently used in the treatment of AD (Wollen, 2010). Thiazole scaffold attracted several research groups working in AD to identify active agent against AD. Zhuang et al. designed new ligands, **6** and **7**, which were novel thiazole derivatives and selective Aβ aggregate ligands (Zhuang et al., 2001). Both compounds had high binding affinity determined in vitro having Kd of 0.13 and 0.06 nM of aggregates of 0.15 and Aβ (1-40) nM of aggregates of Aβ (1-42), respectively. Autoradiography analysis indicated that ligands **6** and **7** were labeled well in the post-mortem brain fragments of patients with Down syndrome in which there are mainly aggregates of Aβ (1-42). Moreover, experiments of biodistribution stated that compounds **6** and **7** were found highly brain uptake (1.89 and 1.57% dose/organ, respectively) and retained in typical mice. To sum up, these thiazole ligands labeled with radioisotopes would also serve as a good-quality biomarker to develop Aβ (1-42) and Aβ (1-40) aggregates of amyloidogenesis in Alzheimer's patients.

Shiradkar et al. used methods of microwave irradiation to produce a library of triazolyl tethered thiazoles and screened them based on their ability to inhibit cdk5/p25 (Shiradkar et al., 2007). Structure–activity relationship (SAR) analysis of the structure showed that change in the structure of screening at C-2 of thiazole ring and thiol plus amino group present in triazole core led to the detection of strong cdk5/p25 inhibitor compounds **8** as well as **9**. Both compounds **8** and **9** exhibited IC_{50} of 42 nM and 30 nM correspondingly compared to screening hit **10** that had an IC_{50} of 48 nM. Further, compounds **8** as well as **9** were selective to cdk2 1.2 and 17-fold.

SAR found that both length and flexibility of the linker played an essential role in binding to GSK-3β and AChE. It is worth noting that compound **11** was a lead compound (IC_{50} = 22.2 nM in GSK-3β and 1.2 nM in AChE) exhibiting good blood–brain barrier (BBB) absorptivity, selective kinase inhibition (especially DYRK1A IC50 = 28.3 nM) and anti-inflammatory effects in neurons. Cyclopropylmethyl ether groups and carboxamide with methyl group in **11** played key roles in its strong dual/triple-target action. Its effectiveness in preventing tau pathology was confirmed by biological assay as a triple-action therapeutic agent that acts on both pathological (Aβ plaques, tau tangles) and symptomatic (cholinergic dysfunction) deficits of AD. Phosphorylation and cognitive enhancement of a mouse model of scopolamine induced by the Morris water maze. These results have suggested the potential of **11** as a triple-action therapeutic agent to treat the pathological hallmarks (tau tangles and Aβ plaques), as well as symptomatic deficits (cholinergic dysfunction) of AD.

A number of derivatives of imidazo[2,1-b]thiazole produced by Andreani et al. were tested with chemiluminescent microassay on their AChE and BuChE inhibitory capacities (Andreani et al., 2008). The findings revealed that compounds **12**, **13** and **14** had selective AChE-inhibitory properties with the IC_{50} of 60, 4.2 and 35 mM, correspondingly. These agents were found to be weak when compared to the conventional AChE inhibitors, tacrine and donepezil.

Cui et al. developed some new benzothiazoles combined with bithiophenes as potential β-amyloid probes (Cui et al., 2010). These compounds had a high binding affinity against the Aβ aggregates with a Ki range of 0.11 and 4.64 nM. The in vitro fluorescent staining experiment shows that these compounds were found to stain the plaques of Aβ in the sections of the brain of APP/PS1 transgenic mice. Two of the strongest compounds, **15** [Ki 1/4 0.61 nM with Aβ (1-42) and 0.11 nM with Aβ (1-40)] and **16** [Ki 1/4 0.31 nM with Aβ (1-42) and 0.25 nM with Aβ (1-40)], were effectively iodinated with radioiodine to continue with the further studies of both advanced in vitro and in vivo research. In the study of in vitro autoradiography of fragments of the brain sections, compounds [^{125}I] **15** and [^{125}I] **16** demonstrated good binding with the plaques of Aβ. The result of the in vivo biodistribution study revealed [^{125}I] **16** displayed strong brain uptake of 3.42% ID/g at 2 min and brain clearance (0.53% ID/g at 60 min), whereas [^{125}I] **15** displayed low brain uptake (0.87% ID/g at 2 min). [^{125}I] **16** may be investigated further as in vivo imaging agent to differentiate plaques of *A*β in patients with AD.

A one-step preparation of 2-aminothiazoles has been reported by Lagoja et al. that demonstrated respectable coverage of neuronal toxicity caused by tau in a neuronal cell line of human beings (Lagoja et al., 2011). SAR study showed that substitution of R_1 (methyl group) with the phenyl ring decreased effectiveness by five times (**17**, IC_{50} 1/4 90 ng/mL and **18**, IC_{50} 1/4 20 ng/mL). It was also observed that the substitution of methyl R_2 with an isopropyl group gave an even stronger inhibitor **19** (IC_{50} 1/4 4 ng/mL).

Thiazole derivatives have also been synthesized in a series of antagonists of RAGE to treat AD (Lee et al., 2012). SAR research indicates that derivatives of alkylamino group exhibited good inhibitory effect with regard to Aβ-RAGE binding. Nevertheless, derivative **20** lacking alkylamino group also responded with favorable inhibiting ability on binding the Aβ-RAGE with IC_{50} value of 1.21 mM (percentage inhibition at 10 mM1/4 54.7%). Moreover, **20** also prevented the Aβ transport across BBB (percentage inhibition 1/4 68.6) and also prevented the Aβ-associated NF-κB activation by inhibiting the Aβ-RAGE binding.

Huang et al. designed, synthesized and tested berberine-phenylbenzothiazole and tacrine-phenylbenzothiazole hybrid molecules as multifunctional anti-Alzheimer agents. The most active inhibitor of AChE was tacrine conjugated with phenylbenzothiazole by the 3-carbon spacer (compound **21**) with an IC_{50} of 0.017 mM and seven times selectivity over BuChE. Compound **21** contained the same Aβ aggregation-inhibiting ability as that of curcumin (51.8 vs 52.1% at 20 mM, respectively). Molecular docking and kinetic experiments indicated that compound **21** bound to gorge, peripheral and central pockets of AChE simultaneously (Huang et al., 2012). According to research studies, the compounds can be used as templates to produce more anti-AD agents with multifunctional properties in the future.

A series of 16 new thiazole-styryl hybrids were reported by Cacciatore which were strategically designed following the combination of cinnamoyl and thiazole pharmacophores in a single molecular framework, targeting several hallmarks of the pathophysiology of AD (Gouleni et al., 2023).

All these compounds were prepared through the use of a multistep pathway that used amide **22** derived from cinnamic acid, thionation through P_2S_5, then thiazole cyclization with α-halo carbonyl intermediates to synthesize thiazole **23**. Structures **24** and **25** were identified as main candidates among the library, which exhibited strong neuroprotective action in SH-SY5Y cells treated with Aβ (1-42) oligomers that improved cell capability by 30–34%. These compounds were strong and dose-dependent inhibitors of AChE and β-secretase (BACE1) with 196 being better at inhibiting AChE (IC50 = 63.6 ug/mL) and **24** being better at inhibiting BACE1 (IC50 = 15.2 ug/mL). Moreover, **25** had the best anti-amyloidogenic activity and inhibited more than 80% of the Aβ aggregation, which was verified through the use of Congo red staining. Hoechst stain and flow cytometry also confirmed that both compounds did not induce genotoxicity but rather lowered Aβ-induced necrosis and apoptosis. Their high binding affinity with AChE and BACE1 was supported by molecular docking. Physicochemical profiling in silico indicated that **25** is a better drug-like molecule than **24**, but they both need optimization to increase BBB absorptivity and aqueous solubility (Chaudhari et al., 2025) (Figure 10.4).

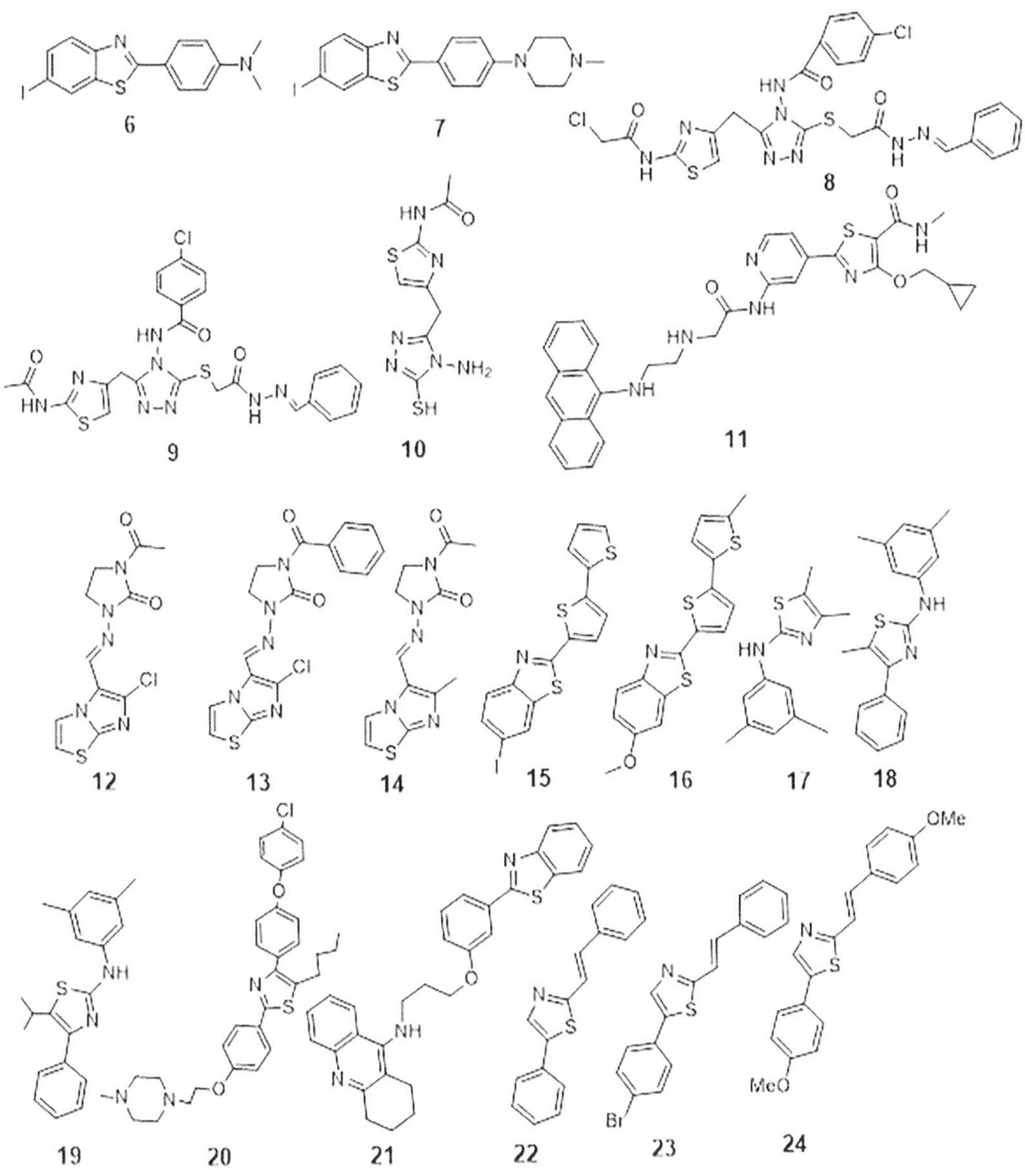

FIGURE 10.4 Structures of thiazole derivative with anti-Alzheimer activities.

10.3.2 1,3-Thiazole Derivatives in Parkinson's Disease (PD)

A number of thiazole-containing scaffolds have been explored as being useful in targeting important enzymes and disease-related pathological procedures in PD.

PD after the AD is the second most common disease, which is a progressive degeneration of dopaminergic neurons located in the substantia nigra pars compacta, as well as pathological deposition of misfolded α-synuclein masses in the form of Lewy bodies (Dorsey et al., 2018; Poewe et al., 2017). The current developments in the PD area have contributed to a more profound knowledge of pathophysiology in this condition, underlining mitochondrial dysfunction, oxidative stress or neuroinflammation as the main factors (Henrich et al., 2023; Savelieff et al., 2018).

The 1,3-thiazole core because of its aromatic nature, hydrogen-bonding ability and the possibility of π–π stacking has consequently become a popular scaffold of privilege in medicinal chemistry (Das et al., 2015; Ruankham et al., 2025). Most of the 1,3-thiazole analogs possess dopaminergic neuroprotective properties, MAO-B inhibitory properties, α-synuclein aggregation inhibitory properties and cellular antioxidant defense activation, including the Nrf2/HO-1 pathway, making them dual symptomatic and disease-modifying. The first thiazole–pyridine–fused compounds were investigated by Azam et al. (2010) to counter oxidative stress and catalepsy triggered by neuroleptics in mice. Structural verification was done on all the synthesized derivatives by spectroscopy and elemental analysis, and anti-Parkinsonian and neuroprotective screening was performed. The majority of compounds exhibited observable activity with the most significant protection in the case of compound **26**. The observations of structure–activity indicated that propyl replacement at position 3 contributed to better anti-Parkinsonian activity than other alkyl and aromatic replacements (Azam et al., 2010).

These findings are consistent with recent findings that thiazole hybrids with benzothiazole, quinoline or chalcone cores can be highly active in MAO-B inhibition and oxidative cascade inhibition involved in dopaminergic cell damage. The MAO-B a proven PD drug target-inhibition assists in the reduction of dopamine breakdown and ROS production, which is further beneficial in the therapeutic importance of early thiazole–pyridine–fused designs.

Subsequently, Yurttaş et al. (2013) prepared a series of piperazine–thiazole hybrids, namely 2-(4-substituted piperazin-1-yl)-N-[4-(2-methylthiazol-4-yl)phenyl]acetamides and used them as acetylcholinesterase (AChE) and butyrylcholinesterase (BChE) inhibitors based on the Ellman method. The assays were performed in the presence of phosphate buffer (25°C, pH 8.0), and the substrates were butyrylthiocholine (BTC) and acetylthiocholine (ATC), and donepezil was used as a positive control. Compounds **27** and **28** were found to be of high potency as AChE inhibitors and similar to donepezil; hence, it was suggested that those were relevant CNS-acting cholinergic modulators (Yurttaş et al., 2013).

The complementary role of cholinergic effect with that of other modern thiazole derivatives is that the latter have been shown to have multitarget profiles, such as antioxidant activity and anti-inflammatory activity. These molecules represent a multitarget directed ligand (MTDL) paradigm, which is more popular in PD drug discovery, as well as modulating neurotransmission and, at the same time, decrease pathological oxidative stress.

Turan-Zitouni et al. (2018) synthesized N-[1-(((3,4-diphenylthiazol-2(3H)-ylidene)amino)methyl) cyclopentyl]acetamide derivatives as substrates to assess their activity against the isoenzymes MAO-A and MAO-B. Compounds **29–32** had comparable activities to the reference drug moclobemide (IC_{50} = 6.061 ± 0.262 μM), with values of 7.06 ± 0.18, 6.56 ± 0.20, 6.78 ± 0.15 and 7.09 ± 0.17 μM, respectively. Compound **30** was the strongest of them and was a reversible and non-competitive inhibitor of both isoforms, which was corroborated by docking findings (Turan-Zitouni et al., 2018).

The reversible and selective inhibition of MAO-B of such type is a key component of PD treatment since it increases the levels of dopamine and decreases oxidative stress. Also, the affinity of MAO-B has been demonstrated to be tuned by structural changes of the thiazole core, including α-aryl and electron-donating substitutions, as well as auxiliary HCs, and offers the chance to maximize potency and BBB penetration.

In another contribution of importance, Secci et al. (2019) evaluated a group of 4-(3-nitrophenyl) thiazol-2-yl hydrazone analogs regarding antioxidant activity and selective hMAO-B inhibition. All products of the syntheses were tested against recombinant hMAO-A and hMAO-B. The strongest hMAO-B inhibition was manifested with compounds **33** (completely reversible) and **34** (partially reversible). It was predicted that the pharmacokinetics would be orally absorbable and exhibit CNS penetration similar to known CNS-active drugs, confirming therapeutic potential (Secci et al., 2019).

These results are consistent with the currently growing body of evidence that thiazole-based antioxidants have the ability to reduce α-synuclein fibrillization, preserve mitochondrial membrane potential and regulate Nrf2 signaling, providing multi-pathway neuroprotection.

Dutta and coworkers also broadened the therapeutic address of the thiazole-linked pharmacophores, making multifunctional dopamine D_2/D_3 receptor agonists, especially the compound **35**. They were able to provide improved selectivity, effectiveness and drug-like properties through the strategic SAR-guided change of the D3 agonist D-264 (**35**). One of them, (–)-**36**, exhibited high D_3 activity (EC_{50} = 0.10 nM) and a D_2/D_3 selectivity ratio of 159 and served as a complete agonist in [^{35}S]GTPγS functional reactions. In vivo, (–)-**36** also reversed Parkinsonian symptoms in reserpine-treated and 6-OHDA-lesioned rat models and also proved to be strong in neuroprotection against MPP^+ toxicity in MN9D cells. This stereochemistry of the molecule was important, whereas the opposite enantiomer was far less potent. Final analogs (**37–40, 36, 41–43**) were readily obtained by the synthetic sequence of t-Boc protection, Suzuki coupling, reductive amination and demethylation. Overall, (–)-**36** is set as a very selective D_3 agonist, which has a high neuroprotective capacity (Modi et al., 2014).

Kamonrat and coworkers (in 2025) prepared a series of 2-sulfonylthiazoles by N-sulfonylation of 2-aminothiazole with substituted benzene sulfonyl chloride to produce **44**, **45** and **46**. These molecules showed strong neuroprotection of SH-SY5Y cells in 6-OHDA by recovering the mitochondrial membrane potential, mitigating LDH leak and inhibiting the production of ROS. Notably, they did not have any toxicity to normal MRC-5 until 50 µg/mL. Mechanistic research showed an increase in SIRT1, which involved mitochondrial resilience and antioxidative mechanisms in their work. Further in silico absorption, distribution, metabolism and excretion (ADME) analyses indicated high levels of CNS drug-likeness and these halogenated thiazole sulfonamides are promising as multitarget neuroprotective agents (Figure 10.5).

10.3.3 1,3-Thiazole Derivatives in Huntington's Disease

This is a neurodegenerative disease that is autosomal dominant and for which there is currently no curative treatment (Ghotbi et al., 2020). Even though plenty of pathogenic pathways have been defined, the complication of the disease implies that various therapeutic targets should be identified. The pathology of Huntington's disease (HD) is caused by mutations in the huntingtin (HTT) protein, which naturally has neuroprotective functionality, but is toxic when the polyglutamine tract expands, thereby causing massive dysfunction in neurons.

A key molecular implication of mutant HTT is the increased genomic binding of RE1-silencing transcription factor (REST/RSF) to RE1/NRSE elements, which suppresses the expression of a number of neuronal genes, including neuromodulatory brain-derived neurotrophic factor (BDNF), which is one of the important neuroprotective mediators and the expression of which is closely associated with HD progression.

Leone et al. (2008) tested 2-aminothiazole derivatives library to reverse this dysregulation. Their SAR analysis presented three active compounds with the most active one being compound **47**. It is suggested that these molecules can have their therapeutic effect on inhibiting the RE1/NRSE-mediated transcriptional repression process, which partially recovers the gene expression (BDNF) and leads to an improved neurodegenerative profile linked with HD (Leone et al., 2008).

To add to these data, Zhang et al. (2005) have constructed a cell-based high-throughput screening system to monitor fused thiazole derivatives in the context of inhibition of polyglutamine (polyQ) aggregation, one of the hallmarks of Huntington pathology. Based on a yeast-based screening platform, a number of thiazole-containing analogs were identified. One of those, analog **48**, was exceptionally potent, exhibiting an IC_{50} = 10 nM in polyQ aggregation inhibition of HD neurons. Moreover, its usefulness in decreasing neurodegeneration was proven in vivo research. The SAR data indicated the extreme importance of the presence of phenyl substitution on the thiazole ring in the retention of a good anti-aggregation effect (Zhang et al., 2005).

Collectively, the data indicate that the differences in substituents in the thiazole core have a significant effect on neuroprotective activity. These results confirm the increment of further targeting thiazole scaffolds as potential therapeutic leads in HD (Figure 10.6).

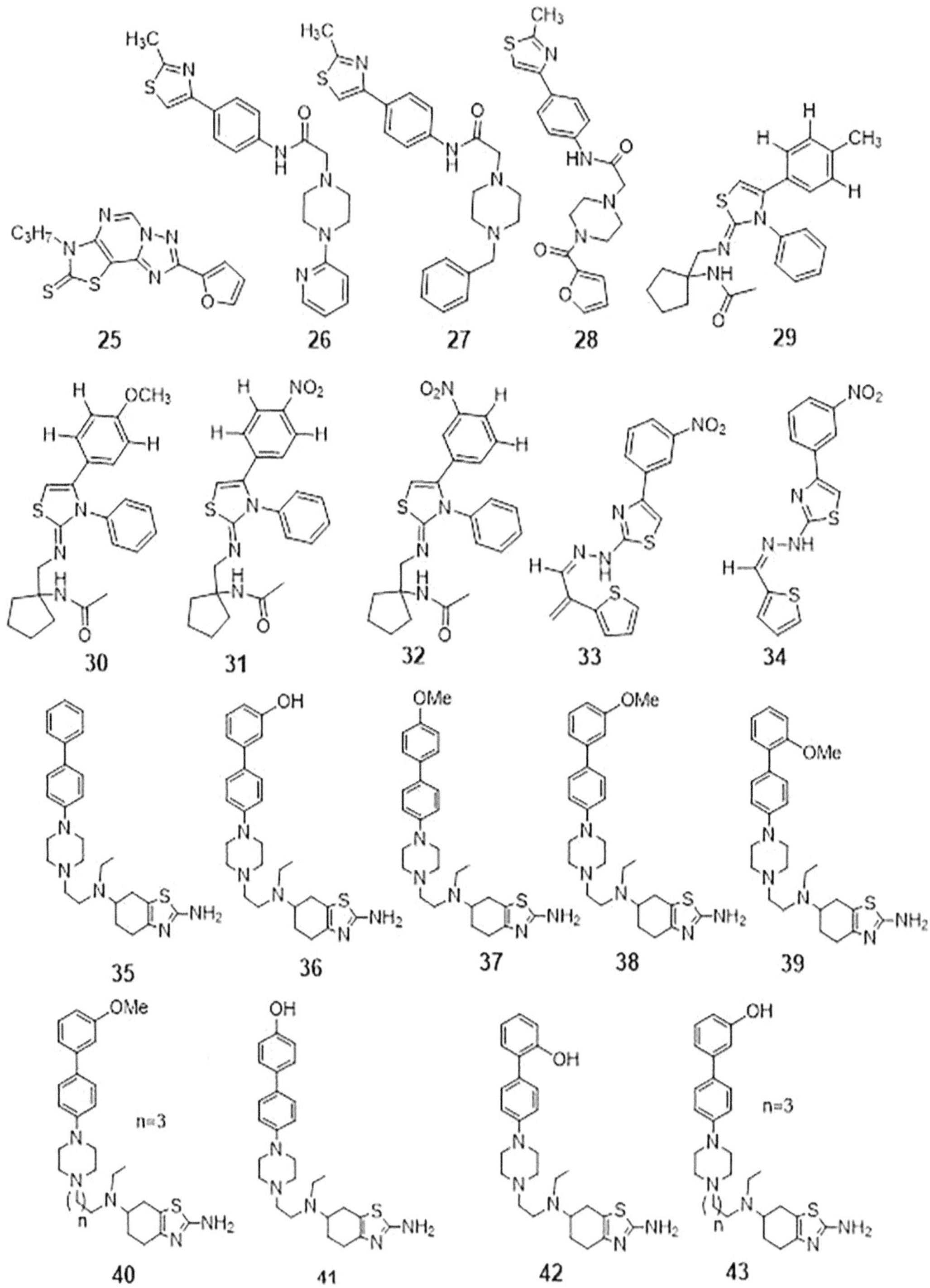

FIGURE 10.5 Structures of thiazole derivative with anti-Parkinson's activities.

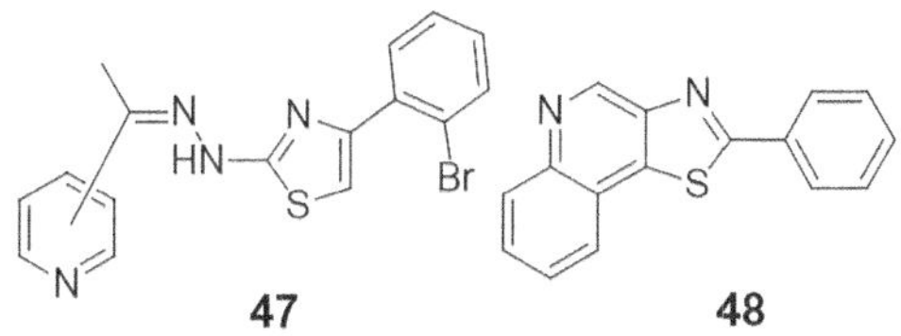

47 48

FIGURE 10.6 Structures of thiazole derivatives with anti-Huntington's activities.

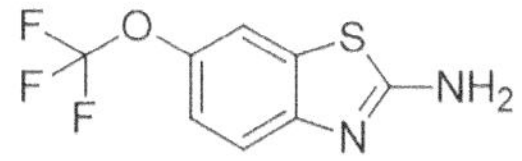

49

FIGURE 10.7 Riluzole (a thiazole derivative) with anti-ALS activities.

10.3.4 1,3-Thiazole Derivatives in Amyotrophic Lateral Sclerosis

Neuroprotection in ALS can be achieved with 1,3-thiazole and benzothiazole scaffolds (Bensimon et al., 1994). The clinically best-defined drug is Riluzole (Figure 10.7 (**49**)) that has a moderately beneficial impact on the survival of ALS patients, as it inhibits the release of glutamates, regulates sodium channels of voltage and stabilizes neuronal excitability (Bensimon et al., 1994; Miller & Moore, 2012). On the basis of Riluzole, a number of derivatives and hybrids such as Riluzole–triazole hybrids and substituted benzothiazoles have been designed in order to increase CNS penetration and neuroprotection (Mignani et al., 2020; Sweeney et al., 2018). In vitro tests suggest that these chemicals have the ability to decrease excitotoxicity and oxidative stress, but direct evidence of these compounds in an ALS model has not been well documented in vivo. Other thiazoles with substitutions, e.g. pyridyl-, phenyl- and methoxy-thiazoles, are also in research as neuroprotectants and could be used to make a scaffold in future ALS therapies (Chaudhary et al., 2023).

10.4 IN VITRO AND IN VIVO STUDIES

10.4.1 In Vitro Studies

10.4.1.1 Cholinesterase Inhibition and Anti-Amyloid Effects

Various thiazole-containing compounds have exhibited robust *in vitro* AChE (acetylcholinesterase) inhibition activities. Ghotbi et al. have aimed pyridinium–thiazole hybrids which inhibited acetylcholinesterase at μM stages while concurrently decreasing Aβ aggregation. SAR revealed that aromatic substituents and substituents with electron-donating ability delay considerably amended dual action (Ghotbi et al., 2020).

Likewise, Hemaida and coworkers prepared novel thiazoles that showed acetylcholinesterase inhibition activities. The IC_{50} values of these compounds were analogous to donepezil, supporting their ability as anti-Alzheimer's agents (Hemaida et al., 2021).

Mirjalili and coworkers recently have studied thiazole–coumarin–based compounds showing strong AChE inhibition (Mirjalili et al., 2024).

10.4.1.2 MAO-B Inhibition in the PD

MAO-B inhibitors are a basis in **PD** treatment, and derivatives of benzothiazole have exhibited a robust inhibitory role. Cao and coworkers synthesized a sulfonamide of benzothiazole that show MAO-B inhibition in the nM levels. These derivatives of thiazoles also decrease oxidative stress and show their neuroprotective activity (Cao et al., 2023).

10.4.1.3 Antioxidant and Cytoprotective Effects

Oxidative stress plays a considerable role in neuronal damage. Numerous derivatives of thiazole have exhibited protecting properties in and PC12 and SH-SY5Y cells exposed to hydrogen peroxide. Studies described the renovation of mitochondrial membrane ability. Hemaida and coworkers have proved that thiazole derivatives reduce the intracellular ROS.

10.4.2 IN VIVO STUDIES

10.4.2.1 In Vivo Role in AD (Alzheimer's Disease)

Thiazole compounds with robust *in vitro* activities have been developed to low-scale *in vivo* studies. Ghotbi and coworkers have proved that pyridinium–thiazole derivatives upgraded memory and learning in behavioral examinations in mice, comprising the Morris water maze. A reduced acetylcholenerase activity and decreased oxidative stress were shown by the brain tissue examination (Ghotbi et al., 2020).

In 2024, report was published by Mirjalili and coworkers which shows that thiazole–coumarin derivatives showed considerable brain penetration and created mental development in scopolamine-mediated amnesia models in mice (Mirjalili et al., 2024).

10.4.2.2 In Vivo Parkinson's Disease Neuroprotection

MAO-B inhibitors of benzothiazole scaffold have been tried in 6-OHDA rodents. Cao et al. in 2023 proved that their lead molecule considerably conserved dopaminergic neurons in the substantia nigra and enhanced motor behavior (Cao et al., 2023).

These outcomes confirm benzothiazole and thiazole frameworks as potential contenders for further mechanistic and behavioral revisions in PD.

10.4.2.3 Limitations and Translational Considerations

The main challenges associated with thiazole-based drugs in this field include variable BBB penetration, limited bioavailability of these drugs and inadequate profiles of toxicology.

10.5 FUTURE PERSPECTIVES AND CHALLENGES

Despite the fact that the clinical data are still scarce, the current technological advancements in the domain of rational drug design and molecular optimization can potentially speed up the process of thiazole derivatives entering the human trial phase. There are already structure-guided medicinal chemistry methods being utilized to improve BBB penetration, metabolic stability and target specificity. Biomarker-driven trials can also be used in clinical development to evaluate disease measures in an objective manner, e.g. Aβ, tau, α-synuclein, oxidative stress parameters. The innovations of drug delivery, which are liposomes, polymeric nanoparticles and other nanocarriers, may enhance brain delivery and minimize peripheral toxicity. Multitarget-directed derivatives that can be used to simultaneously regulate oxidative stress, protein aggregation, neuroinflammation and neurotransmitter imbalance are of particular promise.

The therapeutic ability of thiazole derivatives in neurodegenerative diseases is still being recognized, yet there are still major scientific and developmental challenges. Innovation in this field is being spurred by new technologies in medicinal chemistry, computational modeling and systems biology.

10.6 CONCLUSION

Thiazole derivatives have become one of the most promising multi-purpose therapeutic scaffolds for treating ALS, Parkinson's, Huntington's and ADs. They can lower oxidative stress, inhibit aggregation of proteins (Aβ, tau or α-synuclein), regulate neurotransmitters, stabilize mitochondria, inhibit apoptosis and prevent neuroinflammation. Medicinal chemists can rationally optimize BBB penetration, bioavailability and broad-spectrum activity because of their structural flexibility. Nevertheless, significant issues persist, among them the fact that there are very few thiazoles under human scrutiny, inter-patient variability in BBB permeability and metabolism, and the need to study chronic ND treatment on a long-term basis. The development of the MTDLs, computational drug design, AI-based screening, nanocarriers and personalized medicine provides feasible solutions to these shortcomings. Further combination of preclinical innovations and clinical trial–based optimization will be necessary to make the thiazole derivatives into actual disease-modifying agents. To sum up, thiazoles are one of the most broadly applicable and scientifically appealing scaffolds to discover neurodegenerative drugs in modern times. They have a great potential to help to provide a contribution to next-generation therapy of the complicated pathologies of NDs with further refinement.

REFERENCES

Andreani, Aldo, Silvia Burnelli, Massimiliano Granaiola, Massimo Guardigli, Alberto Leoni, Alessandra Locatelli, Rita Morigi, Mirella Rambaldi, Manuela Rizzoli, and Lucilla Varoli. (2008). Chemiluminescent high-throughput microassay applied to imidazo [2, 1-b] thiazole derivatives as potential acetylcholinesterase and butyrylcholinesterase inhibitors. *European Journal of Medicinal Chemistry, 43*(3), 657–661.

Ayati, Adile, Saeed Emami, Ali Asadipour, Abbas Shafiee, and Alireza Foroumadi. (2015). Recent applications of 1, 3-thiazole core structure in the identification of new lead compounds and drug discovery. *European Journal of Medicinal Chemistry, 97*, 699–718.

Azam, Faizul, Bashir A El-gnidi, Ismail A Alkskas, and Musa A Ahmed. (2010). Design, synthesis and anti-Parkinsonian evaluation of 3-alkyl/aryl-8-(furan-2-yl) thiazolo [5, 4-e][1, 2, 4] triazolo [1, 5-c] pyrimidine-2 (3 H)-thiones against neuroleptic-induced catalepsy and oxidative stress in mice. *Journal of Enzyme Inhibition and Medicinal Chemistry, 25*(6), 818–826.

Aziz, H. (2025). Insights into antimicrobial potential of functionalized thiazoles: In vitro and In silico insights. *Journal of Molecular Liquids,* 424, 12706. https://doi.org/10.1016/j.molliq.2025.127064.

Aziz, H. (2026). Thiazoles inhibit reactive oxygen and nitrogen species mediated diabetes mellitus: integrated in vitro and in silico insights. *RSC Advances,* https://doi.org/10.1039/d6ra01667g

Bensimon, Gilbert, Lucette Lacomblez, Vincent Meininger, and ALS/Riluzole Study Group. (1994). A controlled trial of riluzole in amyotrophic lateral sclerosis. *New England Journal of Medicine, 330*(9), 585–591.

Butterfield, D Allan, and Barry Halliwell. (2019). Oxidative stress, dysfunctional glucose metabolism and Alzheimer disease. *Nature Reviews Neuroscience, 20*(3), 148–160.

Cao, Zhongcheng, Xingyue Wang, Tianlong Zhang, Xianwu Fu, Fan Zhang, and Jiang Zhu. (2023). Discovery of novel 2-(4-(benzyloxy)-5-(hydroxyl) phenyl) benzothiazole derivatives as multifunctional MAO-B inhibitors for the treatment of Parkinson's disease. *Journal of Enzyme Inhibition and Medicinal Chemistry, 38*(1), 2159957.

Castagnolo, Daniele, Mafalda Pagano, Martina Bernardini, and Maurizio Botta. (2009). Domino alkylation-cyclization reaction of propargyl bromides with thioureas/thiopyrimidinones: A new facile synthesis of 2-aminothiazoles and 5H-thiazolo [3, 2-a] pyrimidin-5-ones. *Synlett, 2009*(13), 2093–2096.

Chaudhari, Prakash, Chhanda C Danta, and Adel Nefzi. (2025). Recent synthetic advances in the preparation of 1, 3-thiazole compounds and their therapeutic applications in degenerative diseases. *Organic & Biomolecular Chemistry.*

Chaudhary, Jasmine, Akash Jain, Ashwani Dhingra, Bhawna Chopra, Vishal Sharma, Jatin Gupta, and Aditi Kaushik. (2023). 1, 3-thiazole derivatives: A scaffold with considerable potential in the treatment of neurodegenerative diseases. *Current Topics in Medicinal Chemistry, 23*(23), 2185–2196.

Cui, Meng-Chao, Zi-Jing Li, Rui-Kun Tang, and Bo-Li Liu. (2010). Synthesis and evaluation of novel benzothiazole derivatives based on the bithiophene structure as potential radiotracers for β-amyloid plaques in Alzheimer's disease. *Bioorganic & Medicinal Chemistry, 18*(7), 2777–2784.

Das, Banibrata, Seenuvasan Vedachalam, Dan Luo, Tamara Antonio, Maarten EA Reith, and Aloke K Dutta. (2015). Development of a highly potent D2/D3 agonist and a partial agonist from structure–activity relationship study of N 6-(2-(4-(1 H-indol-5-yl) piperazin-1-yl) ethyl)-N 6-propyl-4, 5, 6, 7-tetrahydrobenzo [d] thiazole-2, 6-diamine analogues: Implication in the treatment of Parkinson's disease. *Journal of Medicinal Chemistry, 58*(23), 9179–9195.

Dorsey, Edward Ray, Todd Sherer, Michael S. Okun, and Bastiaan R. Bloem. (2018). The emerging evidence of the Parkinson pandemic. *J Parkinsons Dis, 8*, S3–S8.

Ghotbi, Golaleh, Mohammad Mahdavi, Zahra Najafi, Farshad Homayouni Moghadam, Maryam Hamzeh-Mivehroud, Soodabeh Davaran, and Siavoush Dastmalchi. (2020). Design, synthesis, biological evaluation, and docking study of novel dual-acting thiazole-pyridiniums inhibiting acetylcholinesterase and β-amyloid aggregation for Alzheimer's disease. *Bioorganic Chemistry, 103*, 104186.

Górecki, S., Kudelko, A., & Olesiejuk, M. (2025). Antimicrobial Activity of 1,3,4-Thiadiazole Derivatives. *Pharmaceuticals*, *18*, 1348. https://doi.org/10.3390/ph18091348

Gouleni, Niki, Annalisa Di Rienzo, Ahmet Yılmaz, Harun Selvitopi, Mehmet Enes Arslan, Adil Mardinoglu, Hasan Turkez, Antonio Di Stefano, Stamatia Vassiliou, and Ivana Cacciatore. (2023). Novel styryl-thiazole hybrids as potential anti-Alzheimer's agents. *RSC Medicinal Chemistry*, *14*(11), 2315–2326.

Gupta, Radha R, Mahendra Kumar, and Vandana Gupta. (2013). *Heterocyclic chemistry: Volume II: Five-membered heterocycles*. Springer Science & Business Media.

Hardiman, Orla, Ammar Al-Chalabi, Adriano Chio, Emma M Corr, Giancarlo Logroscino, Wim Robberecht, Pamela J Shaw, Zachary Simmons, and Leonard H van den Berg. (2017). Erratum: Amyotrophic lateral sclerosis (nature reviews. Disease primers (2017) 3 (17071)). *Nature Reviews. Disease Primers*, *3*, 17085.

Hemaida, Aya Y, Ghada S Hassan, Azza R Maarouf, Jacques Joubert, and Ali A El-Emam. (2021). Synthesis and biological evaluation of thiazole-based derivatives as potential acetylcholinesterase inhibitors. *ACS Omega*, *6*(29), 19202–19211.

Heneka, Michael T, Monica J Carson, Joseph El Khoury, Gary E Landreth, Frederic Brosseron, Douglas L Feinstein, Andreas H Jacobs, Tony Wyss-Coray, Javier Vitorica, and Richard M Ransohoff. (2015). Neuroinflammation in Alzheimer's disease. *The Lancet Neurology*, *14*(4), 388–405.

Henrich, Martin T, Wolfgang H Oertel, D James Surmeier, and Fanni F Geibl. (2023). Mitochondrial dysfunction in Parkinson's disease–a key disease hallmark with therapeutic potential. *Molecular Neurodegeneration*, *18*(1), 83.

Huang, Ling, Tao Su, Wenjun Shan, Zonghua Luo, Yang Sun, Feng He, and Xingshu Li. (2012). Inhibition of cholinesterase activity and amyloid aggregation by berberine-phenyl-benzoheterocyclic and tacrine-phenyl-benzoheterocyclic hybrids. *Bioorganic & Medicinal Chemistry*, *20*(9), 3038–3048.

Hutchinson, Ian, Sharon A Jennings, B Rao Vishnuvajjala, Andrew D Westwell, and Malcolm FG Stevens. (2002). Antitumor benzothiazoles. 16. Synthesis and pharmaceutical properties of antitumor 2-(4-aminophenyl) benzothiazole amino acid prodrugs. *Journal of Medicinal Chemistry*, *45*(3), 744–747.

Lagoja, Irene, Christophe Pannecouque, Gerard Griffioen, Stefaan Wera, Veronica Maria Rojasdelaparra, and Arthur Van Aerschot. (2011). Substituted 2-aminothiazoles are exceptional inhibitors of neuronal degeneration in tau-driven models of Alzheimer's disease. *European Journal of Pharmaceutical Sciences*, *43*(5), 386–392.

Lee, Yun Suk, Hee Kim, Young-Ho Kim, Eun Joo Roh, Hogyu Han, and Kye Jung Shin (2012). Synthesis and structure-activity relationships of tri-substituted thiazoles as RAGE antagonists for the treatment of Alzheimer's disease. *Bioorganic & Medicinal Chemistry Letters*, *22*, 7555–61.

Leone, Samantha, Cesare Mutti, Aleksey Kazantsev, Mattia Sturlese, Stefano Moro, Elena Cattaneo, Dorotea Rigamonti, and Alessandro Contini. (2008). SAR and QSAR study on 2-aminothiazole derivatives, modulators of transcriptional repression in Huntington's disease. *Bioorganic & Medicinal Chemistry*, *16*(10), 5695–5703.

Lingaraju, Gejjalagere S, Toreshettahally R Swaroop, Ajjampura C Vinayaka, Kothanahally S Sharath Kumar, Marilinganadoddi P Sadashiva, and Kanchugarakoppal S Rangappa. (2012). An easy access to 4, 5-disubstituted thiazoles via base-induced click reaction of active methylene isocyanides with methyl dithiocarboxylates. *Synthesis*, *44*(09), 1373–1379.

Maradiya, Hari Raghav, and Vithal Soma Patel. (2003). Synthesis and application of disperse dyes based on 2-aminothiazole derivatives. *Chemistry of Heterocyclic Compounds*, *39*(3), 357–363.

Mignani, Serge, Jean-Pierre Majoral, Jean-François Desaphy, and Giovanni Lentini. (2020). From riluzole to dexpramipexole via substituted-benzothiazole derivatives for amyotrophic lateral sclerosis disease treatment: Case studies. *Molecules*, *25*(15), 3320.

Miller, Robert G, and Dan H Moore. (2012). Riluzole for amyotrophic lateral sclerosis (ALS)/motor neuron disease (MND). *Cochrane Database of Systematic Reviews* (3).

Mirjalili, Bi Bi Fatemeh, Seyedeh Azita Fazeli Attar, and Fereshteh Shiri. (2024). Synthesis, biological evaluation and in silico study of 4-(benzo [d] thiazole-2-yl) phenols based on 4-hydroxy coumarin as acetylcholinesterase inhibitors. *Scientific Reports*, *14*(1), 26459.

Mishra, Chandra Bhushan, Shikha Kumari, and Manisha Tiwari. (2015). Thiazole: A promising heterocycle for the development of potent CNS active agents. *European Journal of Medicinal Chemistry*, *92*, 1–34.

Modi, Gyan, Tamara Antonio, Maarten Reith, and Aloke Dutta. (2014). Structural modifications of neuroprotective anti-Parkinsonian (–)-N 6-(2-(4-(Biphenyl-4-yl) piperazin-1-yl)-ethyl)-N 6-propyl-4, 5, 6, 7-tetrahydrobenzo [d] thiazole-2, 6-diamine (D-264): An effort toward the improvement of in vivo efficacy of the parent molecule. *Journal of Medicinal Chemistry*, *57*(4), 1557–1572.

Mrowicka, M., Mrowicki, J., Dragan, G., & Majsterek, I. (2023). The importance of thiamine (vitamin B1) in humans. *Bioscience Reports*, *43*(10), BSR20230374. https://doi.org/10.1042/BSR20230374

Poewe, Werner, Klaus Seppi, Caroline M Tanner, Glenda M Halliday, Patrik Brundin, Jens Volkmann, Anette-Eleonore Schrag, and Anthony E Lang. (2017). Parkinson disease. *Nature Reviews Disease Primers*, *3*(1), 1–21.

Querfurth, Henry W, and Frank M LaFerla. (2010). Mechanisms of disease. *The New England Journal of Medicine*, *362*(4), 329–344.

Ross, Christopher A, and Sarah J Tabrizi. (2011). Huntington's disease: From molecular pathogenesis to clinical treatment. *The Lancet Neurology*, *10*(1), 83–98.

Ruankham, Waralee, Ratchanok Pingaew, Veda Prachayasittikul, Apilak Worachartcheewan, Suphissara Sathuphong, Setthawut Apiraksattayakul, Tanawut Tantimongcolwat, Virapong Prachayasittikul, Supaluk Prachayasittikul, and Kamonrat Phopin. (2025). Neuroprotective thiazole sulfonamides against 6-OHDA-induced Parkinsonian model: In vitro biological and in silico pharmacokinetic assessments. *RSC Advances*, *15*(6), 4281–4295.

Savelieff, Masha G, Geewoo Nam, Juhye Kang, Hyuck Jin Lee, Misun Lee, and Mi Hee Lim. (2018). Development of multifunctional molecules as potential therapeutic candidates for Alzheimer's disease, Parkinson's disease, and amyotrophic lateral sclerosis in the last decade. *Chemical Reviews*, *119*(2), 1221–1322.

Schapira, Anthony H, and Peter Jenner. (2011). Etiology and pathogenesis of Parkinson's disease. *Movement Disorders*, *26*(6), 1049–1055.

Secci, Daniela, Simone Carradori, Anél Petzer, Paolo Guglielmi, Melissa D'Ascenzio, Paola Chimenti, Donatella Bagetta, Stefano Alcaro, Gokhan Zengin, and Jacobus P Petzer. (2019). 4-(3-Nitrophenyl) thiazol-2-ylhydrazone derivatives as antioxidants and selective hMAO-B inhibitors: Synthesis, biological activity and computational analysis. *Journal of Enzyme Inhibition and Medicinal Chemistry*, *34*(1), 597–612.

Selkoe, Dennis J, and John Hardy. (2016). The amyloid hypothesis of Alzheimer's disease at 25 years. *EMBO Molecular Medicine*, *8*(6), 595–608.

Shiradkar, Mahendra Ramesh, Kalyan Chakravarthy Akula, Varaprasad Dasari, Vijayakumar Baru, Bhoomeshwar Chiningiri, Santosh Gandhi, and Ranjit Kaur. (2007). Clubbed thiazoles by MAOS: A novel approach to cyclin-dependent kinase 5/p25 inhibitors as a potential treatment for Alzheimer's disease. *Bioorganic & Medicinal Chemistry*, *15*(7), 2601–2610.

Sweeney, Joseph B, Marcus Rattray, Victoria Pugh, and Lucy A Powell. (2018). Riluzole–triazole hybrids as novel chemical probes for neuroprotection in amyotrophic lateral sclerosis. *ACS Medicinal Chemistry Letters*, *9*(6), 552–556.

Todorova, A.K., Juettner, F., Linden, A., Pluess, T., & von Philipsborn, W. (1995). Nostocyclamide: A new macrocyclic, thiazole-containing allelochemical from Nostoc sp. 31 (cyanobacteria). *The Journal of Organic Chemistry*, *60*(24), 7891–7895. https://doi.org/10.1021/jo00129a032)

Turan-Zitouni, Gülhan, Aouatef Tabbi, Weiam Hussein, Abdullah Burak Karaduman, Begüm Nurpelin Sağlık, and Yusuf Özkay. (2018). Synthesis and evaluation of N-[1-(((3, 4-diphenylthiazol-2 (3H)-ylidene) amino) methyl) cyclopentyl] acetamide derivatives for the treatment of diseases belonging to MAOs. *Journal of Chemistry 2018*(1), 3547942.

Wollen, Keith A. (2010). Alzheimer's disease: The pros and cons of pharmaceutical, nutritional, botanical, and stimulatory therapies, with a discussion of treatment strategies from the perspective of patients and practitioners. *Alternative Medicine Review*, *15*(3), 223–44.

Yurttaş, Leyla, Zafer Asım Kaplancıklı, and Yusuf Özkay. (2013). Design, synthesis and evaluation of new thiazole-piperazines as acetylcholinesterase inhibitors. *Journal of Enzyme Inhibition and Medicinal Chemistry*, *28*(5), 1040–1047.

Zhang, Linda, Raymond Chuen-Chung Chang, Leung-Wing Chu, and Henry Ka-Fung Mak. (2012). Current neuroimaging techniques in Alzheimer's disease and applications in animal models. *American Journal of Nuclear Medicine and Molecular Imaging*, *2*(3), 386.

Zhang, Xiaoqian, Donna L Smith, Anatoli B Meriin, Sabine Engemann, Deborah E Russel, Margo Roark, Shetia L Washington, Michele M Maxwell, J Lawrence Marsh, and Leslie Michels Thompson. (2005). A potent small molecule inhibits polyglutamine aggregation in Huntington's disease neurons and suppresses neurodegeneration in vivo. *Proceedings of the National Academy of Sciences*, *102*(3), 892–897.

Zhuang, Zhi-Ping, Mei-Ping Kung, Catherine Hou, Daniel M. Skovronsky, Tamar L. Gur, Karl Plössl, John Q. Trojanowski, Virginia M.-Y. Lee, and Hank F. Kung. (2001). Radioiodinated styrylbenzenes and thioflavins as probes for amyloid aggregates. *Journal of Medicinal Chemistry*, *44*(12), 1905–1914.

11 Nitrogen-Containing Heterocyclic Natural Products as Leads for Neurodegenerative Disorders

Bhupender Nehra, Manoj Kumar, Viney Chawla, and Pooja A. Chawla

11.1 INTRODUCTION

Neurodegenerative disorders comprise a diverse range of long-term, progressive illnesses, i.e., characterized by selective loss of neurons followed by cognitive, behavioral and motor deficits. Irreversible neuronal dysfunction within particular brain regions is a prevalent feature of diseases like Alzheimer's disease (AD), Parkinson's disease (PD), Huntington's disease, amyotrophic lateral sclerosis and other dementias as mentioned in Figure 11.1. Due to their lengthy disease course, lack of curative treatments and rising global frequency triggered by aging populations, these disorders place a heavy social and economic burden on society (Nishit et al., 2022). Whereas each condition has unique clinical characteristics, such as cholinergic deficiencies in AD or dopaminergic neuronal loss in

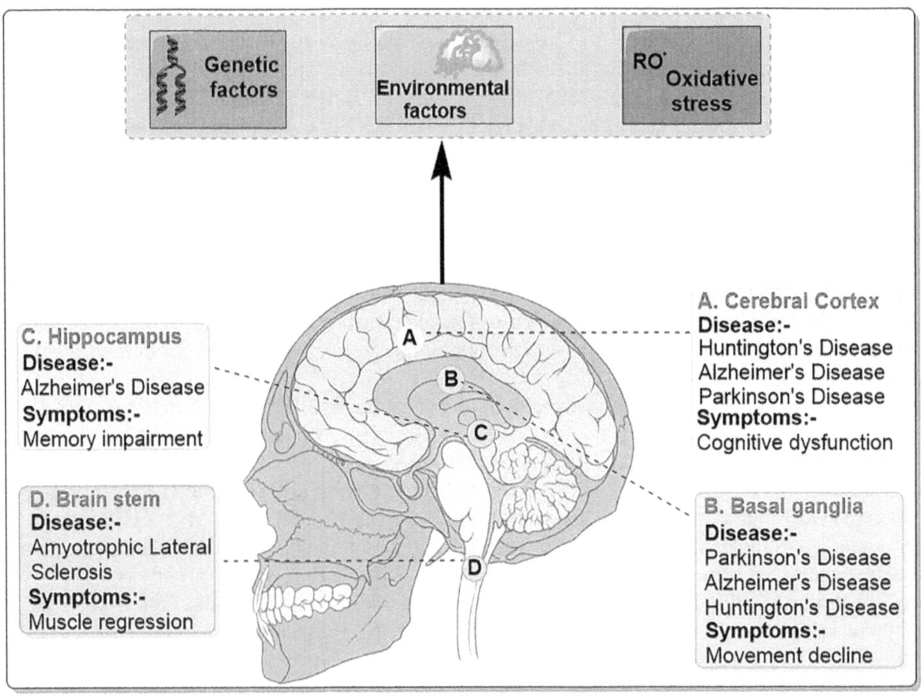

FIGURE 11.1 Several neurodegenerative diseases and their symptoms inked with their brain regions.

DOI: 10.1201/9781003772019-11

PD, all of these have a common theme of disturbed neuronal homeostasis that gradually overwhelms intrinsic repair mechanisms (Ni et al., 2024). Moreover, requirement for therapeutic drugs that can modify several cellular pathways involved in neuronal survival is highlighted through complexity of their multifactorial disease. A number of interlinked metabolic mechanisms that progressively impair neuronal survival are thought to be responsible for the molecular pathophysiology of neurodegenerative diseases (Sumien et al., 2021). Chronic oxidative stress driven by mitochondrial malfunction, compromised redox balance and an excessive build-up of reactive oxygen and nitrogen species simultaneously. These disruptions lead to neuronal damage *via* promotion of protein oxidation, DNA damage and lipid peroxidation. Furthermore, chronic activation of microglia and astrocytes is another keen element in neuroinflammation which releases pro-inflammatory cytokines, complement factors and reactive species in order to spread neurodegeneration (Bido et al., 2021). Protein misfolding and aggregation such as polyglutamine expansions in HD (Harding et al., 2021), α-synuclein aggregates in PD and β-amyloid plaques in AD are crucial to disrupt proteostasis and causes synaptic dysfunction. Neuronal susceptibility is further exacerbated by impairment of neurotransmitter systems, decreased axonal transport, calcium dyshomeostasis and deregulation of apoptotic pathways (Chau & Kim, 2022). In sum, these interrelated pathophysiological processes show the role of multi-target approaches and natural product-based leads to attract increased attention while mono-target drug regimen possesses little success.

Most of the available treatments in case of severe neurodegenerative diseases are symptomatic to afford temporary relief. Herein, cholinesterase inhibitors and NMDA receptor modulators used in AD to provide some cognitive improvements (Ordoñez et al., 2025). Further, Levodopa replacement therapy reduces motor symptoms in PD while it eventually loses its effectiveness and is linked to long-term side effects such dyskinesias (Wiesen & Atlas, 2022). In addition, efforts to create disease-modifying treatments that target tau, α-synuclein, amyloid-β and neuroinflammatory pathways have produced uneven clinical results which reflects challenge to interfere with intricate and advanced pathogenic cascades. Other challenges include inadequate knowledge of early pathogenic triggers, variability in patient genetics and illness stages along with poor CNS penetration of therapeutic options (Webster et al., 2025). Novel therapeutic leads with favorable pharmacokinetic profiles, better safety margins and ability to simultaneously target several molecular processes are therefore critically required. These characteristics are frequently present in structurally varied natural compounds. Natural products have historically been a rich source of pharmacologically active compounds and they have special benefits in field of neurodegenerative illnesses due to their structural diversity and potential for polypharmacology. Numerous metabolites that originate from plants, microbes and marine environments have been refined by evolutionary selection to interact with a variety of biological targets (Silva et al., 2020). These are appealing options for multi-target drug development due to their capacity to regulate oxidative stress, inflammation, protein aggregation, mitochondrial function and neurotransmitter systems (Caruana et al., 2021). Furthermore, natural N-containing heterocycles often exhibit favorable CNS activity and excellent target selectivity because of their intrinsic heteroatom-driven interactions with enzymes, receptors and ion channels. In this context, galantamine and physostigmine are two authorized medications for neurological conditions that come straight from natural sources to demonstrate their translational potential (Safir et al., 2025). Natural products continue to be essential sources of novel lead structures that can address the complex nature of neurodegenerative illnesses as research progresses as depicted in Figure 11.2.

11.2 NITROGENOUS HETEROCYCLES: CHEMICAL AND BIOLOGICAL SIGNIFICANCE

11.2.1 Structural Diversity and Biosynthetic Origins

All bacteria, fungi and marine organisms use different biosynthetic processes to produce N-containing heterocycles which show remarkable structural variety. Their structures vary from

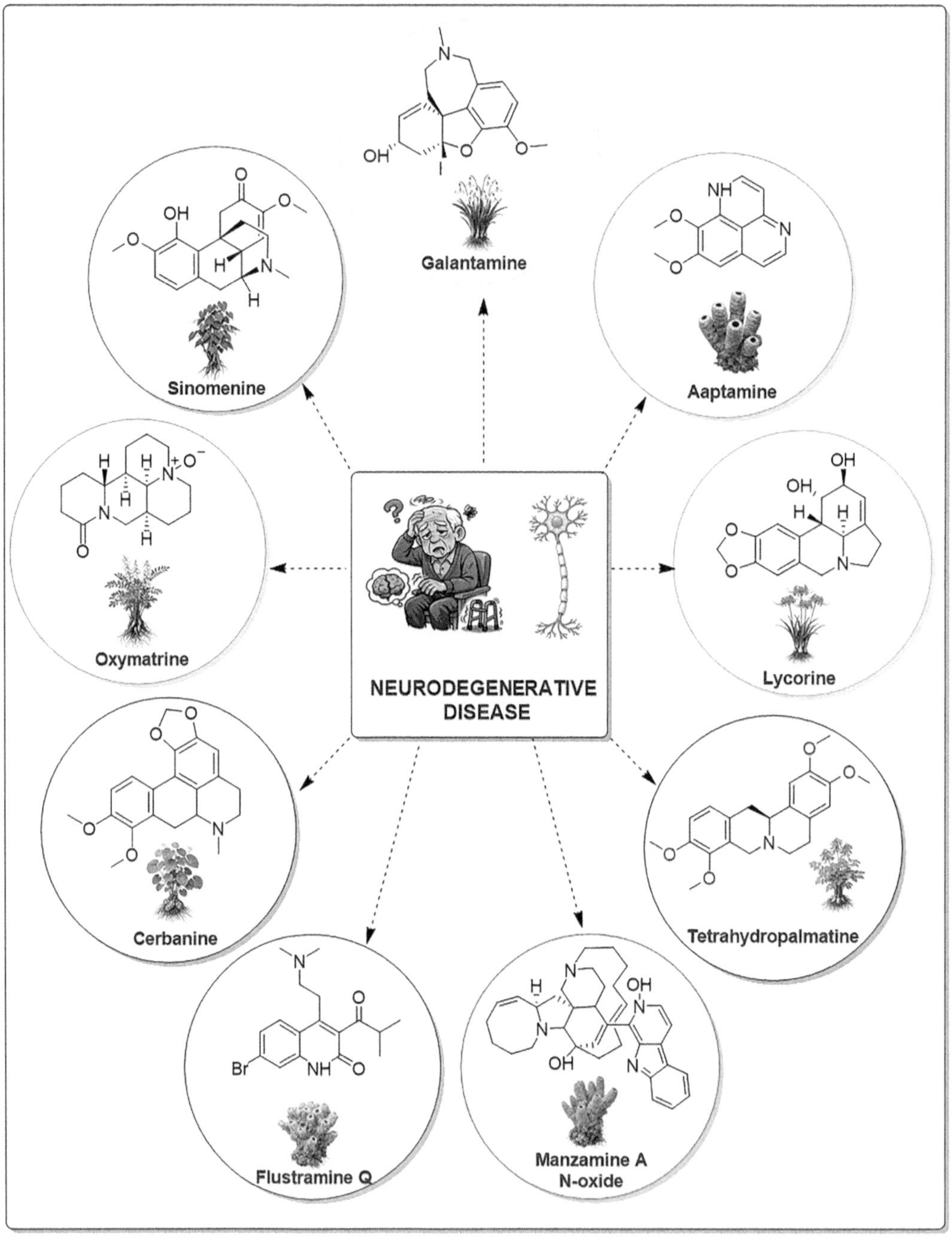

FIGURE 11.2 Various N-heterocyclic natural products as effective therapeutics against neurodegenerative illnesses.

basic monocyclic scaffolds like imidazole and pyridine to complicated fused polycyclic systems such as marine alkaloids, indoles, β-carbolines, quinolines and isoquinolines (Ismail et al., 2023). These scaffolds are frequently produced biosynthetically *via* enzymatic cyclizations, oxidations, methylations and rearrangements of amino acid precursors such as tryptophan, tyrosine, lysine and ornithine (Andrés et al., 2022; Lee et al., 2023). Whereas isoquinoline alkaloids are generated when dopamine condenses with aldehydes while indole alkaloids are produced when tryptophan is decarboxylated followed by Pictet–Spengler or Mannich-type reactions (Berger et al., 2022; Ismail et al., 2023). Unlike terrestrial species, marine organisms including sponges and tunicates produce distinctive nitrogenous alkaloids with unique ring systems and halogenation patterns. Their great functional diversity and frequent discovery as bioactive compounds with CNS relevance are both attributed to this biosynthetic richness (Ayyappan & Ekambaram, 2025).

11.2.2 Pharmacophoric Role of Nitrogen in Drug–Target Interactions

As N-atoms can participate in H-bonding, protonation–deprotonation balances and electronic modulation, they have a major influence on molecular recognition processes. By taking part in ionic interactions, H-bond donation or acceptance and coordination with metal-containing enzyme active sites, nitrogen heteroatoms frequently act as crucial pharmacophoric centers that improve binding affinity and selectivity in drug–target interactions relevant to central nervous system Catanesi et al. (2021). Basic or neutral nitrogen also affects physicochemical characteristics like lipophilicity and pKa which consequently affect membrane permeability and CNS penetration. Additionally, N-atoms are often involved in stabilization of desired conformation, molecular orientation inside a binding pocket or engagement with catalytic residues in enzymes (e.g., kinases, acetylcholinesterase (AChE) and monoamine oxidases (MAOs)) (Krishnan et al., 2023). Nitrogen-containing heterocycles are essential to create neuroprotective compounds with optimal target engagement by virtue of their diverse contributions.

11.2.3 Importance of Heterocyclic Scaffolds in CNS Drug Design

Heterocyclic scaffolds predominate in CNS-active pharmacological libraries because they offer compact structures with excellent binding effectiveness, metabolic stability and customizable pharmacokinetics. In particular, nitrogenous heterocycles improve CNS drug design through precise control of lipophilicity, topological polar surface area and molecular weight, i.e., parameters essential to penetrate blood–brain barrier. In addition to reduce entropic costs upon binding, their rigid or semi-rigid structures promote robust receptor connections (Shrestha et al., 2025). Furthermore, many heterocyclic cores have intrinsic affinities for ion channels, neurotransmitter receptors and enzymes essential to brain function that afford precise control of glutamatergic, dopaminergic, cholinergic and serotonergic pathways. Because of their synthetic adaptability, substituents that improve bioavailability, lessen toxicity or provide multi-target activity can be included (Laurencin et al., 2023). Therefore, N-containing heterocycles remain fundamental in development of neuroprotective medicines derived from natural products that can tackle intricate pathophysiology of neurodegeneration.

11.3 CLASSES OF N-CONTAINING NATURAL HETEROCYCLES IN NEUROPROTECTION

11.3.1 Indole Alkaloids

Due to their structural similarity toward important neurotransmitters like serotonin and melatonin, indole alkaloids constitute a significant class of naturally occurring neuroactive chemicals (Table 11.1). In this context, reserpine is derived from *Rauwolfia serpentina* and affects central monoaminergic transmission by permanently VMAT2 blockade. It has been used traditionally to treat hypertension but is essential to comprehend biology of catecholamines (Aljarba et al., 2023). In next, a β-carboline alkaloid from *Banisteriopsis caapi,* i.e., harmine has antioxidant qualities, neurogenic potential through DYRK1A regulation and MAO-A inhibition which contribute to neuroprotective advantages (Xie et al., 2025). Further, *Vinca minor* is a source of vincamine and

TABLE 11.1
Classification of N-Containing Natural Heterocycles with Neuroprotective Potential

Entry	Heterocyclic Class	Representative Natural Products	Neuroprotective Activities	Key Neurodegenerative Targets and Associated Mechanisms
1.	Indole alkaloids	Reserpine, harmine, harmaline, yohimbine, vincamine, ibogaine	• MAO-A/B inhibition, serotonergic and dopaminergic modulation, neurogenesis stimulation, inhibition of neuroinflammation, improved cerebral blood flow, Nrf2-mediated antioxidant activity	• MAO-A/B inhibition reduces oxidative deamination • Harmine modulates DYRK1A and enhances neurogenesis • Vincamine improves cerebral perfusion and mitochondrial oxygenation
2.	Quinoline/ isoquinoline alkaloids	Berberine, palmatine, tetrandrine, sanguinarine, chelerythrine	• Strong anti-amyloidogenic activity, mitochondrial protection, AMPK activation, anti-inflammatory and anti-apoptotic actions • Modulation of cholinergic transmission	• Berberine inhibits Aβ fibril formation, reduces tau phosphorylation, activates AMPK to restore mitochondrial respiration • Palmatine suppresses NF-κB signaling and reduces neuroinflammatory cascades
3.	Pyridine/ piperidine alkaloids	Nicotine, galantamine, anabasine, lobeline, cytisine	• Enhancement of cholinergic neurotransmission, nAChR agonism or partial agonism, AChE inhibition, upregulation of neurotrophic factors, modulation of dopaminergic signaling	• Galantamine acts as both AChE inhibitor and nAChR allosteric potentiator • Nicotine stimulates neurotrophic pathways (BDNF/NGF) and reduces microglial reactivity
4.	Imidazole/ pyrrolidine alkaloids	Histamine analogs, proline-based alkaloids (e.g., kainic acid derivatives), loline alkaloids	• Histaminergic modulation (H1–H4 receptors), ROS scavenging, microglial deactivation, neurovascular protection, involvement in synaptic plasticity	• Histamine activates H3/H4 receptors to reduce excitotoxicity • Proline derivatives exhibit antioxidant and mitochondrial-protective functions
5.	Marine-derived nitrogen heterocycles	Manzamine A, fascaplysin, axinelline A, dragmacidin D, pseudopteroxazole	• Anti-amyloidogenic activity, kinase inhibition (GSK-3β, CDKs), suppression of microglial activation, mitochondrial protection, inhibition of tau hyperphosphorylation	• Manzamine A inhibits GSK-3β and attenuates tau phosphorylation • Fascaplysin inhibits CDKs and prevents Aβ fibril maturation; dragmacidin D disrupts glutamate-induced excitotoxicity

its semisynthetic analogues which are useful to treat cerebrovascular insufficiency and cognitive impairment because they promote neuronal metabolism, maintain mitochondrial activity and increase cerebral blood flow (da Motta et al., 2024). These indole-derived scaffolds have multi-target actions to address neurotransmission, oxidative stress and mitochondrial health.

11.3.2 Quinoline and Isoquinoline Alkaloids

Quinoline and isoquinoline alkaloids are structurally different aromatic heterocycles, i.e., widely found in therapeutic plants. Further, Berberine is a protoberberine alkaloid that possess neuroprotective benefits *via* modulation of AMPK signaling, lowering oxidative stress, blocking cholinesterases and reducing neuroinflammation through suppression of pro-inflammatory cytokines. Also, cognitive resilience is further improved by its capacity to affect glucose metabolism and mitochondrial dynamics (Akbarizadeh-Mashkani et al., 2025). Palmatine is also structurally similar and exhibits antioxidant activity, inhibits β-amyloid formation and attenuates microglial activation by which it became useful in AD. Due to their polypharmacological character, these alkaloids also show positive interactions with AChE, MAO enzymes and other kinases. Their profound penetration into hydrophobic regions of CNS targets is made possible by their aromatic N-containing frameworks which support their ongoing investigation as neuroprotective lead agents (Wang et al., 2023).

11.3.3 Pyridine and Piperidine Derivatives

The great affinity of pyridine and piperidine derivatives for nicotinic and cholinergic receptors affords them a unique status in neuropharmacology. Nicotine has been crucial to demonstrate neuroprotective potential of nicotinic acetylcholine receptor (nAChR) regulation, despite its potential addiction liabilities. Nicotine improves attention, increases synaptic plasticity and reduces inflammation *via* activation of α4β2 and α7 nAChRs through cholinergic signaling pathways (Rivera-García et al., 2022). Moreover, galantamine is an approved medication for AD since it inhibits AChE and allosterically potentiates nAChRs. In addition, it is a naturally occurring piperidine-based alkaloid from *Galanthus* species. Both synaptic resilience as well as cholinergic neurotransmission are enhanced through this dual mechanism (Nozaki et al., 2024). Hence, pyridine/piperidine scaffold encompasses best physicochemical characteristics for CNS penetration and is a flexible template to produce neuroprotective drugs with many targets.

11.3.4 Imidazole and Pyrrolidine Derivatives

Natural metabolites that contain imidazole and pyrrolidine have important functions in brain signaling regulation and preservation of cellular homeostasis. Imidazole-based substances affect arousal, cognition and neuroinflammation *via* acting on histamine receptors in CNS. These substances include histamine analogues, i.e., derived from amino acid metabolic pathways. These could potentially be used as therapeutic molecules for inflammatory neurodegeneration and cognitive disorders due to their capacity to modulate H1, H2 and H3 receptors (Katsube et al., 2024). Furthermore, pyrrolidine derivatives, i.e., often produced from proline or ornithine biosynthesis may contribute to neuroprotection. Natural products based on proline can also help to prevent excitotoxic stress and preserve integrity of mitochondria. Their relevance in CNS-focused medicinal chemistry is supported by compact heterocycles which facilitate close interactions with enzyme active sites (Figure 11.1) (Shalaby et al., 2022).

11.3.5 Marine-Derived N-Heterocycles

Nitrogenous heterocycles with extraordinary structural characteristics and powerful bioactivities can be found in abundance in marine environments. Manzamine A, a complex alkaloid from marine sponges that contains β-carboline exhibits strong neuroprotective benefits *via* suppressing inflammatory pathways, promotion of mitochondrial function and GSK-3β signaling modification (Llorach-Pares et al., 2020). Further, fascaplysin is a planar nitrogenous pigment from *Fascaplysinopsis* species and possesses neuroprotective effect (Chung et al., 2024). In next, axinelline A is another metabolite generated from sponges which has proven capacity to lower neuronal oxidative stress and suppress neuroinflammation. Due to their structural distinctiveness and increased lipophilicity, marine nitrogenous heterocycles can effectively penetrate CNS and provide prospects for discovery of novel neuroprotective mechanisms (Li & Sun, 2025). Table 11.2 depicts various marine-derived N-heterocyclic molecules with potent neurodegenerative effects.

TABLE 11.2
Marine-Derived Nitrogenous Heterocycles with Reported Anti-Neurodegenerative Activities

S. No.	Compound	Source Organism	Core N-Heterocycle	Neuroprotective Potential	Mechanistic Insights
1.	Manzamine A	*Haliclona* spp., *Acanthostrongylophora* spp.	β-carboline fused to macrocyclic structure	Anti-inflammatory, anti-amyloidogenic, mitochondrial protective, anti-tau activity	• Inhibits GSK-3β, reduces tau phosphorylation at Ser202/Thr205 • Suppresses microglial NF-κB signaling • Restores mitochondrial complex I and IV activity
2.	Fascaplysin	*Fascaplysinopsis* spp.	Indoloquinolinium, planar cationic scaffold	Strong anti-amyloid effect, kinase inhibition, ROS modulation	• Inhibits CDK4 and GSK-3β, reducing tau pathologies • Binds strongly to Aβ aggregates, interrupting β-sheet formation • Modulates redox enzymes
3.	Axinelline A	*Axinella* spp.	Imidazole-containing bicyclic structure	Anti-amyloidogenic, moderate antioxidant, synaptic protective	Binds Aβ oligomerization interfaces • Prevents protofibril elongation • Stabilizes synaptic proteins under oxidative stress
4.	Dragmacidin D	*Dragmacidon* sp., sponges from family Dendrophylliidae	Bis-indole alkaloid with guanidine/imino functionality	Anti-excitotoxic, antioxidant, neuroinflammatory suppression	• Blocks glutamate-induced Ca^{2+} overload • Restores mitochondrial $\Delta\psi m$ • Reduces COX-2 and IL-6 expression
5.	Pseudo pteroxazole	*Pseudopterogorgia elisabethae*	Oxazole-containing nitrogen heterocycle	Anti-inflammatory, neuroprotective under oxidative challenge	Inhibits TLR-mediated microglial activation • Scavenges superoxide radicals • Enhances neuronal survival pathways (Akt/CREB)

11.4 MECHANISTIC INSIGHTS INTO NEUROPROTECTIVE ACTION

11.4.1 Antioxidant and Free Radical Scavenging Mechanisms

Numerous N-containing heterocyclic natural compounds have neuroprotective benefits through effective antioxidant processes (see Figure 11.3). By directly scavenging ROS and RNS, stabilizing free radicals by resonance and inhibiting spread of lipid peroxidation in neuronal membranes, these substances reduce oxidative damage. Further, superoxide anions, hydroxyl radicals and peroxynitrite are neutralized by redox cycling which is facilitated by their heteroatom-rich scaffolds that allow electron donation or acceptance (Iova et al., 2023). Furthermore, several alkaloids enhance glutathione production, restore catalase, superoxide dismutase and glutathione peroxidase activity. Also, they upregulate phase II detoxifying enzymes *via* Nrf2–ARE signaling to modify endogenous antioxidant

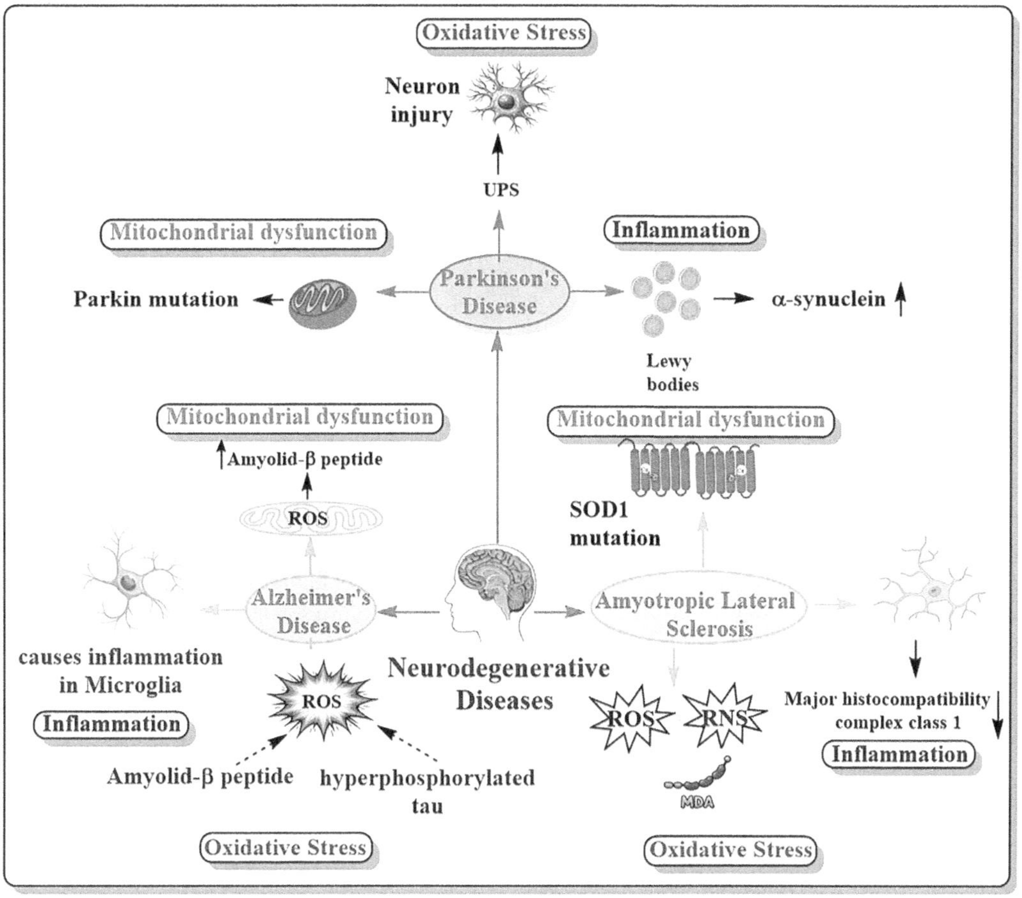

FIGURE 11.3 Major neuronal dysfunction mechanisms associated with neurodegenerative diseases.

defense systems. These heterocyclic metabolites counteract one of the first and most damaging processes in neurodegenerative pathogenesis by protection of neuronal architecture and preservation of synaptic efficacy through reduced oxidative alterations of proteins, DNA and lipids (Waheeb et al., 2025).

11.4.2 Anti-Inflammatory and Microglial Modulation

The development of neurodegenerative diseases is greatly aided by neuroinflammation which is mainly caused by activated microglia and reactive astrocytes. By inhibiting important signaling pathways including NF-κB, MAPKs and JAK/STAT, nitrogenous heterocyclic natural products influence these inflammatory cascades by lowering transcription of pro-inflammatory mediators like TNF-α, IL-1β, IL-6, iNOS and COX-2 (Zuppe & Reed, 2024). By restricting excessive cytokine production and avoiding transition to a neurotoxic M1 phenotype while concurrently supporting pro-repair M2 phenotype (support neuronal survival), many alkaloids reduce microglial overactivation. Together, these substances lessen inflammatory damage by reducing nitric oxide and prostaglandin E2 synthesis in order to decrease complement protein formation and attenuate astrocytic reactivity (Xie, 2025).

11.4.3 Inhibition of Protein Aggregation and Amyloidogenesis

Misfolded protein aggregates such as β-amyloid and tau clumps in AD, α-synuclein fibrils in PD and huntingtin polyglutamine expansions in Huntington's disease are characteristic clinical manifestations of numerous neurodegenerative illnesses. By interacting with aggregation-prone proteins, stabilization of their native conformations and interference with oligomerization processes, N-rich natural heterocycles demonstrate strong anti-amyloidogenic effects (Rupreo et al., 2025). Their conjugated aromatic structures limit nucleation and fibril elongation *via* facilitation of hydrophobic contacts, H-bonding and π–π stacking with amyloidogenic sequences. A number of alkaloids also trigger cellular clearance systems by autophagy promotion and proteasomal breakdown of misfolded proteins. These metabolites help to maintain synaptic function and lessen cytotoxicity related to protein aggregation through regulation of chaperone proteins, β-secretase and γ-secretase inhibition and limiting production of neurotoxic oligomers (Onifade et al., 2025).

11.4.4 Mitochondrial Protection and Anti-Apoptotic Pathways

Neuronal vulnerability is largely caused due to mitochondrial failure, and nitrogenous natural products protect neurons by maintaining mitochondrial integrity and halting apoptosis. These substances maintain ATP production which is necessary for brain function by optimizing electron transport chain activity, lowering mitochondrial ROS production and stabilizing mitochondrial membrane potential. Numerous heterocyclic metabolites reduce cytochrome c release and downstream caspase activation by blockage of mitochondrial permeability transition pore (Duan et al., 2020). They prevent cell death in response to oxidative or metabolic stressors by shifting balance toward anti-apoptotic signaling through modulation of Bcl-2 family proteins. Additionally, some metabolites improve overall neural resilience by promotion of mitochondrial biogenesis through activation of PGC-1α and SIRT pathways. Hereby, N-containing heterocycles reduce early and late mitochondrial processes linked to development of neurodegenerative diseases through these pathways (Shaker et al., 2021).

11.4.5 Modulation of Neurotransmitter Systems and Enzymatic Targets (AChE, MAO-B, GSK-3β)

Numerous heterocycles that include nitrogen have a direct impact on neurotransmitter pathways that help rebuild disrupted synaptic equilibrium in neurodegenerative illnesses. In this concern, AChE inhibition by a significant group improves cholinergic transmission which is essential for memory and cognition. Others reduce dopamine oxidative deamination and load of free radicals in Parkinsonian pathology by modulation of MAO-B (Bhat & Kumar, 2022). A number of β-carbolines and indole alkaloids have potential to change AD *via* inhibition of GSK-3β, i.e., a kinase involved in tau hyperphosphorylation, neuroinflammation and apoptosis. nAChRs, serotonergic receptors, glutamatergic signaling and dopaminergic pathways are some other neuroactive processes. Herein, nitrogenous heterocycles exhibit multi-target pharmacology required to address intricate network of dysregulated neurotransmission in neurodegenerative illnesses through their interactions with several enzyme and receptor targets (van Onselen & Downing, 2021).

11.5 MOLECULAR MECHANISMS ASSOCIATED WITH NEUROPROTECTION

Nitrogen-containing heterocyclic natural compounds have neuroprotective effects that arises from a complicated network of molecular processes concurrently to support cellular survival and resilience. These substances control intracellular signaling cascades such as PI3K/Akt, ERK/MAPK, AMPK and CREB pathways which together support synaptic maintenance, neuronal growth and stress survival (Shoaib et al., 2023). Numerous alkaloids assist both neurogenesis as well as synaptogenesis *via* increasing expression of neurotrophic factors including BDNF and NGF. By preventing excessive activation of voltage-gated calcium channels and NMDA receptors, their capacity to regulate calcium homeostasis prevents excitotoxicity (Jain et al., 2020). These metabolites preserve intracellular protein quality control by lowering ER stress, re-establishing proteostasis

TABLE 11.3
Molecular Targets Modulated by N-Containing Natural Heterocycles

Entry	Target	Biological Role in Neurodegeneration	Nitrogenous Natural Products	Mechanistic Effect
1.	Acetylcholinesterase (AChE)	• Hydrolyzes acetylcholine • Excessive activity leads to cholinergic deficits in AD	Galantamine, berberine, huperzine A (structurally related bicyclic alkaloid)	• Galantamine binds to catalytic anionic site and allosterically potentiates nAChRs • Berberine fits into AChE active site to enhance cholinergic tone
2.	Monoamine oxidase-B (MAO-B)	• Catalyzes oxidative deamination of dopamine • Major source of H_2O_2 and oxidative stress in PD	Harmine, β-carbolines (harmaline), peganine	• Harmine selectively inhibits MAO-A/B *via* reversible binding toward FAD cofactor site and hence decreases oxidative dopamine catabolism
3.	GSK-3β	Maintain tau phosphorylation, neuronal apoptosis and inflammatory gene expression	Manzamine A, fascaplysin	• Manzamine A acts as a potent non-ATP competitive inhibitor decreasing tau hyperphosphorylation • Fascaplysin binds ATP pocket to inhibit CDKs and indirectly downregulate GSK-3β activity
4.	nAChRs (α4β2, α7 subtypes)	Regulate learning, memory, neurotrophic factor release and dopaminergic pathways	Nicotine, cytisine, galantamine	• Nicotine activates α7-nAChR to enhance neurotrophic signaling • Cytisine acts as a partial agonist to promote neuroprotection without strong addictive potential
5.	NF-κB/MAPK signaling pathways	Control inflammation, microglial activation and cytokine gene expression	Palmatine, berberine, tetrandrine	• Palmatine suppresses IκBα degradation preventing NF-κB activation • Tetrandrine blocks MAPK phosphorylation limiting inflammatory cascades
6.	Amyloidogenic pathways (Aβ, BACE1)	It leads to Aβ production and plaque formation	Berberine, axinelline A	• Berberine downregulates BACE1 expression reducing Aβ42 levels • Axinelline A interferes with Aβ oligomer assembly

and encourage autophagic flux. In addition, they have an impact on epigenetic regulators such as histone acetyltransferases and deacetylases which modifies gene expression patterns associated with neural healing. Nitrogenous heterocycles are positioned as adaptable neuroprotective agents that can operate at several nodes within degenerative cascade due to combination of their chemical actions (Li et al., 2021). Various molecular targets that effectively encountered by nitrogen-based natural heterocyclic compounds are represented in Table 11.3.

11.6 SYNTHETIC MODIFICATIONS AND SEMISYNTHETIC DERIVATIVES

Enhancement in potency, selectivity, pharmacokinetics and safety while maintaining fundamental heterocyclic structure responsible for biological activity is ultimate objective of structural optimization of N-containing natural compounds. Here, modification in substituent patterns, adding EDGs or EWGs to change electrical distribution and increasing conformational rigidity to boost receptor binding are standard medicinal chemistry techniques. Target interactions can be improved or off-target effects can be decreased through modification of steric bulk, H-bonding capacity or ring fusion patterns (Wan et al., 2021). Furthermore, increasing CNS penetration frequently requires improved lipophilicity or decreased polar surface area. Moreover, metabolic stability and resistance to enzymatic degradation can be adjusted through selective alteration of functional groups like hydroxyls, methoxy or amines. Researchers can transform natural scaffolds into more potent neuroprotective leads with the help of these optimization techniques (Anastassova et al., 2022).

In order to improve activity, stability or pharmacokinetics, bioisosteric replacement is essential to afford more drug-like natural heterocycles *via* replacement of atoms or groups with structurally or electrically comparable counterparts. In this, binding orientation and metabolic robustness can be significantly changed by substituting fluorine for hydroxyl groups, switching amide linkages for bioisosteric heterocycles or adding azoles and diazines (Andrade et al., 2023). Further, integration of different pharmacophores into a single molecule or scaffold hybridization has become a potent method to achieve multi-target activity in complicated neurodegenerative illnesses. For instance, indole–quinoline, piperidine–coumarin or isoquinoline–flavonoid scaffolds can be combined to create hybrid molecules with improved antioxidant, anti-amyloid and enzyme-inhibitory abilities (Viegas et al., 2022). A promising area in development of natural product-based neuroprotective drugs is demonstrated by this hybridization which produces molecular entities that can simultaneously modulate several pathogenic pathways while preserving advantageous pharmacokinetics. Numerous completely synthetic and semisynthetic analogues developed from naturally occurring N-heterocycles have demonstrated enhanced neuroprotective potential (Wan et al., 2021). In order to improve MAO-A/B selectivity and decrease cytotoxicity, modified β-carboline derivatives have been synthesized that can shield dopaminergic neurons in PD models. In comparison to their parent molecule, structural derivatives of berberine such as 8-alkyl and 9-O-substituted analogues have greater cholinesterase inhibitory action and better BBB permeability (Shome et al., 2024). In this context, galantamine-based semisynthetic analogues with lipophilic or aromatic extensions exhibit enhanced AChE inhibition and nicotinic receptor allosteric potentiation (Stavrakov et al., 2020). In further, analogues generated from marine alkaloids such as modified manzamines and fascaplysin derivatives have proven increased GSK-3β inhibition and decreased toxicity that indicates possibility of rational modification to convert intricate natural scaffolding into neuroprotective medicines coupled with clinical relevance (Priyadharshini et al., 2024). These case studies illustrate the role of systematic structural modifications to preserve inherent biological benefits of natural heterocycles while significantly enhancing therapeutic characteristics.

11.7 CURRENT PROGRESS IN NEUROPROTECTIVE POTENTIAL OF N-CONTAINING NATURAL PRODUCTS AND THEIR DERIVATIVES

Ali and their team (2025) disclosed the study of sinomenine hydrochloride (Figure 11.4) as a protective agent against AD. Sinomenine hydrochloride significantly reduced cadmium-induced oxidative stress *via* lowering ROS/LPO and restoring GSH along with upregulation of Nrf2/HO-1 antioxidant pathway. Further, it suppressed major inflammatory markers such as TLR4, GFAP, Iba-1, p-NF-κB and TNF-α while reversing apoptosis by decreasing Bax and increasing Bcl-2 with improved neuronal integrity. Further, SH also restored synaptic proteins (PSD-95, SNAP-23) by reversing cadmium-induced synaptic dysfunction. Behavioral tests (MWM and Y-maze) confirmed improved learning and memory that indicate its strong neuroprotective effect (Ali et al., 2025).

Nguyen et al. (2025) explored *in silico* investigation of marine alkaloid aaptamine (Figure 11.4) as potent molecule for neurodegenerative disease. This study showed aaptamine as potent inhibitor, as compared to reference drugs safinamide (MAO-B), tolcapone (COMT) and donepezil (AChE). It was also shown that aaptamine binds AChE more strongly than donepezil. In addition, COMT and MAO-B contribute to neurodegeneration through dopamine breakdown and oxidative stress while

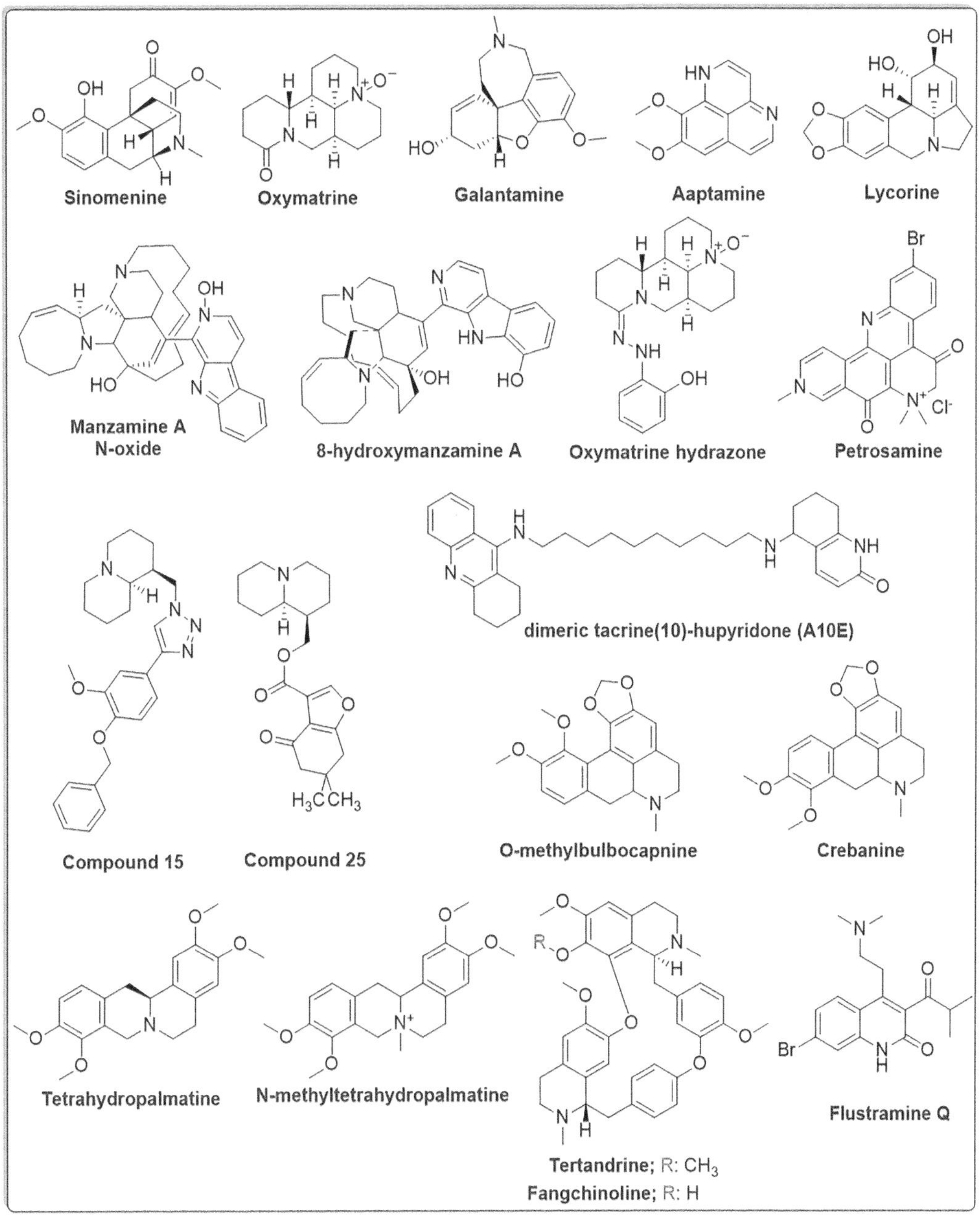

FIGURE 11.4 Structures of N-heterocyclic natural molecules or their derivatives against neurodegenerative diseases.

aaptamine demonstrated stable 500-ns interactions with these targets. It also showed strong van der Waals-driven binding within catalytic pockets similar to their reference drugs (Nguyen et al., 2025).

In the year 2024, a study of petrosamine (*Petrosia* sp.,) was conducted by Riberio and colleagues which showed strong neuroprotection against aluminum-induced toxicity in both *in vitro* (SH-SY5Y neuroblastoma cells) and *in vivo* (zebrafish) AD models. An *in vitro* study demonstrated that petrosamine (Figure 11.4) restored cell viability, reduced apoptosis and suppressed oxidative stress while strong inhibition toward AChE activity. Further, *in vivo* activity revealed reduced neuronal apoptosis, improved developmental defects and significantly lowered elevated AChE activity in zebrafish. Petrosamine exhibited a very potent AChE inhibitory potential with IC_{50} value of 0.091 μM, which was almost 60 times stronger than Rivastigmine (IC_{50} = 5.5 μM) and more potent than galantamine. These results establish petrosamine as a highly promising marine-derived therapeutic lead for AD (Ribeiro et al., 2024).

In another article, Kowal et al. (2023) reported evaluation of 25 marine alkaloids from *Flustra foliacea* for AChE inhibitory profile with the help of Ellman's *in vitro* assay. Flustramine Q (Figure 11.4) showed highest inhibition (82%) with a potent IC_{50} value of 9.6 μM, as compared to reference drug physostigmine. Further, molecular docking demonstrated that flustramine Q showed strong binding toward human AChE with a docking score of −9.38 kcal/mol due to interaction with key residues Trp86, Tyr337, Tyr133 and Asp74 through hydrophobic, hydrogen and cation–π interactions (Kowal et al., 2023).

Maharajan and his team (2023) discussed improvement of oxidative stress induced in HT22 cells by oxymatrine (Figure 11.4). They worked on HT22 neuronal cells in which oxymatrine markedly reduced H_2O_2-induced oxidative stress *via* lowering intracellular ROS. In addition, AMPK phosphorylation restored which in turn suppressed key senescence markers p53, p21 and p16. Furthermore, oxymatrine protected neuronal viability in a dose-dependent manner that demonstrate direct anti-senescence and antioxidant actions. In D-galactose-induced aging mice, oxymatrine-enhanced hippocampal AMPK activity, reduced p53 expression and significantly lowered RAGE levels which indicates decrease of neuroinflammation Maharajan et al. (2023).

In year 2023, Alanzi and Alsalhi screened 388 marine sponge metabolites through virtual docking against five Alzheimer-related protein targets including COX-1 (PDB: 1egq), TACE (PDB: 2fv5), beta-secretase (PDB: 2wjo), OTUD1 (PDB: 4bop) and BACE1 (PDB: 4dju) by employing glide scores, ADMET filters and Lipinski parameters to identify drug-like candidates. From this, five top inhibitors were selected viz. manzamine A N-oxide, isonaamine E, 32,33-dihydro-31-hydroxymanzamine A (for both beta-secretase and OTUD1) and 8-hydroxymanzamine A (Figure 11.4). In addition, molecular dynamics and MM-PBSA analyses confirmed stable binding, favorable free-energy profiles and strong van der Waals/electrostatic interactions within catalytic pockets of each protein (Alanzi & Alsalhi, 2023).

In this research work, comparative investigation of two natural alkaloids (galantamine and lycorine; Figure 11.4) regarding their pharmacological potential against with amyloid β and to treat AD was described by Kola et al. (2023). Herein, NMR and CD spectroscopy study was performed which concluded that structural changes in Aβ40 and Aβ42 peptides occurred upon addition of GAL and LYC that helps in modification of secondary structure and may influence amyloid aggregation. Moreover, antioxidant study revealed that LYC able to delay ascorbate oxidation by inhibiting ROS and decrease oxidative stress in AD as compared to GAL (Kola et al., 2023).

Schepetkin and their research group (2023) synthesized new lupinine triazole derivatives and evaluated them along with 50 lupinine-based esters for AChE inhibition. Among all, **15** (Figure 11.4) was most potent derivative with IC_{50} value of 7.2 μM, i.e., comparable to galantamine (IC_{50} = 8.2 μM). Among ester series, compound **25** (Figure 11.4) showed strongest activity by having IC_{50} value of 24.4 ± 3.4 μM. Kinetic analysis demonstrated that compound **15** follows a mixed-type inhibition mechanism and docking study on AChE enzyme (PDB: 4EY7) revealed strong binding interactions (interface energy = −24.05 kcal/mol) through H-bonds with Tyr337, Tyr124 and Phe295. Additionally, all derivatives were found non-cytotoxic up to 50 μM which highlighted their potential as safe AChE inhibitor leads for Alzheimer's therapeutic development (Schepetkin et al., 2023).

Yang et al. (2022) synthesized oxymatrine hydrazone (Figure 11.4) and evaluated against sevoflurane-induced neuron damage. In this work, synthesized compound protects neurons from 3% sevoflurane toxicity in which viability normally dropped to 62% at 72 h but was almost fully restored by 5.0 μM pretreatment. Sevoflurane increased apoptosis from 0.99% (control) to 39.08% that was markedly reversed by oxymatrine hydrazone *via* suppressing Bax and cleaved caspase-3 and restoring anti-apoptotic BCL-2. It also rescued strong decline in ERK1/2 phosphorylation, bringing p-ERK1/2 levels close to control while the MEK inhibitor U0126 abolished this effect (Yang et al., 2022).

In this literature, inhibitory action of aaptamine (Figure 11.4) toward two cholinesterase (AChE and BuChE) as an anti-Alzheimer's agent was explored by Miao and their team (2022). An *in vivo* study was performed with zebrafish AD model resulted that recovery rate for dyskenia by aaptamine were 34.8, 58.8 and 60.0% at doses of 5, 10 and 20 μM, respectively, as than that of donepezil (63.7% at 8 μM dose) (Miao et al., 2022).

Tong et al. (2022) investigate tetrandrine (Figure 11.4) as a therapeutic agent for tauopathies that focuses on its ability to reduce tau aggregation and cognitive deficits. An *in vitro* study concluded with tau-P301L expressing SH-SY5Y neuronal cells, tetrandrine-restored lysosomal acidification, enhanced autophagy-lysosomal degradation and significantly reduced hyperphosphorylated and aggregated tau by inhibiting TPC2-mediated Ca^{2+} dysregulation. Further, *in vivo* study in P301S transgenic tau mice demonstrated that tetrandrine treatment reduced tau tangles, neuroinflammation and synaptic dysfunction which significantly improved learning and memory performance (Tong et al., 2022).

In year 2021, Bi *et al.*, reported effect of sinomenine (Figure 11.4) in cerebral injury and inflammation in mouse model. Herein, SINO at 20 mg/kg markedly reduced middle cerebral artery occlusion-induced cerebral damage that lowers brain water content and restore antioxidant enzyme activity. In microglia, SINO (50–200 μM) robustly activated Nrf2 within 4 h and elevated HO-1, NQO1 protein and mRNA levels in a dose- and time-dependent manner. SINO also reduced MCAO-induced inflammation *via* decreasing TNF-α, IL-1β and NOS2 expression while antioxidant balance was restored (Bi et al., 2021).

In the same year, Liu and his co-worker showed the potency of tetrahydropalmatine (Figure 11.4) alkaloid against neurotoxicity induced by methamphetamine. It was demonstrated that MA caused neurodegeneration, reduced BDNF and TrkB levels while increased calmodulin and induced apoptosis that leads to behavioral deficits. Also, THP reverses these effects by upregulation of BDNF, activated TrkB, decreased CAM and inhibited apoptotic pathways which improved neuronal structure and function (Liu et al., 2021).

Development of dimeric tacrine(10)-hupyridone (A10E) (Figure 11.4) was carried out by Xuan et al. (2021) which markedly improved cognition in APP/PS1 and Aβ-oligomer mouse models. It lowered hippocampal $A\beta_{1-40}$ and $A\beta_{1-42}$ by more than 40–50%, reduced plaque load ($p < 0.01$) and suppressed neuroinflammation through decreasing IL-6 and TNF-α levels by ~35–45%. A10E also normalized cholinergic dysfunction by reducing elevated AChE activity ($p < 0.05$) and restored ChAT protein/activity while nanomolar concentrations (1–3 nM) activated BDNF/TrkB/Akt pathway to counter Aβ-induced neuronal toxicity (Xuan et al., 2021).

Daowtak and their team (2021) evaluated four alkaloids (Figure 11.4) from *Stephania venosa* viz. crebanine, O-methylbulbocapnine, tetrahydropalmatine and N-methyltetrahydropalmatine for their ability to inhibit amyloid formation by using human insulin and Aβ40/Aβ42 peptides as models. Hereby, crebanine and tetrahydropalmatine showed the strongest anti-amyloid effects by significantly increased fibrillation half-time ($t_{0.5}^{ins}$ from 0.55 to 0.81 h and 0.69 h) and reducing ThT fluorescence (crebanine = 0.44 ± 0.06; tetrahydropalmatine = 0.57 ± 0.07) (Daowtak et al., 2021).

According to research work performed by Ren et al. (2020), tetrandrine (Figure 11.4) was used to treat AD through suppressing inflammatory activation of microglia in 5XFAD mouse model. This study was focused on neurodegeneration-related brain changes particularly amyloid plaque burden, microglial-mediated inflammation and neuronal apoptosis. Here, tetrandrine treatment significantly reduced amyloid plaque deposition and lowered pro-inflammatory cytokines such as TNF-α, IL-1β and IL-6 while also decreasing COX-2, iNOS and NF-κB expression. It was also noticed that 40 mg/kg tetrandrine markedly improved learning and memory performance in Morris water maze that restored escape latency close to wild-type levels (Ren et al., 2020).

In same year, Kong and co-workers assessed cholinesterase inhibitory potential of Stephaniae tetrandrae Radix to support brain health by targeting cholinesterase enzymes relevant to AD. This study aimed to identify active herbal components and their ability evaluation to inhibit AChE and BChE enzymes linked to cognitive decline. STR extracts rich in alkaloids showed strong AChE inhibition with ~82% AChE inhibition and ~41% BChE inhibition at 200 μg/mL. Moreover, fangchinoline emerged as most potent neuroprotective candidate by targeting AChE and showed IC_{50} value of 2.58 ± 0.28 μM. Further, docking study was performed on AChE (PDB ID: 4EY7) enzyme with fangchinoline and cyclanoline that experienced binding energy of −7.11 and −7.00 kJ/mol, respectively (Kong et al., 2020).

Gan et al. (2020) disclosed oxymatrine (Figure 11.4)-mediated reduction in dopaminergic neuronal damage and microglia-mediated neuroinflammation in MPTP/MPP^+-induced models of PD. This study showed that MPTP caused severe motor loss (decreased rotarod time from 135.63 ± 14.03 to 42.23 ± 5.60 s) and reduced striatal dopamine (1,625.69 ± 175.6–567.78 ± 45.99) while oxymatrine restored these deficits and increased DA and DOPAC levels at 5, 10 and 20 mg/kg dose. Moreover, oxymatrine also suppressed microglial activation and reduced TNF-α, IL-1β, IL-6 and NO levels (Gan et al., 2020).

11.8 FUTURE DIRECTIONS AND CHALLENGES

Single-target medications frequently lack adequate modulation of various pathogenic processes implicated in neurodegenerative illnesses due to their complex and interrelated character. The development of compounds with concurrent regulation capabilities toward oxidative stress, neuroinflammation, protein aggregation, mitochondrial malfunction and neurotransmitter imbalances is known as multi-target drug design. Because of their chemical diversity, capacity to engage various binding pockets and compatibility with hybridization techniques, N-containing heterocyclic scaffolds are particularly well-suited for MTDL creation (Wang et al., 2023). In addition, integration of cholinesterase inhibition with antioxidant activity, kinase modulation with anti-amyloid capabilities or MAO-B inhibition with anti-inflammatory effects within a single molecular entity became focus of rational design efforts. Despite the significant potential of MTDLs, there are still issues regarding safety, polypharmacological toxicity prevention and attaining balanced potency across targets which necessitate rigorous optimization supported by biological validation and *in silico* modeling (Keller et al., 2020).

Natural product-based neuroprotective drug discovery is being revolutionized by developments in artificial intelligence (AI), machine learning and multi-omics technologies. Further, optimization of heterocyclic scaffolds is accelerated by quick prediction of bioactivity, BBB permeability and metabolic liabilities afforded by AI-driven molecular modeling. Herein, metabolomics, transcriptomics and proteomics are some examples of omics platforms that offer a thorough understanding about molecular networks influenced by natural alkaloids related to neuroprotection (Cui et al., 2025). By using these approaches, SAR analysis is improved, synergistic interactions between natural and synthetic derivatives are identified while logical design of semisynthetic analogues is encouraged. Furthermore, AI-supported biosynthetic route reconstruction assists to identify uncommon heterocycles by anticipating enzymatic steps and microbial synthesis promotion. This combination of computational and systems biology with natural product chemistry offers more accurate targeting of neurodegenerative pathways along with quick increase in number of available lead structures (Maiese, 2024). Many nitrogen-based natural products come from plants or marine species, i.e., difficult to grow, harvest sustainably or extract in large amounts for industrial use. By facilitating heterocycle creation through modified microbial hosts like *Escherichia coli*, *Streptomyces* and yeast strains optimized for alkaloid biosynthesis, synthetic biology provides novel alternatives. Researchers can create complex nitrogenous scaffolds in scalable, ecological systems by reconstructing biosynthetic gene clusters, altering enzymatic selectivity or adding new metabolic pathways (Urui et al., 2021). Furthermore, sustainability is improved by using renewable feedstocks, biocatalysis and green chemistry concepts. These initiatives are crucial to maintain a steady supply and to reduce negative effects on environment and enhancement in economic development of neuroprotective heterocyclic lead compounds (Zhao et al., 2025).

11.9 CONCLUSIONS

Heterocyclic natural products that include nitrogen are a highly effective source of bioactive compounds with great potential for treatment of neurodegenerative diseases. They can address the intricate biochemical landscape to drive neuronal degeneration due to their structural diversity, evolutionary optimization and ability to interact with different molecular pathways. These substances have complex neuroprotective mechanisms that go beyond the boundaries of traditional single-target treatments ranging from reduction of oxidative stress and neuroinflammation to prevent protein aggregation alongside regulation of mitochondrial integrity. Although integration with synthetic biology allows sustainable access to rare and important scaffolds, advances in synthetic modification, in silico modeling and delivery methods continue to extend their therapeutic potential. Despite ongoing interdisciplinary research challenges in optimization of pharmacokinetics, safety and translational efficacy, nitrogenous heterocyclic natural products are promising lead candidates for next-generation neurotherapeutics, i.e., capable to reduce rising global burden of neurodegenerative diseases.

REFERENCES

Akbarizadeh-Mashkani, M. H., Afshinmajd, S., Iranzadeh, S., & Roghani, M. (2025). Involvement of synaptophysin and microtubule-associated protein 2 in the neuroprotective effect of berberine in an amyloid β-induced rat model of Alzheimer's disease. *Avicenna Journal of Phytomedicine*, *15*(6), 1741.

Alanzi, A. R., & Alsalhi, M. S. (2023). Exploring marine sponge metabolites as promising leads for Alzheimer's disease: Virtual screening, ADMET analysis, and molecular dynamics simulations. *Biomedical and Biotechnology Research Journal (BBRJ)*, *7*(3), 484–496.

Ali, W., Choe, K., Rehman, I. U., Park, H. Y., Jang, S., Ullah, S., ..., & Kim, M. O. (2025). Sinomenine hydrochloride impedes memory impairments via Nrf2/HO-1-mediated inhibition of oxidative stress, neuroinflammation and apoptosis in mice brain. *Journal of Neuroimmune Pharmacology*, *20*(1), 85.

Aljarba, N. H., Qais, F. A., Alqahtani, R. A., & Alkahtani, S. (2023). Multi-targeted virtual screening of phytocompounds of *Rauwolfia serpentina* against caspase-8, BACE, and AChE for the treatment of neurodegenerative diseases. *Journal of Disability Research*, *2*(3), 70–85.

Anastassova, N., Aluani, D., Hristova-Avakumova, N., Tzankova, V., Kondeva-Burdina, M., Rangelov, M., ..., & Yancheva, D. (2022). Study on the neuroprotective, radical-scavenging and MAO-B inhibiting properties of new benzimidazole arylhydrazones as potential multi-target drugs for the treatment of Parkinson's disease. *Antioxidants*, *11*(5), 884.

Andrade, V. H. D., Rodrigues, E. Y. M., Dias, N. A., Ferreira, G. F., Carvalho, D. B. D., das Neves, A. R., ..., & Lagatta, D. C. (2023). Neuroprotective profile of triazole grandisin analogue against amyloid-beta oligomer-induced cognitive impairment. *ACS Chemical Neuroscience*, *14*(24), 4298–4310.

Andrés, C. M. C., Pérez de la Lastra, J. M., Andrés Juan, C., Plou, F. J., & Pérez-Lebeña, E. (2022). Impact of reactive species on amino acids—Biological relevance in proteins and induced pathologies. *International Journal of Molecular Sciences*, *23*(22), 14049.

Ayyappan, V., & Ekambaram, Y. (2025). Analyzing the role of natural marine compounds in developing biomedical applications and pharmaceuticals. *Natural and Engineering Sciences*, *10*(2), 37–47.

Berger, A., Valant-Vetschera, K., Schinnerl, J., & Brecker, L. (2022). Alkaloid diversification in the genus Palicourea (Rubiaceae: Palicoureeae) viewed from a (retro-) biogenetic perspective. *Phytochemistry Reviews*, *21*(3), 915–939.

Bhat, J. A., & Kumar, M. (2022). Neuroprotective effects of theobromine in permanent bilateral common carotid artery occlusion rat model of cerebral hypoperfusion. *Metabolic Brain Disease*, *37*(6), 1787–1801.

Bi, F., Zhang, Y., Liu, W., & Xie, K. (2021). Sinomenine activation of Nrf2 signaling prevents inflammation and cerebral injury in a mouse model of ischemic stroke. *Experimental and Therapeutic Medicine*, *21*(6), 647.

Bido, S., Muggeo, S., Massimino, L., Marzi, M. J., Giannelli, S. G., Melacini, E., ..., & Broccoli, V. (2021). Microglia-specific overexpression of α-synuclein leads to severe dopaminergic neurodegeneration by phagocytic exhaustion and oxidative toxicity. *Nature Communications*, *12*(1), 6237.

Caruana, M., Camilleri, A., Farrugia, M. Y., Ghio, S., Jakubíčková, M., Cauchi, R. J., & Vassallo, N. (2021). Extract from the marine seaweed Padina pavonica protects mitochondrial biomembranes from damage by amyloidogenic peptides. *Molecules*, *26*(5), 1444.

Catanesi, M., Caioni, G., Castelli, V., Benedetti, E., d'Angelo, M., & Cimini, A. (2021). Benefits under the sea: The role of marine compounds in neurodegenerative disorders. *Marine Drugs*, *19*(1), 24.

Chau, E., & Kim, J. R. (2022). α-synuclein-assisted oligomerization of β-amyloid (1–42). *Archives of Biochemistry and Biophysics*, *717*, 109120.

Chung, H. J., Kim, J. E., Huh, Y., San Lee, J., Kim, S. W., Na, K., …, & Jeong, N. Y. (2024). The multi-targeted effect of fascaplysin on the proliferation and dedifferentiation of schwann cells inhibits peripheral nerve degeneration by blocking CDK4/6 and androgen receptor. *Experimental Neurobiology*, *33*(6), 266.

Cui, G., Li, M., Guo, W., Gao, M., Zhu, Q., & Liao, J. (2025). AI driven network pharmacology: Multi-scale mechanisms of traditional Chinese medicine from molecular to patient analysis. *Computational and Structural Biotechnology Journal*, *27*, 5087.

da Motta, K. P., Martins, C. C., da Rocha, V. M., Soares, M. P., Mesko, M. F., Luchese, C., & Wilhelm, E. A. (2024). Insights into vincristine-induced peripheral neuropathy in aged rats: Wallerian degeneration, oxidative damage, and alterations in ATPase enzymes. *ACS Chemical Neuroscience*, *15*(21), 3954–3969.

Daowtak, K., Pompimon, W., Salerno, M., & Udomtanakunchai, C. (2021). *Decelerate of amyloid fibrillation by the alkaloids extracted from Stephania venosa*.

Duan, W. J., Liang, L., Pan, M. H., Lu, D. H., Wang, T. M., Li, S. B., …, & He, R. R. (2020). Theacrine, a purine alkaloid from kucha, protects against Parkinson's disease through SIRT3 activation. *Phytomedicine*, *77*, 153281.

Gan, P., Ding, L., Hang, G., Xia, Q., Huang, Z., Ye, X., & Qian, X. (2020). Oxymatrine attenuates dopaminergic neuronal damage and microglia-mediated neuroinflammation through cathepsin D-dependent HMGB1/TLR4/NF-κB pathway in Parkinson's disease. *Frontiers in Pharmacology*, *11*, 776.

Harding, R. J., Deme, J. C., Hevler, J. F., Tamara, S., Lemak, A., Cantle, J. P., …, & Arrowsmith, C. H. (2021). Huntingtin structure is orchestrated by HAP40 and shows a polyglutamine expansion-specific interaction with exon 1. *Communications Biology*, *4*(1), 1374.

Iova, O. M., Marin, G. E., Lazar, I., Stanescu, I., Dogaru, G., Nicula, C. A., & Bulboacă, A. E. (2023). Nitric oxide/nitric oxide synthase system in the pathogenesis of neurodegenerative disorders—An overview. *Antioxidants*, *12*(3), 753.

Ismail, H., Khalid, D., Waseem, D., Ijaz, M. U., Dilshad, E., Haq, I. U., …, & Saleem, S. (2023). Bioassays guided isolation of berberine from *Berberis lycium* and its neuroprotective role in aluminium chloride induced rat model of Alzheimer's disease combined with insilico molecular docking. *PLoS One*, *18*(11), e0286349.

Jain, S., Kim, E. S., Kim, D., Burrows, D., De Felice, M., Kim, M., …, & Majid, A. (2020). Comparative cerebroprotective potential of d-and l-carnosine following ischemic stroke in mice. *International Journal of Molecular Sciences*, *21*(9), 3053.

Katsube, M., Ishimoto, T., Fukushima, Y., Kagami, A., Shuto, T., & Kato, Y. (2024). Ergothioneine promotes longevity and healthy aging in male mice. *GeroScience*, *46*(4), 3889–3909.

Keller, S., Polanski, W. H., Enzensperger, C., Reichmann, H., Hermann, A., & Gille, G. (2020). 9-Methyl-β-carboline inhibits monoamine oxidase activity and stimulates the expression of neurotrophic factors by astrocytes. *Journal of Neural Transmission*, *127*(7), 999–1012.

Kola, A., Lamponi, S., Currò, F., & Valensin, D. (2023). A comparative study between lycorine and galantamine abilities to interact with AMYLOID β and reduce in vitro neurotoxicity. *International Journal of Molecular Sciences*, *24*(3), 2500.

Kong, X. P., Ren, H. Q., Liu, E. Y., Leung, K. W., Guo, S. C., Duan, R., …, & Tsim, K. W. (2020). The cholinesterase inhibitory properties of stephaniae tetrandrae radix. *Molecules*, *25*(24), 5914.

Kowal, N. M., Di, X., Omarsdottir, S., & Olafsdottir, E. S. (2023). Flustramine Q, a novel marine origin acetylcholinesterase inhibitor from *Flustra foliacea*. *Future Pharmacology*, *3*(1), 38–47.

Krishnan, N., Mariappanadar, V., Dhanabalan, A. K., Devadasan, V., Gopinath, S. C., & Raman, P. (2023). Purification, identification and in silico models of alkaloids from *Nardostachys jatamansi*—bioactive compounds for neurodegenerative diseases. *Biomass Conversion and Biorefinery*, *13*(16), 14889–14900.

Laurencin, C., Lancelot, S., Merida, I., Costes, N., Redouté, J., Le Bars, D., …, & Ballanger, B. (2023). Distribution of α2-adrenergic receptors in the living human brain using [11C] yohimbine PET. *Biomolecules*, *13*(5), 843.

Lee, T. Y., Yang, W., Cha, D. S., & Han, Y. T. (2023). Synthesis of a natural quinoline alkaloid isolated from the deep-sea-derived fungus and its potential as a therapeutic for Parkinson's disease. *Journal of Asian Natural Products Research*, *25*(5), 446–455.

Li, B., Yang, Y., Wang, Y., Zhang, J., Ding, J., Liu, X., …, & Sun, C. (2021). Acetylation of NDUFV1 induced by a newly synthesized HDAC6 inhibitor HGC rescues dopaminergic neuron loss in Parkinson models. *Iscience*, *24*(4).

Li, Q., & Sun, X. (2025). Pharmacological interventions for pain and fatigue management in elderly leukemia patients: A comparative study of analgesic and antidepressant efficacy. *Pakistan Journal of Pharmaceutical Sciences, 38*(6).

Liu, L., Liu, M., Zhao, W., Zhao, Y. L., & Wang, Y. (2021). Tetrahydropalmatine regulates BDNF through TrkB/CAM interaction to alleviate the neurotoxicity induced by methamphetamine. *ACS Chemical Neuroscience, 12*(18), 3373–3386.

Llorach-Pares, L., Rodriguez-Urgelles, E., Nonell-Canals, A., Alberch, J., Avila, C., Sanchez-Martinez, M., & Giralt, A. (2020). Meridianins and lignarenone B as potential GSK3β inhibitors and inductors of structural neuronal plasticity. *Biomolecules, 10*(4), 639.

Maharajan, N., Lee, C. M., Vijayakumar, K. A., & Cho, G. W. (2023). Oxymatrine improves oxidative stress-induced senescence in HT22 cells and mice via the activation of AMP-activated protein kinase. *Antioxidants, 12*(12), 2078.

Maiese, K. (2024). Artificial intelligence and disease signature pathways: Driving innovation to elucidate underlying pathogenic mechanisms. *Current Neurovascular Research, 21*(3), 229–233.

Miao, S., He, Q., Li, C., Wu, Y., Liu, M., Chen, Y., ..., & Gong, K. (2022). Aaptamine–a dual acetyl–and butyrylcholinesterase inhibitor as potential anti-Alzheimer's disease agent. *Pharmaceutical Biology, 60*(1), 1502–1510.

Nguyen, T. L. A., Nguyen, H. L., Pham, T. M., Truong, D. H., Ngo, T. C., Li, M. S., & Dao, D. Q. (2025). Enzyme inhibitory activity of marine alkaloid aaptamines for neurodegenerative diseases: An in silico study. *Royal Society Open Science, 12*(7).

Ni, H., Liu, M., Cao, M., Zhang, L., Zhao, Y., Yi, L., ..., & Dong, Y. (2024). Sinomenine regulates the cholinergic anti-inflammatory pathway to inhibit TLR4/NF-κB pathway and protect the homeostasis in brain and gut in scopolamine-induced Alzheimer's disease mice. *Biomedicine & Pharmacotherapy, 171*, 116190.

Nishit, P., Kumar, V. S., Ishi, T., Lokesh, A., Cao, H., & Sanjib, B. (2022). Neurodegenerative disorders of Alzheimer, Parkinsonism, Amyotrophic lateral sclerosis and multiple sclerosis: An early diagnostic approach for precision treatment. *Metabolic Brain Disease, 37*(1), 67–104.

Nozaki, S., Hijioka, M., Wen, X., Iwashita, N., Namba, J., Nomura, Y., ..., & Kitamura, Y. (2024). Galantamine suppresses α-synuclein aggregation by inducing autophagy via the activation of α7 nicotinic acetylcholine receptors. *Journal of Pharmacological Sciences, 156*(2), 102–114.

Onifade, I. A., Umar, H. I., Aborode, A. T., Awaji, A. A., Jegede, I. D., Adeleye, B. H., ..., & Idowu, N. (2025). In silico study of selected alkaloids as dual inhibitors of β-and γ-secretases for Alzheimer's disease. *Journal of Alzheimer's Disease, 103*(4), 1191–1215.

Ordoñez, W. O. C., Palomino, N. V., Varela, P. E. V., Martínez, I. B., Alves, L. B., & Giuliatti, S. (2025). Alkaloids from *Caliphruria subedentata* (Amaryllidaceae) as regulators of AChE, BuChE, NMDA and GSK3 activity: An in vitro and in silico approach for mimicking Alzheimer's disease. *Neurochemical Research, 50*(2), 1–19.

Priyadharshini, G. B., Hassan, S., Meenatchi, R., Kiran, G. S., Jayanthy, C., Veera Bramhachari, P., & Selvin, J. (2024). Neuroactive peptides and neuroprotective molecules from marine sponges and associated bacteria: An untapped resource for systemic drug development. In *Marine bioactive molecules for biomedical and pharmacotherapeutic applications* (pp. 283–323). Springer Nature Singapore.

Ren, D., Fu, Y., Wang, L., Zhong, X., Yuan, J., Jiang, C., ..., & Li, Z. (2020). *Tetrandrine ameliorated Alzheimer's Disease through suppressing the inflammatory activation of microglia in 5XFAD mouse.*

Ribeiro, J., Araújo-Silva, H., Fernandes, M., da Silva, J. A., Pinto, F. D. C. L., Pessoa, O. D. L., ..., & Gomes, A. C. (2024). Petrosamine isolated from marine sponge *Petrosia* sp. demonstrates protection against neurotoxicity in vitro and in vivo. *Natural Products and Bioprospecting, 14*(1), 16.

Rivera-García, L. G., Francis-Malavé, A. M., Wilson, T. D., Ferchmin, P. A., Eterovic, V., & Carrasquillo, Y. (2022). 4R-Tobacco-Cembranoid is a Positive Allosteric Modulator of the Alpha7 nicotinic acetylcholine receptor that modulates inflammation-induced thermal hypersensitivity in mice. *bioRxiv*, 2022-01.

Rupreo, V., Saha, R., Bhattacharyya, J., & Mitra, R. K. (2025). Molecular mechanisms of Lobeline-mediated inhibition of lysozyme amyloidogenesis: A synergistic approach using biophysical and cheminformatics techniques. *Biophysical Chemistry*, 107502.

Safir, W., Khan, Y., Uddin, N., Malik, A., & Maqbool, A. (2025). Unlocking therapeutic potential of alkaloids inhibiting acetylcholine esterase: A comparative molecular docking and MD simulation analysis with known drugs of galantamine and physostigmine. *Trends in Medical and Biological Sciences, 1*(1), 1–16.

Schepetkin, I. A., Nurmaganbetov, Z. S., Fazylov, S. D., Nurkenov, O. A., Khlebnikov, A. I., Seilkhanov, T. M., ..., & Quinn, M. T. (2023). Inhibition of acetylcholinesterase by novel lupinine derivatives. *Molecules, 28*(8), 3357.

Shaker, F. H., El-Derany, M. O., Wahdan, S. A., El-Demerdash, E., & El-Mesallamy, H. O. (2021). Berberine ameliorates doxorubicin-induced cognitive impairment (chemobrain) in rats. *Life sciences*, *269*, 119078.

Shalaby, A. M., Sharaf Eldin, H. E., Abdelsameea, A. A., Abdelnour, H. M., Alabiad, M. A., Elkholy, M. R., & Aboregela, A. M. (2022). Betahistine attenuates seizures, neurodegeneration, apoptosis, and gliosis in the cerebral cortex and hippocampus in a mouse model of epilepsy: A histological, immunohistochemical, and biochemical study. *Microscopy and Microanalysis*, *28*(5), 1734–1748.

Shoaib, S., Ansari, M. A., Fatease, A. A., Safhi, A. Y., Hani, U., Jahan, R., ..., & Islam, N. (2023). Plant-derived bioactive compounds in the management of neurodegenerative disorders: Challenges, future directions and molecular mechanisms involved in neuroprotection. *Pharmaceutics*, *15*(3), 749.

Shome, A., Chahat, Chawla, V., & Chawla, P. A. (2024). Neuroprotective effect of natural indole and β-carboline alkaloids against Parkinson's disease: An overview. *Current Medicinal Chemistry*, *31*(38), 6251–6271.

Shrestha, R. L., MC, S., Tamang, A., Poudel, M., Parajuli, N., Shrestha, A., ..., & Adhikari Subin, J. (2025). Evaluating the inhibitory efficacy of Oxalis phytocompounds on monoamine oxidase B: An integrated approach targeting age related neurodegenerative diseases through molecular docking and dynamics simulations. *PLos One*, *20*(7), e0329168.

Silva, J., Martins, A., Alves, C., Pinteus, S., Gaspar, H., Alfonso, A., & Pedrosa, R. (2020). Natural approaches for neurological disorders—The neuroprotective potential of *Codium tomentosum*. *Molecules*, *25*(22), 5478.

Stavrakov, G., Philipova, I., Lukarski, A., Atanasova, M., Zheleva, D., Zhivkova, Z. D., ..., & Doytchinova, I. (2020). Galantamine-curcumin hybrids as dual-site binding acetylcholinesterase inhibitors. *Molecules*, *25*(15), 3341.

Sumien, N., Cunningham, J. T., Davis, D. L., Engelland, R., Fadeyibi, O., Farmer, G. E., ..., & Cunningham, R. L. (2021). Neurodegenerative disease: Roles for sex, hormones, and oxidative stress. *Endocrinology*, *162*(11), bqab185.

Tong, B. C. K., Huang, A. S., Wu, A. J., Iyaswamy, A., Ho, O. K. Y., Kong, A. H. Y., ..., & Cheung, K. H. (2022). Tetrandrine ameliorates cognitive deficits and mitigates tau aggregation in cell and animal models of tauopathies. *Journal of Biomedical Science*, *29*(1), 85.

Urui, M., Yamada, Y., Ikeda, Y., Nakagawa, A., Sato, F., Minami, H., & Shitan, N. (2021). Establishment of a co-culture system using *Escherichia coli* and Pichia pastoris (*Komagataella phaffii*) for valuable alkaloid production. *Microbial Cell Factories*, *20*(1), 200.

van Onselen, R., & Downing, T. G. (2021). Neonatal reserpine administration produces widespread neuronal losses and α-Synuclein inclusions in a rat model. *Neurotoxicity Research*, *39*(6), 1762–1770.

Viegas, F. V. P. D., Gontijo, V. S., de Freitas Silva, M., Ortiz, C. J., dos Reis Rosa Franco, G., Ernesto, J. T., ..., & Viegas, C. (2022). Curcumin, resveratrol and cannabidiol as natural key prototypes in drug design for neuroprotective agents. *Current Neuropharmacology*, *20*(7), 1297–1328.

Waheeb, T. S., Abdulkader, M. A., Ghareeb, D. A., & Moustafa, M. E. (2025). Neuroprotective efficacy of berberine and caffeine against rotenone-induced neuroinflammatory and oxidative disturbances associated with Parkinson's disease via inhibiting α-synuclein aggregation and boosting dopamine release. *Inflammopharmacology*, 1–22.

Wan, L. X., Zhang, J. F., Zhen, Y. Q., Zhang, L., Li, X., Gao, F., & Zhou, X. L. (2021). Isolation, structure elucidation, semi-synthesis, and structural modification of C19-diterpenoid alkaloids from *Aconitum apetalum* and their neuroprotective activities. *Journal of Natural Products*, *84*(4), 1067–1077.

Wang, Y., Pei, H., Chen, W., Du, R., Li, J., & He, Z. (2023). Palmatine protects PC12 cells and mice from Aβ25-35-induced oxidative stress and neuroinflammation via the Nrf2/HO-1 pathway. *Molecules*, *28*(24), 7955.

Webster, J. M., Yang, Y. T., Miller, A. T., Zane, A., Scholz, K., Stone, W. J., ..., & Harms, A. S. (2025). Tau, amyloid-β and α-synuclein co-pathologies synergistically enhance neuroinflammation and neuropathology. *bioRxiv*, 2024-10.

Wiesen, T., & Atlas, D. (2022). Novel anti-apoptotic L-DOPA precursors SuperDopa and SuperDopamide as potential neuroprotective agents for halting/delaying progression of Parkinson's disease. *Cell Death & Disease*, *13*(3), 227.

Xie, J. (2025). Sanguinarine alleviates post-stroke cognitive impairment in rats by modulating the p38 MAPK signaling pathway: Sanguinarine improves post-stroke cognitive impairment. *Indian Journal of Experimental Biology (IJEB)*, *63*(09), 726–737.

Xie, Z., Liu, W., Dang, R., Hu, X., Cai, F., Xiang, Z., ..., & Wang, C. (2025). Effects and mechanisms of harmine on ameliorating ethanol-induced memory impairment. *Journal of Ethnopharmacology*, *337*, 118789.

Xuan, Z., Gu, X., Yan, S., Xie, Y., Zhou, Y., Zhang, H., …, & Cui, W. (2021). *Dimeric tacrine (10)-hupyridone effectively combats Alzheimer's Disease as A multi-target-directed ligand.*

Yang, J., Zhang, Y., Zhang, G., Zhao, W., & Yang, H. (2022). Synthesis of oxymatrine hydrazone and its preventive action against sevoflurane induced neuron damage through ERK pathway up-regulation. *Acta Biochimica Polonica, 69*(3), 537–542.

Zhao, M., Wu, X., Xu, D., Zhang, B., & Li, F. (2025). Efficient synthesis of tetrahydroisoquinoline alkaloids from dopamine and natural phenolic acids by whole-cell biocatalysis. *ACS Sustainable Chemistry & Engineering.*

Zuppe, H., & Reed, E. (2024). Common cytokine receptor gamma chain family cytokines activate MAPK, PI3K, and JAK/STAT pathways in microglia to influence Alzheimer's disease. *Frontiers in Molecular Neuroscience, 17*, 1441691.

12 Pharmacokinetic and Pharmacodynamics of Nitrogen Heterocycles in the Brain

Shraddha M. Gupta, Dinesh D. Rishipathak, Pratiksha N. Sonwane, and Kapil Kumar

12.1 INTRODUCTION

Nitrogen-containing heterocycles are a crucial structural class in contemporary medicinal chemistry due to their significant presence in bioactive natural products and approved pharmaceuticals. Their structural versatility, capacity to modulate electronic distribution, and ability to engage in hydrogen bonding render them crucial scaffolds for high-affinity interactions with biological targets, thereby significantly influencing potency and selectivity in drug molecules (Hossain et al., 2023; Kerru et al., 2020; Rizzo et al., 2023. Recent analyses indicate that most newly approved small-molecule drugs contain at least one nitrogen heterocycle, highlighting their importance in current drug design and discovery processes (de la Torre & Albericio, 2024; Jampílek & Kráľová, 2024; Pibiri, 2024).

Nitrogen heterocycles significantly influence pharmacodynamic properties and are essential for enhancing pharmacokinetic and physicochemical characteristics, including aqueous solubility, membrane permeability, metabolic stability, and oral bioavailability (Gao et al., 2022; Ohshima & Nakahara, 2024; Pal & Singh, 2023). Their synthetic accessibility, enhanced by developments in transition-metal catalysis, heterocycle annulation strategies, and environmentally friendly synthetic methodologies, has significantly broadened the chemical space for medicinal chemists, facilitating swift analog development and structure–activity relationship (SAR) exploration (Dwivedi et al., 2025; Kumar et al., 2023; Ward & O'Boyle, 2025). Recent literature emphasizes notable advancements in the construction of heterocyclic scaffolds, including pyridines, indoles, imidazoles, triazoles, and fused bicyclic systems, which are increasingly utilized in lead optimization programs (Ebenezer et al., 2022; Mo et al., 2024).

Nitrogen heterocycles are present in nearly all therapeutic areas, including oncology, infectious diseases, neurological disorders, and cardiovascular diseases. Their presence in biologically active natural products, including alkaloids, nucleobases, and amino-acid-derived frameworks, underscores their biological significance and appropriateness as drug-like motifs (Jadala et al., 2026; Nehra et al., 2025). Their electronic tunability, favorable pharmacokinetic properties, and extensive pharmacological potential position nitrogen heterocycles as essential components in contemporary drug discovery.

12.2 PHARMACOKINETICS OF NITROGEN HETEROCYCLES

12.2.1 Absorption

The physicochemical characteristics of nitrogen-containing heterocycles, particularly their ionization state (pKa), lipophilicity (log P/log D), molecular weight, and topological polar surface area (TPSA), have a significant impact on their absorption. The degree of passive diffusion and involvement in carrier-mediated uptake are determined by these characteristics taken together. Permeability

DOI: 10.1201/9781003772019-12

across the gastrointestinal (GI) epithelium is generally better for heterocycles with low hydrogen-bond donors, TPSA < 90 Å^2, and moderate lipophilicity (log P 1–3) (Stellnberger et al., 2024). For example, because of its tiny size and capacity to exist partially in an unprotonated, membrane-permeable form at physiological pH, nicotine, a bicyclic nitrogen heterocycle, exhibits fast oral absorption.

Complex heterocycles like clozapine and fluoxetine depend on their balanced lipophilicity and partial ionization for efficient diffusion across enterocytes. In contrast, strongly basic heterocycles with pKa> 10, like ranitidine (a di-substituted heterocyclic amine), demonstrate diminished membrane permeability owing to nearly complete protonation in the GI tract. This condition requires transporter-mediated uptake or elevated doses to attain therapeutic plasma concentrations.

Experimental methods for determining pKa and log D are crucial for predicting the absorption of central nervous system (CNS)-directed heterocycles, as computational predictions may be unreliable for structures with multiple nitrogen atoms or fused ring systems (Stellnberger et al., 2024). The protonation states significantly affect the fraction of a drug that can be absorbed; therefore, medicinal chemists frequently modify heterocycle substitutions to optimize pKa and permeability during the lead optimization process.

12.2.2 Distribution

The distribution of nitrogen heterocycles is affected by factors, including tissue perfusion, plasma protein binding, lipophilicity, and active transport mechanisms across tissues like the blood–brain barrier (BBB). Lipophilic heterocycles, especially those characterized by low TPSA and few hydrogen bond donors, generally exhibit increased volumes of distribution. Diazepam, characterized by a diazepine ring, exhibits significant distribution into adipose tissue, resulting in an extended elimination half-life.

12.2.2.1 Blood–Brain Barrier Penetration

The capacity of CNS-active nitrogen heterocycles to penetrate the BBB is a critical determinant of their therapeutic potential. Molecules like sertraline, clozapine, and lamotrigine effectively penetrate the BBB due to an optimal balance of lipophilicity and pKa, which ensures they remain adequately un-ionized at physiological pH (Banks et al., 2024; Spielvogel et al., 2025). Absorption and distribution of some drugs containing nitrogen heterocycles are shown in Table 12.1. Structural characteristics include the following:

- TPSA is less than 70 Å^2.
- Less than two hydrogen bond donors.
- Molecular weight is less than 450 Da.
- Adjustment of pKa to prevent excessive positive charge.
- Collectively improve BBB permeability (Spielvogel et al., 2025). Minor deviations, including the addition of an extra nitrogen atom, may decrease permeability as a result of heightened polarity or enhanced P-glycoprotein (P-gp) efflux.

12.2.2.2 Transporter Interactions

Transporters play a crucial role in distribution dynamics. Basic heterocycles, including quinine and diphenhydramine, serve as substrates for P-glycoprotein, thereby restricting their penetration into the brain. Heterocycles that resemble endogenous amines, such as selegiline, an MAO-B inhibitor, may engage organic cation transporters (OCTs) to facilitate CNS uptake(Huang et al., 2022).

12.2.3 Metabolism

Nitrogen heterocycles are subject to significant Phase I and Phase II metabolic processes, primarily involving cytochrome P450 (CYP) enzymes, monoamine oxidases (MAOs), and various conjugation pathways.

TABLE 12.1
Absorption and Distribution of Drugs Containing Nitrogen Heterocycles

Drug (Class) **Key Heterocycle/Scaffold**	**Absorption (F, Tmax)**	**Distribution (Vd, PPB, BBB)**
Nicotine (smoking cessation) 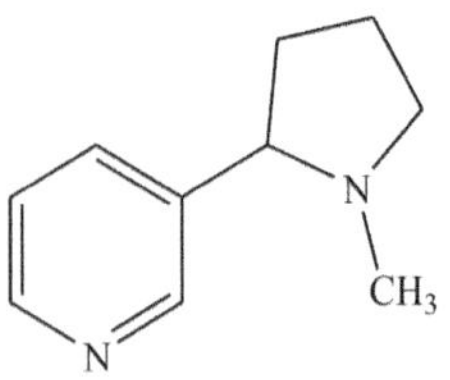 Pyridine, pyrrolidine	High F via buccal/transdermal/oral forms Tmax ~10–30 min (inhaled faster)	Vd moderate Low PPB Rapid CNS uptake (readily penetrant) (Goldenson et al., 2021)
Clozapine (antipsychotic) 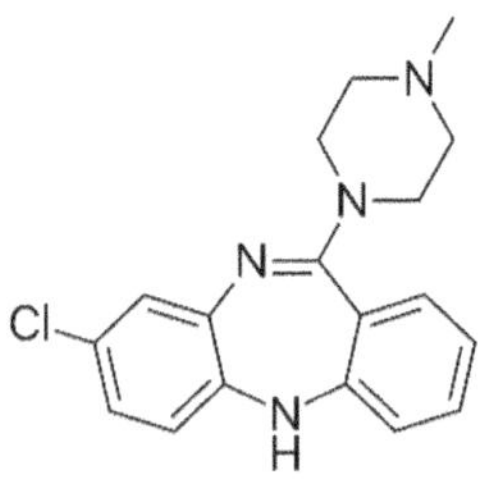 Dibenzodiazepine	Good oral absorption (F variable due to first-pass) Tmax ~2–6 h	Large Vd PPB ~95% BBB penetrant (CNS active) (Freibüchler & Seifert, 2024)
Fluoxetine (SSRI) 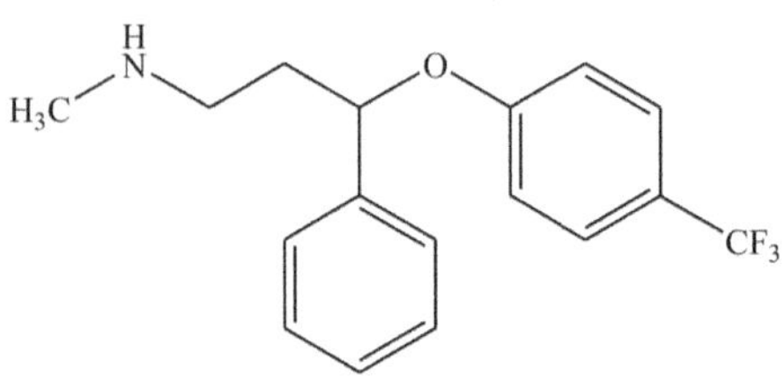 Arylamine/secondary amine heterocycle	F 70–90% Tmax 6–8 h	Very large Vd (tissue accumulation) PPB ~95% BBB penetrant (Sohel et al., 2024)
Lamotrigine (antiepileptic) 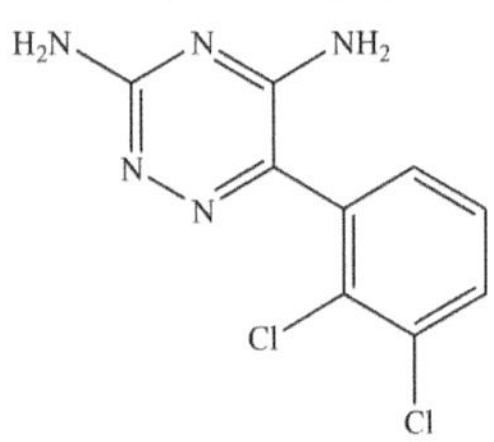 Triazine	Good oral F (~>90% for some formulations) Tmax 1–3 h (IR)	Moderate Vd Low to moderate PPB (~55%) Crosses BBB (antiseizure effect) (Li et al., 2025)
Selegiline (MAO-B inhibitor) 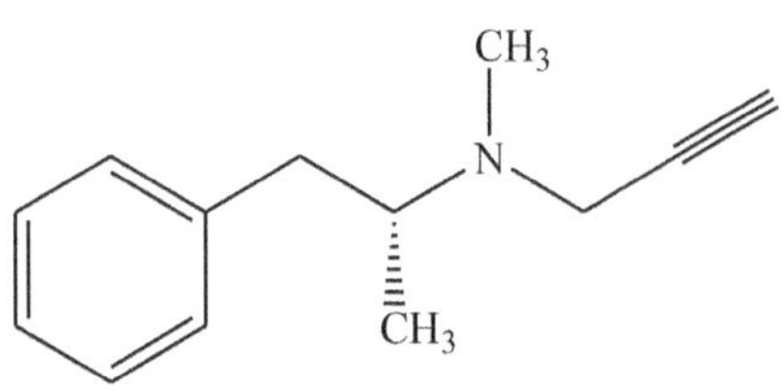 Propargylamine, phenethylamine	Good oral F (but high first-pass for oral) Transdermal form bypasses first-pass	Moderate Vd BBB penetrant (CNS target) (Moore & Saadabadi, 2023)

(Continued)

TABLE 12.1 *(Continued)*
Absorption and Distribution of Drugs Containing Nitrogen Heterocycles

Drug (Class) **Key Heterocycle/Scaffold**	**Absorption (F, Tmax)**	**Distribution (Vd, PPB, BBB)**
Imipramine (tricyclic antidepressant) 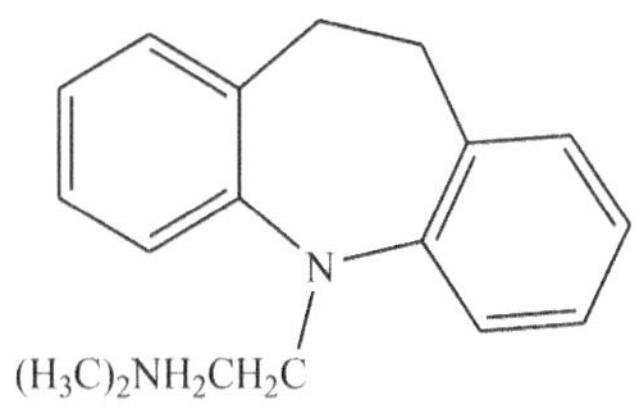 Dibenzoazepine	Well absorbed orally but extensive first-pass Tmax ~2 h	Large Vd High PPB BBB penetrant (Fayez & Gupta, 2023)
Buspirone (anxiolytic) 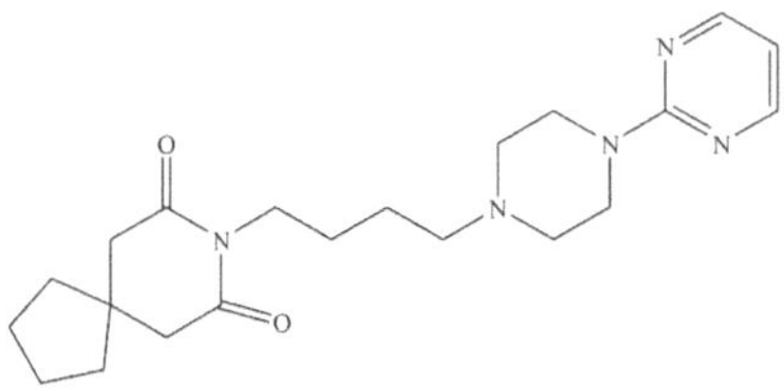 Pyrimidinylpiperazine	Low F due to extensive first-pass metabolism Tmax 40–90 min	Moderate Vd PPB ~86% Limited CNS exposure relative to other anxiolytics but clinically active (Wilson & Tripp, 2023)
Memantine (NMDA antagonist) 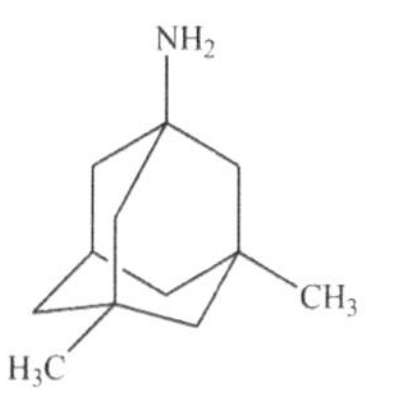 Adamantane	Good oral F (~100% for immediate release) Tmax 3–8 h.	Moderate Vd Low-moderate PPB BBB penetrant (Kuns et al., 2026)
Diazepam (benzodiazepine) 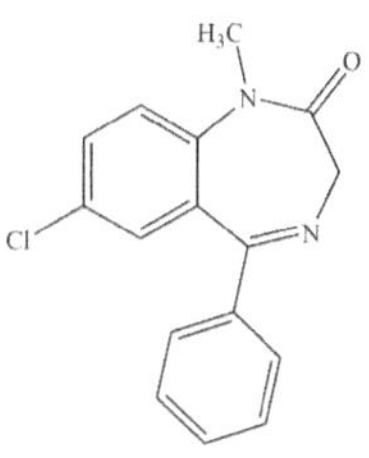 Benzodiazepine	High oral F Tmax 0.5–1.5 h (oral) Rapid onset IV/rectal	Very large Vd High PPB Readily penetrates BBB (Dhaliwal et al., 2023)
Morphine (opioid agonist) 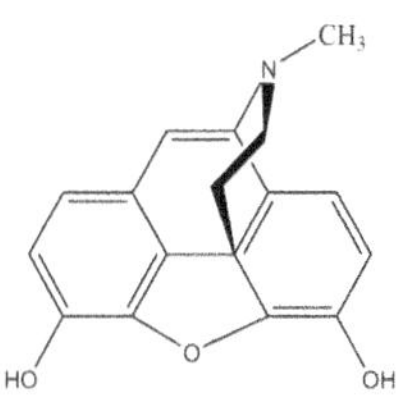 Phenanthrene	Oral F low (~20–40%) due to first-pass Tmax ~ 60 min oral	Vd variable Moderate PPB Limited direct BBB penetration but active metabolites contribute to CNS effect (Murphy et al., 2025)

BBB, blood–brain barrier penetration; F, oral bioavailability; PPB, plasma protein binding; Vd, apparent volume of distribution.

12.2.3.1 CYP450-Mediated Oxidation

CYP450 oxidation enhances polarity to promote excretion. CYP3A4, CYP2D6, and CYP1A2 play a crucial role in the metabolism of heterocyclic drugs.

- **Clozapine** is metabolized through N-oxidation and aromatic ring hydroxylation, predominantly by CYP1A2 and CYP3A4 enzymes.
- **Imipramine,** a tricyclic heterocycle, is metabolized through N-demethylation by CYP2D6, resulting in the formation of the active metabolite desipramine.
- **Buspirone,** a pyrimidinyl–piperazine heterocycle, undergoes rapid oxidation by CYP3A4, leading to low oral bioavailability.

CYP-mediated metabolism may result in detoxification or the production of active or toxic metabolites. PAINS-like heterocycles can experience aromatic oxidation, leading to the formation of reactive intermediates that may contribute to idiosyncratic hepatotoxicity (Abdelmonem et al., 2024; Chai et al., 2022; Zhao et al., 2021).

12.2.3.2 Monoamine Oxidase Metabolism

MAO enzymes are essential for the metabolic conversion of heterocyclic compounds that structurally mimic natural neurotransmitters. MAO-B predominantly facilitates the oxidative deamination of pharmaceuticals such as selegiline, rasagiline, and tranylcypromine, resulting in the production of inactive or partially active metabolites. Conversely, MAO-A swiftly metabolizes indole-containing neurotransmitters, such as serotonin analogs, thereby markedly diminishing their duration of action (Hâncu et al., 2024; Shahwan et al., 2024).

Consequently, medicines formulated as either MAO-B inhibitors or MAO substrates necessitate structural alterations that impart resistance to rapid MAO-mediated oxidative deamination. This is typically accomplished by including bulky or sterically hindered heterocyclic substituents, which diminish enzyme accessibility and improve metabolic stability (Khalid et al., 2021; Luo et al., 2024).

12.2.3.3 Phase II Conjugation

Following Phase I oxidation, the majority of heterocyclic drugs primarily undergo glucuronidation or sulfation as key Phase II metabolic pathways.

Lamotrigine, a triazine heterocycle, is primarily excreted via UGT-mediated glucuronidation, resulting in the formation of water-soluble metabolites.

Morpholine-containing drugs typically experience significant sulfate or glucuronide conjugation, leading to the formation of highly polar metabolites that are effectively excreted through the kidneys. Major metabolism pathways and possible metabolites of some drugs are depicted in Table 12.2.

12.2.4 Elimination

Nitrogen-containing heterocycles are primarily eliminated via renal and hepatic pathways. The physicochemical properties of these compounds significantly affect their clearance rates. Lipophilic heterocycles characterized by a high volume of distribution are prone to tissue accumulation, leading to prolonged elimination times. Polar heterocycles exhibit higher water solubility and restricted tissue distribution, resulting in their rapid clearance, primarily through renal excretion.

12.2.4.1 Renal Excretion

Heterocyclic compounds that are polar and water-soluble are primarily excreted via renal pathways. Clearance is contingent upon processes, including filtration, active secretion, and reabsorption, all of which are affected by the ionization state of the metabolite.

TABLE 12.2
Metabolism (Major Pathways and Metabolites)

2AQ1AZ	Major Phase I Enzymes/Pathways	Notable Metabolites (Active/Toxic)
Nicotine	Hepatic CYP2A6 primarily Multiple oxidation pathways	Cotinine (major metabolite, longer t½)
Clozapine	CYP1A2, CYP3A4, CYP2D6 (minor); N-oxidation, demethylation	Norclozapine (desmethyl)—active; clozapine N-oxide.
Fluoxetine	CYP2D6 (major), CYP3A4 N-demethylation	Norfluoxetine—active, long t½ → prolonged effect
Lamotrigine	Minimal CYP Predominantly UGT1A4-mediated N-glucuronidation (Phase II)	Lamotrigine-2-N-glucuronide (major inactive)
Selegiline	Extensive first-pass CYP and MAO metabolism Oxidation/dealkylation	Desmethylselegiline and amphetamine metabolites (some active)
Imipramine	CYP2D6, CYP1A2 N-dealkylation and hydroxylation	Desipramine (active metabolite)
Buspirone	CYP3A4 oxidation to hydroxy-buspirone and dealkylated metabolites 1-PP (1-pyrimidinylpiperazine)—pharmacologically active	1-PP (active, anxiolytic properties secondary)
Memantine	Minimal hepatic metabolism Some hepatic oxidation Largely excreted unchanged	Parent drug largely responsible for activity
Diazepam	CYP2C19, CYP3A4 N-demethylation and hydroxylation	Nordiazepam (desmethyldiazepam), temazepam, oxazepam—active (contribute to long effects)
Morphine	UGT2B7 glucuronidation (Phase II) >> minor CYP	Morphine-6-glucuronide (M6G)—active (potent analgesic) M3G (inactive/toxic potential)

- Glomerular filtration plays a crucial role in the preliminary elimination of polar metabolites.
- Active tubular secretion, frequently facilitated by transporters like OCT2, improves renal excretion.
- Reabsorption is influenced by ionization, with ionized compounds exhibiting limited tubular reabsorption.

The protonated form of memantine, an adamantyl–amine heterocycle, is primarily cleared renally through active secretion, demonstrating the significance of transporter-mediated mechanisms in the elimination of heterocycles(Muschong et al., 2024; Podgorski et al., 2023).

12.2.4.2 Hepatic Excretion

Nitrogen-containing drugs that possess high molecular weight or form bulky Phase II conjugates frequently undergo biliary elimination rather than renal clearance. Their size, lipophilicity, and conjugation profile facilitate transport into bile through hepatic efflux pathways.

- **Macrolide–heterocycle hybrids**, such as certain azithromycin derivatives, are commonly excreted via the **biliary route**.
- **Clozapine N-oxide**, formed through hepatic metabolism, can also undergo **biliary excretion**, reflecting the role of metabolite size and polarity in determining elimination pathways.

12.2.5 Influence of Lipinski and Beyond-Ro5 Space

Lipinski's rule of five establishes a fundamental framework for predicting the absorption and elimination of drugs. Many heterocyclic compounds, particularly macrocycles and polyheterocycles, are classified as beyond-Rule-of-Five (bRo5) compounds. These molecules frequently contravene conventional Ro5 criteria while still attaining sufficient therapeutic exposure, attributable to factors like improved membrane permeability or dependence on active transport mechanisms (Amin et al., 2022; Ibeyaima et al., 2024; Luo et al., 2024).

- Macrocyclic heterocycles often surpass the Ro5 thresholds while preserving bioavailability due to their conformational flexibility.
- Polyheterocycles can employ carrier-mediated transport to offset inadequate passive diffusion(Muschong et al., 2024).

Due to structural complexity and deviation from classical guidelines, PBPK (physiologically based pharmacokinetic) modeling is essential for accurately predicting human clearance, half-life, and overall pharmacokinetic behavior(Khalid et al., 2021; Ozbek et al., 2024; Tsakalozou et al., 2022). The primary route of elimination and half-life for some drugs containing nitrogen heterocycles is indicated in Table 12.3.

TABLE 12.3
Elimination (t½, Clearance, Primary Route, Clinical Implications)

Drug	t½ (Typical) Elimination Half-Life	Clearance/Principal Elimination Route	Clinical PK Considerations
Nicotine	Short (1–2 h) for nicotine Cotinine t½ ~16–20 h	Hepatic metabolism → renal excretion of metabolites	Formulation-dependent delivery (inhaled vs transdermal) strongly changes Tmax/Cmax
Clozapine	t½ ~8–12 h (varies widely)	Hepatic metabolism; renal and biliary elimination of metabolites	Therapeutic drug monitoring recommended due to variable clearance and toxicity risk (agranulocytosis)
Fluoxetine	Parent t½ 1–3 days Norfluoxetine t½ 7–15 days (very long)	Hepatic metabolism Slow clearance due to active metabolite	Long washout and drug-interaction potential (CYP2D6 inhibition)
Lamotrigine	t½ 24–35 h (adult) Decreased by enzyme inducers, increased by valproate	Hepatic glucuronidation Renal excretion of glucuronide	Drug interactions (valproate increases t½, inducers decrease it)
Selegiline	t½ oral ~1–3 h (but effects are longer due to metabolites) Transdermal formulation extends exposure	Extensive hepatic metabolism Renal elimination of metabolites	Route (transdermal vs oral) changes exposure and dietary restrictions (MAO effects)
Imipramine	t½ parent ~12–24 h Active metabolites extend effect [38]	Hepatic metabolism Renal elimination of metabolites [38]	CYP2D6 polymorphisms → large interindividual variability [38]
Buspirone	t½ ~2–3 h (but active metabolites may contribute)	Extensive hepatic metabolism (CYP3A4) Renal/fecal elimination of metabolites	Low oral F Many formulations and DDIs via CYP3A4 matter clinically
Memantine	t½ ~60–100 h (long)	Renal excretion of unchanged drug (major)	Dosing adjustments needed in renal impairment
Diazepam	t½ variable (20–50 h) due to active metabolites May be longer in elderly	Hepatic metabolism to active metabolites → renal excretion	Prolonged effects, accumulation Caution in elderly/hepatic impairment
Morphine	t½ 2–4 h (parent) M6G longer in renal impairment	UGT2B7 glucuronidation → renal elimination of glucuronides	*Renal impairment*: accumulation of M6G → increased opioid effect/toxicity

12.3 PHARMACODYNAMICS OF NITROGEN HETEROCYCLES

Nitrogen-containing heterocycles demonstrate a wide range of pharmacodynamic activities. Nitrogen atoms significantly influence receptor-binding affinity, ionization behavior, and hydrogen-bonding capacity through their presence, position, and electronic characteristics. The structural characteristics enable effective interaction with various biological targets, such as neurotransmitter receptors, ion channels, membrane transporters, and essential metabolic enzymes (Kerru et al., 2020; Wang et al., 2023).

Nitrogen-containing heterocycles influence pharmacological effects through the modulation of essential neurotransmitter pathways, such as the dopaminergic, serotonergic, adrenergic, and GABAergic systems. Many clinically relevant antidepressants and anxiolytics possess tertiary or cyclic amine groups that structurally mimic endogenous neurotransmitters, facilitating high-affinity interactions with serotonin transporters (SERTs) and serotonin receptors, including 5-HT_1A and 5-HT_2A (Bhosale et al., 2024; Rajkumar & Parkes, 2021). Several heterocyclic antipsychotics, especially those with pyridazinone or piperazine scaffolds, act as antagonists at dopamine D_2 and serotonin 5-HT_2A receptors, which contributes to their antipsychotic and mood-stabilizing effects (Nowak et al., 2023).

Nitrogen-containing heterocycles can act as both hydrogen bond donors and acceptors, allowing them to achieve optimal binding orientations in receptor pockets and enzyme active sites, which improves potency and selectivity (Hughes et al., 2022; Petrovic et al., 2023). The versatility of hydrogen bonding is crucial for ligands that interact with ion channels, including voltage-gated sodium and calcium channels. Heterocyclic scaffolds, such as imidazoles, triazoles, and pyridines, contribute to the stabilization of specific channel conformations and the modulation of cellular excitability (Zheng et al., 2021). Drugs like lamotrigine and carbamazepine, which contain nitrogen-rich heterocycles, demonstrate therapeutic efficacy in epilepsy by employing a use-dependent blockade of sodium channels, thereby diminishing abnormal neuronal firing (Pal et al., 2021).

Nitrogen-containing heterocycles interact with enzymes by emulating the structural and electrical characteristics of natural substrates. Quinazoline- and pyrimidine-derived tyrosine kinase inhibitors have great specificity for the ATP-binding sites of kinases, as the heterocyclic nitrogen atoms accurately mimic the hydrogen-bonding configurations of native nucleotide bases (Giri et al., 2025). Similarly, other heterocyclic MAO inhibitors, such as propargylamine and hydrazine derivatives, create covalent or quasi-irreversible complexes with MAO-A and MAO-B, resulting in enduring enzyme inhibition and protracted neurotransmitter upregulation (Oliveira et al., 2022).

In addition to their traditional interactions with receptors and enzymes, numerous nitrogen-containing heterocyclic medicines significantly influence transporter systems, including P-glycoprotein (P-gp), OCTs, and solute carrier (SLC) transporters. By modulating these transporters, heterocyclic compounds can modify the uptake, efflux, and distribution of endogenous neurotransmitters and co-administered drugs, thereby affecting synaptic neurotransmission, intracellular concentrations, and the duration and intensity of pharmacological effects (Patel et al., 2023). Selective serotonin reuptake inhibitors (SSRIs) frequently incorporate cyclic amine heterocycles that inhibit SERTs and may also engage with OCTs to modulate neuronal uptake kinetics. Likewise, specific heterocyclic anticancer drugs can circumvent P-gp-mediated efflux, hence increasing intracellular drug concentration and therapeutic efficacy.

The structural attributes of heterocycles, including aromaticity, planarity, and protonation behavior, are crucial in facilitating noncovalent interactions with biological targets. These encompass π–π stacking, cation–π interactions, hydrogen bonding, and electrostatic complementarity, which collectively augment binding affinity, selectivity, and overall efficacy (Xu & Fang, 2020). Planar heterocycles, including imidazoles, indoles, and pyridines, can intercalate or stack with aromatic residues in receptor or enzyme binding pockets, therefore stabilizing the ligand–target complex. The protonation states of heterocyclic nitrogen compounds affect the intensity of electrostatic interactions with acidic or polar residues, which can be crucial for enzyme inhibition and receptor regulation.

Furthermore, the synergistic effects of transporter modification and structural complementarity enable heterocyclic medicines to concurrently optimize pharmacokinetics and pharmacodynamics. This dual capability elucidates why numerous heterocyclic compounds exhibit substantial potency despite low bioavailability or fast metabolic elimination. The multifunctional characteristics of nitrogen-containing heterocycles—encompassing receptor binding, enzyme inhibition, transporter modulation, and structural complementarity—render them highly versatile scaffolds for contemporary drug design and therapeutic applications (Yamada et al., 2025).

The notable pharmacologic diversity of nitrogen-containing heterocycles arises from their structural variability, electronic adaptability, and capacity to engage in many noncovalent interactions. The presence, placement, and electrical characteristics of nitrogen atoms enable these compounds to accurately regulate hydrogen bonding, π–π stacking, cation–π interactions, and electrostatic complementarity across various biological targets. This versatility allows heterocycles to interact with receptors, ion channels, transporters, and enzymes with significant affinity and selectivity, frequently affecting multiple biochemical pathways concurrently. Nitrogen heterocycles work as versatile scaffolding that can precisely modulate pharmacological responses, enhance therapeutic windows, and increase target selectivity. These properties are fundamental to the development of essential drugs across various therapeutic domains, including neurology, where they influence neurotransmitter systems; oncology, through enzyme inhibition and receptor antagonism; infectious diseases, via antimicrobial and antiparasitic effects; and cardiovascular medicine, by engaging with ion channels and enzymatic regulators of vascular tone. The integration of chemical versatility and biological accuracy renders nitrogen heterocycles highly advantageous in contemporary drug discovery and rational pharmacotherapy.

12.4 CASE STUDIES OF CLINICALLY SIGNIFICANT NITROGEN HETEROCYCLES

12.4.1 Pyridine and Pyrazine

Pyridine scaffolds are extensively employed heterocyclic frameworks in CNS therapies, attributed to their aromatic stability, moderate basicity, and capacity for π–π stacking interactions with aromatic residues in receptor binding sites. These characteristics enable pyridine-containing compounds to attain advantageous binding orientations, enhancing both potency and selectivity. Natural pyridine alkaloids, like cytisine (Figure 12.1a), further illustrate the medicinal potential of this framework. Cytisine, extracted from the plant *Cytisus laburnum*, functions as a partial agonist at α4β2 nAChRs and has been utilized in Eastern Europe as a smoking cessation aid. The molecular framework has inspired synthetic derivatives, such as varenicline, illustrating how natural pyridine alkaloids act as lead molecules for CNS medication development. Additional pyridine-based CNS pharmaceuticals encompass nicotine, which engages with nAChRs, and riluzole, a pyridine derivative employed in the treatment of amyotrophic lateral sclerosis (ALS) for its sodium channel modulation and reduction of glutamate release. An exemplary instance is varenicline (Figure 12.1b) that acts as a partial agonist at α4β2 nicotinic acetylcholine receptors (nAChRs). Varenicline partially stimulates these receptors, diminishing nicotine-induced dopamine release in the mesolimbic reward circuit, hence alleviating cravings and withdrawal symptoms in smokers (Jordan et al., 2018; Lin et al., 2020; Singh & Saadabadi, 2024). The pharmacological characteristics of varenicline highlight how modifications in the pyridine ring, including the incorporation of fused rings or electron-withdrawing groups, can influence agonist potency, receptor subtype selectivity, and total CNS activity.

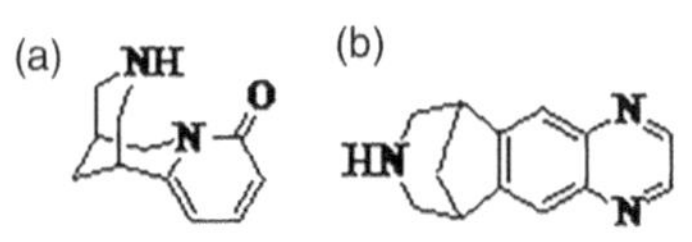

FIGURE 12.1 (a) Cytisine and (b) Varenicline.

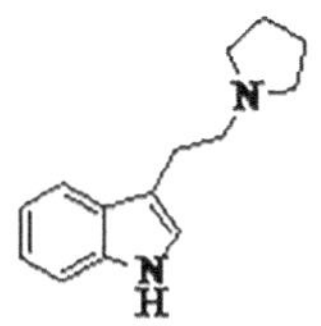

FIGURE 12.2 3-[2-(pyrrolidin-1-yl)ethyl]indole.

Thorough evaluations of pyridine-based CNS agents demonstrate that even slight structural alterations—such as heteroatom relocation, ring fusion, or the addition of substituents at the 2-, 3-, or4-positions of the pyridine ring—can markedly affect receptor affinity, pharmacokinetics, and safety profiles (Lin et al., 2020). The incorporation of electron-withdrawing groups can improve metabolic stability, whereas substituents influencing lipophilicity can alter BBB permeability. These discoveries demonstrate that the strategic modulation of the electrical environment of pyridine nitrogen and neighboring positions enables medicinal chemists to precisely adjust potency, receptor selectivity, and therapeutic safety, rendering pyridine scaffolds exceptionally adaptable in CNS drug discovery.

12.4.2 Indoles

The indole nucleus, which bears structural resemblance to serotonin (5-HT), serves as a favored scaffold for serotonergic agents. A significant case pertains to 3-[2-(pyrrolidin-1-yl)ethyl]indole (Figure 12.2), which were designed as selective 5-HT_1D receptor agonists through strategic substitutions that preferentially enhanced 5-HT_1D interactions over 5-HT_1B interactions (Chambers et al., 1999). These SAR insights demonstrate that indole orientation and substituent patterning can produce significant receptor subtype specificity.

Indole-based 5-HT_3 antagonists, such as ondansetron (Figure 12.3a), granisetron (Figure 12.3b), and tropisetron (Figure 12.3c), are commonly employed in the management of chemotherapy-induced nausea through the inhibition of ligand-gated 5-HT_3 ion channels (Machu, 2011). The indole-like core facilitates high-affinity binding within the receptor's aromatic cage, thereby inhibiting serotonin-mediated neuronal excitation.

Recent studies indicate that indole amide derivatives exhibit potential for Alzheimer's disease, demonstrating multi-target effects such as AChE inhibition and the prevention of amyloid-β aggregation (Świderska-Mosiężko et al., 2025). Docking studies of indole ligands, such as the D2AAK series, indicate that protonated nitrogen atoms exhibit strong interactions with a conserved Asp 3.32 residue, thereby stabilizing receptor binding and affecting GPCR signaling.

The findings highlight the indole scaffold's versatility as a pharmacophore that can modulate CNS receptors and disease-related enzymes.

12.4.3 Imidazole and Piperazines

Imidazole (Figure 12.4a) and piperazine (Figure 12.4b) scaffolds are frequently integrated in medicinal chemistry to improve aqueous solubility, refine receptor targeting, and reduce off-target

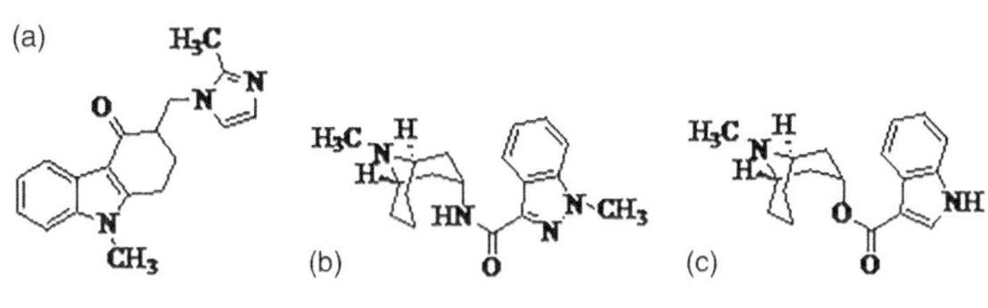

FIGURE 12.3 (a) Ondansetron, (b) Granisetron, and (c) Tropisetron.

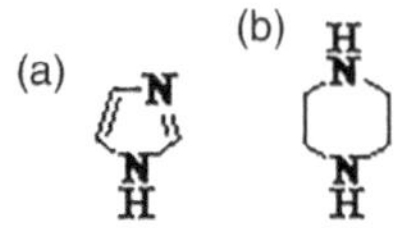

FIGURE 12.4 (a) Imidazole (b) Piperazine.

interactions. A notable example includes aryl-piperazine derivatives featuring an imidazole core, developed as 5-HT$_2$A/2C receptor antagonists with concurrent SERT inhibitory activity, suggesting a potential serotonin antagonist–reuptake inhibitor (SARI) mechanism for antidepressant treatment (Kowalska et al., 2023). In vivo studies using rodent models have shown antidepressant-like behavioral effects, underscoring the functional CNS activity attainable via dual scaffold design.

Studies on SARs indicate that minor alterations to the imidazole ring, particularly substitutions at the N-1 or C-4 positions, can markedly improve receptor selectivity, diminish side-effect liabilities, and optimize pharmacodynamic results(Seo et al., 2011). Molecular docking and computational modeling suggest that imidazole nitrogen atoms play a significant role in hydrogen-bonding interactions within serotonergic receptor binding pockets, whereas the piperazine moiety enhances pharmacokinetic properties, such as increased membrane permeability and metabolic stability (Goel et al., 2023).

The imidazole–piperazine hybrid demonstrates versatility that surpasses CNS applications. For example, when combined with other heterocycles like oxadiazoles, these hybrids exhibit significant anticancer activity by inhibiting kinase pathways and modulating apoptotic signaling (Lin et al., 2017). The dual functionality highlights the adaptability of imidazole–piperazine motifs, demonstrating how rational scaffold combination can be utilized to target various biological systems, enhance drug-like properties, and broaden therapeutic applicability in CNS, oncology, and potentially other areas.

12.4.4 Triazole

Triazole-containing heterocycles are extensively utilized in CNS drug design owing to their metabolic stability, hydrogen-bonding capacity, and moderate polarity, which enhance BBB penetration and target interaction. Triazolobenzodiazepines, including alprazolam (Figure 12.5), function as positive allosteric modulators of GABA *A* receptors, thereby enhancing inhibitory neurotransmission and resulting in anxiolytic and sedative effects. The triazole ring enhances receptor affinity via hydrogen bonding and π–π stacking interactions with aromatic residues in the receptor binding site.

Studies on SARs indicate that modifications to the triazole ring, including N-alkylation or aryl substitutions, can optimize potency, selectivity, and pharmacokinetics. Triazole derivatives featuring electron-withdrawing substituents typically exhibit enhanced metabolic stability and diminished off-target sedation. In addition to traditional benzodiazepine structures, triazole-linked ligands are being explored as serotonin 5-HT 1A agonists and D2 receptor modulators, demonstrating the scaffold's adaptability in influencing various neurotransmitter systems (Abdel-Mohsen et al., 2024).

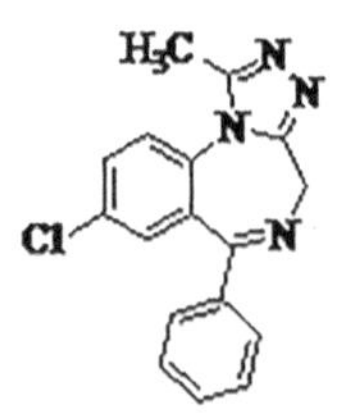

FIGURE 12.5 Alprazolam.

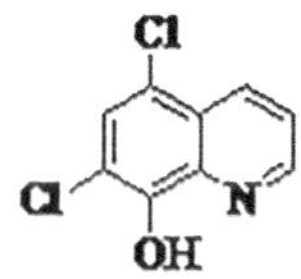

FIGURE 12.6 Clioquinol.

12.4.5 Quinoline

Quinoline-based heterocycles are valued for their rigid aromatic structure, nitrogen basicity, and capacity for π–π and hydrogen-bonding interactions, facilitating selective binding to CNS targets. Quinolines are primarily recognized for their antimalarial and antimicrobial properties; however, CNS-active quinoline derivatives have been formulated as MAO inhibitors, cholinesterase inhibitors, and NMDA receptor modulators.

Clioquinol, a hydroxyquinoline derivative (Figure 12.6), exhibits the ability to chelate metal ions and modulate neuronal oxidative stress, indicating its neuroprotective potential. Modifications to the quinoline ring, including halogenation or N-alkylation, enhance CNS penetration, selectivity for target enzymes, and pharmacokinetic profiles. Quinolone derivatives function as scaffolds for multi-target-directed ligands (MTDLs) in the study of neurodegenerative diseases, demonstrating the scaffold's adaptability in tackling intricate CNS pathologies.

12.5 RATIONAL DESIGN OF INNOVATIVE CNS-ACTIVE PHARMACEUTICALS

The rational design of CNS-active small molecules necessitates the systematic optimization of various molecular properties to ensure sufficient brain exposure, effective target engagement, and an acceptable safety profile. Two primary objectives generally inform early-stage medicinal chemistry decisions: (a) ensuring adequate permeability across the BBB while avoiding excessive nonspecific accumulation in non-target tissues, and (b) attaining high affinity and selectivity at the intended molecular target while reducing susceptibility to rapid metabolic clearance.

Medicinal chemists achieve these objectives by manipulating structural features, including ring size, heterocycle fusion, and substitution patterns, which serve as key factors in modulating physicochemical and pharmacokinetic properties. Adjusting lipophilicity affects passive diffusion across the BBB, whereas polar surface area and hydrogen-bonding capacity influence transporter recognition and aqueous solubility. Altering the pKa of heterocyclic nitrogen compounds can improve target engagement by promoting the suitable protonation state at physiological pH, while three-dimensional shape or conformational rigidity can enhance receptor complementarity and selectivity (Gupta et al., 2024; Urbina et al., 2021; Wu et al., 2023).

The design principles are illustrated in CNS-active scaffolds, including pyridines, indoles, imidazoles, and triazoles. Subtle modifications, such as ring fusion, N-alkylation, or the addition of electron-withdrawing groups, can significantly affect BBB penetration, receptor subtype selectivity, metabolic stability, and overall pharmacodynamic effects. Integrating computational modeling, SAR studies, and early pharmacokinetic profiling enables researchers to prioritize compounds that effectively balance brain exposure, efficacy, and safety, thereby establishing a solid framework for the development of next-generation CNS therapeutics.

12.5.1 Ring Size and Scaffold Topology

Modifying ring size, whether through five- versus six-membered rings, medium rings, or macrocycles, influences conformational flexibility and three-dimensional structure, subsequently impacting target affinity and BBB permeability. Compact, planar five- and six-membered heterocycles, such as pyrroles, pyridines, and indoles, typically achieve an effective balance of lipophilicity and low polar surface area, facilitating passive BBB entry. In contrast, medium rings or constrained bicyclic systems may enhance affinity by pre-organizing the ligand into a bioactive conformation and minimizing entropic penalties during binding (Shearer et al., 2022 Yuan et al., 2024). In contrast, the strategic implementation of ring expansion or macrocyclization facilitates access to "beyond-Rule-of-Five" (bRo5) chemical space for challenging CNS targets. Macrocycles can exhibit unexpectedly favorable CNS exposure when conformationally constrained to obscure polar groups and minimize exposed TPSA (Du et al., 2025). Recent synthetic advancements, such as single atom insertions and late-stage ring editing, enable iterative modifications in ring size during later phases of lead optimization, thereby expediting the testing of SAR hypotheses (Shearer & Ertl, 2022).

12.5.2 Substituents and Electronic Tuning (pKa, Lipophilicity, H-bonding)

Minor modifications to substituents, such as methylation, fluorination, and O-alkylation, significantly influence metabolic stability, pKa, and membrane permeability. Fluorine substitution is a well-established method to enhance metabolic stability by inhibiting oxidative metabolism, fine-tune lipophilicity, and influence conformational preferences. Careful fluorination has preserved numerous CNS leads from swift clearance while maintaining receptor potency (Ong et al., 2024). Modifying pKa through the strategic placement of heteroatoms, such as relocating a basic nitrogen or incorporating an electron-withdrawing group, influences the unprotonated fraction at physiological pH. This, in turn, affects the amount of drug capable of passive diffusion across the BBB. For numerous CNS targets, achieving an optimal balance of moderate basicity is crucial to ensure target engagement while preventing complete protonation, which hinders BBB permeability (Gupta et al., 2024; Urbina et al., 2021). Hydrogen-bond donors and acceptors are minimized or geometrically shielded to decrease TPSA and prevent P-gp recognition, while being maintained in specific positions to facilitate directional interactions that enhance selectivity at the target (Wu et al., 2023; Yuan et al., 2024).

12.5.3 Heterocycle Fusion and Three-Dimensionality

The combination of heterocyclic and carbocyclic rings in structures like indoles, benzodiazepines, and triazolobenzodiazepines typically expands the aromatic surface area and establishes a rigid molecular framework. This configuration improves binding affinity via distinct π–π stacking, cation–π, and hydrophobic interactions with protein residues. This rigidity enables ligands to assume a preorganized conformation that aligns with the geometry of receptor binding pockets, thereby enhancing both potency and selectivity.

Excessive planarity may present certain disadvantages. Highly planar molecules demonstrate increased susceptibility to P-glycoprotein (P-gp)-mediated efflux, potentially restricting brain penetration and exhibiting off-target promiscuity, thereby elevating the risk of side effects. Modern CNS drug design addresses these challenges by incorporating small sp^3-rich linkers, saturated rings, or stereocenters. This approach introduces three-dimensional character that minimizes undesired interactions while preserving favorable target engagement. This strategy has been effectively implemented in molecules, including riluzole derivatives, spirocyclic GABA modulators, and conformationally constrained benzodiazepines, where nuanced three-dimensional features enhance the equilibrium between potency and pharmacokinetic/ADME characteristics (Ding et al., 2024; Du et al., 2025).

Cheminformatics and multiparameter analyses underpin these design principles. Research indicates that the most effective CNS drugs reside within a limited physicochemical space characterized by a balance of aromaticity, rigidity, and three-dimensional structure. This insight has been formalized into CNS-focused multiparameter optimization (MPO) filters, which are extensively utilized in the industry to prioritize compounds exhibiting favorable BBB permeability, target engagement, metabolic stability, and overall drug-likeness (Agarwal et al., 2024; Urbina et al., 2021). These filters incorporate parameters, including lipophilicity, polar surface area, aromatic ring count, and fraction of sp^3 carbons, assisting medicinal chemists in the systematic design of advanced CNS therapeutics.

12.5.4 Multiparameter Optimization (MPO) and in Silico Workflows

Current CNS drug discovery initiatives acknowledge that enhancing a singular molecular characteristic, like lipophilicity or molecular weight, is inadequate to meet the intricate demands for brain penetration, target engagement, and safety. Consequently, MPO frameworks, such as CNS MPO and TEMPO variants, have emerged as essential instruments in early medicinal chemistry. The MPO scores amalgamate various molecular descriptors, including lipophilicity (clogP), molecular weight, TPSA hydrogen-bond donor count, and heterocyclic pKa, into a singular heuristic metric that forecasts the probability of BBB penetration and overall CNS drug-likeness (Ghose et al., 2017 Urbina et al., 2021).

MPO scores are commonly employed in an iterative manner, assisting chemists in the prioritization of analogs, the refinement of substitution patterns, and the balancing of conflicting physicochemical requirements. These methods are frequently utilized alongside physiologically based pharmacokinetic (PBPK) modeling, machine learning (ML)-driven ADME predictions, and preliminary in vitro transporter or permeability assays. This integration enables researchers to predict human exposure, brain distribution, and potential metabolic liabilities prior to conducting extensive in vivo studies. Compounds exhibiting moderate clogP, low TPSA, and a limited number of hydrogen-bond donors are likely to attain high CNS penetration while minimizing P-gp efflux. In contrast, increased aromaticity or polarity is associated with decreased predicted brain exposure.

The integration of MPO with experimental DMPK data, including microsomal stability, Caco-2 permeability, and transporter interactions, facilitates the swift evaluation of lead series, permitting the deprioritization of poorly performing analogs prior to late-stage development. This integrated computational and experimental approach has markedly decreased attrition rates in CNS programs, allowing medicinal chemists to allocate resources to compounds with the greatest likelihood of success while concurrently optimizing potency, selectivity, and safety.

12.5.5 Metabolic Stability and Targeted Blocking Strategies

Medicinal chemists utilize various targeted structural modifications to enhance the metabolic stability of CNS-active heterocycles while preserving their target potency. Common strategies involve deuteration at metabolically sensitive C–H sites, which decelerates oxidative metabolism while preserving receptor interactions, and the incorporation of small electron-withdrawing groups, such as fluorine (F) or trifluoromethyl (CF_3) groups, near established metabolic hotspots. Modifications can decrease Phase I oxidative clearance, enhance oral bioavailability, and reduce the generation of reactive or toxic metabolites (Niu et al., 2024; Ong et al., 2024). Furthermore, local steric shielding, such as that provided by gem-dimethyl groups, can obstruct the access of cytochrome P450 enzymes to susceptible sites, thereby increasing stability.

Prodrug strategies are extensively employed to enhance CNS delivery. In these designs, polar heterocycles are temporarily masked using lipophilic esters or carrier-mediated promoieties to enhance penetration of the BBB. Upon entering the CNS, enzymatic cleavage facilitates the release of the active heterocyclic drug, thereby allowing central pharmacological activity while maintaining systemic stability (Lillethorup et al., 2025). Examples include L-DOPA methyl esters for dopamine replacement therapy and lipophilic carbamate derivatives of GABAergic drugs, which utilize BBB permeability to attain therapeutic concentrations in the CNS.

The integration of chemical and prodrug strategies enables medicinal chemists to optimize the interplay between metabolic stability, BBB delivery, and target engagement, thereby improving CNS exposure and minimizing systemic toxicity—an essential factor in the development of next-generation neurotherapeutics.

12.5.6 Transporter-Aware Design and Efflux Avoidance

Active efflux transporters, notably P-glycoprotein (P-gp) and breast cancer resistance protein (BCRP), constitute significant obstacles to attaining effective drug concentrations in the CNS. As a result, lead series are evaluated early for transporter substrate potential to identify structural liabilities that could restrict brain penetration. Structural characteristics linked to elevated P-gp or BCRP recognition, including highly polar cationic regions, planar polyaromatic surfaces, and extended hydrophobic patches, are systematically altered to diminish efflux. Medicinal chemists can implement polar-shielding groups, decrease exposed polar surface area, or incorporate steric bulk near anticipated transporter recognition elements to reduce active extrusion while maintaining target engagement.

Structural modifications are generally implemented alongside MPO guidance, initial in vitro transporter assays (such as MDCK-MDR1 or Caco-2 cells), and computational predictions of substrate affinity. This integrated strategy facilitates the design of molecules capable of crossing the BBB, evading efflux, and ensuring prolonged residence in the CNS, thereby improving therapeutic efficacy and pharmacodynamic consistency (Agarwal et al., 2024; Wu et al., 2023). Optimization of specific aryl-piperazine and indole derivatives to reduce P-gp recognition has enhanced CNS exposure in preclinical models while maintaining receptor selectivity, underscoring the effectiveness of this strategy in contemporary CNS drug design.

12.5.7 Case-Driven Examples and Workflow

Current CNS drug discovery initiatives demonstrate the effectiveness of rational, evidence-based approaches in transforming metabolic or physicochemical challenges into viable, brain-penetrant candidates. Teams have utilized various complementary strategies to enhance CNS exposure and target engagement.

a. Strategic fluorination involves the introduction of a fluorine atom at metabolically labile positions, which inhibits CYP-mediated oxidation while maintaining the hydrogen-bonding geometry necessary for high-affinity binding to the target. This approach improves metabolic stability while maintaining pharmacodynamics.
b. Heterocycle nitrogen repositioning: Modifying the location of a heterocyclic nitrogen can decrease basicity, thereby enhancing the proportion of unprotonated species at physiological pH (~7.4). This modification enhances passive BBB penetration by augmenting lipophilicity in the neutral form, while preserving receptor interactions.
c. Macrocyclization of flexible leads involves the incorporation of flexible molecules into macrocyclic structures, which minimizes conformational entropy losses during binding and can protect polar surface area, thereby reducing the apparent total polar surface area (TPSA). This improves BBB permeability and target specificity concurrently.

Modifications are generally informed by CNS-focused MPO scoring, physiologically based pharmacokinetic (PBPK) simulations, and in vitro transporter profiling before conducting in vivo studies (Du et al., 2025; Ghose & Herbertz, 2022; Gupta et al., 2024; Lillethorup & Hansen, 2025; Niu & Patel, 2024; Shearer & Ertl, 2022; Yuan et al., 2024). Iterative cycles of design, in silico triage, focused synthesis, and experimental DMPK evaluation form the foundation of contemporary CNS lead optimization, enabling medicinal chemists to systematically balance potency, selectivity, brain penetration, metabolic stability, and safety. These integrated strategies demonstrate how modern drug discovery utilizes computational, chemical, and pharmacological insights to generate clinically viable CNS candidates.

12.6 CONCLUSION

Nitrogen-containing heterocycles are a crucial class of scaffolds in contemporary medicinal chemistry, especially in neuropharmacology and CNS drug design. Their widespread use in therapeutics is due to the distinctive electronic, steric, and physicochemical properties associated with nitrogen atoms, which directly affect drug behavior from absorption to clinical response.

Nitrogen heterocycles exhibit precisely adjustable ADME properties across pharmacokinetic dimensions. The ionization patterns, hydrogen-bonding capacity, and lipophilicity influence oral absorption, membrane permeability, and BBB transport. Additionally, metabolic transformations facilitated by cytochrome P450s, MAOs, and flavin-containing monooxygenases determine clearance rates and half-life. Structural determinants allow medicinal chemists to optimize heterocycles for predictable distribution profiles, minimized metabolic liabilities, and improved CNS exposure.

At the pharmacodynamic level, nitrogen heterocycles engage in various biological interactions, including receptor binding, ion-channel modulation, enzyme inhibition, and regulation of neurotransmitter pathways. Their capacity to function as both hydrogen-bond donors and acceptors enables accurate imitation of endogenous ligands, leading to enhanced receptor affinity, target selectivity, and therapeutic efficacy. This versatility accounts for their widespread use in antipsychotics, antidepressants, anticonvulsants, kinase inhibitors, and antibiotics, where interactions driven by heterocycles mediate clinical efficacy.

Case studies of clinically significant motifs, such as pyridines, indoles, imidazoles, and piperazines, illustrate how minor structural variations influence pharmacological outcomes. Pyridines enhance CNS bioavailability and metabolic stability in neuropsychiatric drugs. Indoles allow for serotonergic pathway targeting due to their structural similarity to tryptamine. Imidazoles facilitate selective receptor modulation. Piperazines improve aqueous solubility and receptor–ligand complementarity. The practical applications highlight the significant clinical importance of scaffold engineering in heterocyclic chemistry.

The combination of structural insights with computational modeling, BBB-permeability prediction, and ligand-based design facilitates a more systematic approach to the development of next-generation CNS-active therapeutics. Medicinal chemists can systematically enhance target engagement and reduce off-target interactions and metabolic degradation by optimizing ring systems, steric bulk, electronic distribution, and fusion patterns. Rational design strategies are enhancing drug discovery processes and broadening therapeutic options for neurological and psychiatric disorders.

Nitrogen heterocycles remain essential building blocks in the development of CNS drugs. Their diverse contributions to pharmacokinetics, pharmacodynamics, molecular recognition, and clinical performance underscore their essential role in modern neurotherapeutics. With advancements in synthetic methodologies, computational tools, and mechanistic understanding, nitrogen heterocycle–based drug design will continue to be fundamental to innovative, effective, and safer pharmacotherapy for the CNS.

REFERENCES

Abdel-Mohsen, H. T., Hashem, A. M., & El-Ashmawy, I. M. (2024). Recent advances in imidazole–heterocycle hybrids for anticancer therapy. *Bioorganic Chemistry, 149*, 105995.

Abdelmonem, B. H., El-Shenawy, H. A., & El-Koussy, A. M. (2024). Decoding the role of CYP450 enzymes in metabolism and pharmacokinetics of small molecules: Implications for drug discovery. *Biomedicines, 12*(7), 1467.

Agarwal, P., Whitaker, K., & Roberts, S. (2024). CNS delivery of targeted protein degraders and transport-aware design principles. *European Journal of Pharmaceutical Sciences, 187*, 106300.

Amin, A., Khan, S., & Verma, R. (2022). Medicinal and industrial applications of nitrogen heterocycles. *The Open Medicinal Chemistry Journal, 16*, 45–68.

Banks, W. A., Rhea, E. M., Reed, M. J., & Erickson, M. A. (2024). The penetration of therapeutics across the blood-brain barrier: Classic case studies and clinical implications. *Cell Reports Medicine, 5*(11), 101760.

Bhosale, S., et al. (2024). Serotonin receptor–ligand interactions: Structural and pharmacological overview. *Frontiers in Pharmacology, 15*, 1345632.

Chai, A. B.et al. (2022). Regulation of P-glycoprotein in the brain: Implications for CNS drug disposition. *International Journal of Molecular Sciences, 23*(23), 14667.

Chambers, M. S., Street, L. J., Goodacre, S., et al. (1999). Synthesis and serotonergic activity of 3-[2-(pyrrolidin-1-yl)ethyl]indoles: Potent agonists for the h5-HT1D receptor with high selectivity. *Journal of Medicinal Chemistry, 42*(4), 691–700.

de la Torre, B. G., & Albericio, F. (2024). The pharmaceutical industry in 2023: An analysis of FDA drug approvals from the perspective of molecules. *Molecules (Basel, Switzerland)*, *29*(3), 585.

Dhaliwal, J. S., Rosani, A., & Saadabadi, A. (2023, Aug 28). Diazepam. In *StatPearls* [Internet]. StatPearls Publishing. https://www.ncbi.nlm.nih.gov/books/NBK537022/

Ding, Y., Qiu, Y., & Zhang, H. (2024). Medicinal chemistry strategies for modifying marketed drug scaffolds: Improving stability, pharmacokinetics and target selectivity. *Pharmaceutics*, *16*(3), 556.

Du, Y., Kleiner, A., &Park, S. (2025). Macrocycles and medium ring systems in contemporary medicinal chemistry: Lessons for CNS drug design. *Chem Med Chem*, *20*(2), e202400788.

Dwivedi, A. R., Jaiswal, S., Kukkar, D., Kumar, R., Singh, T. G., Singh, M. P., Gaidhane, A. M., Lakhanpal, S., Prasad, K. N., & Kumar, B. (2025). A decade of pyridine-containing heterocycles in US FDA approved drugs: A medicinal chemistry-based analysis. *RSC Medicinal Chemistry*, 16, 12–36.

Ebenezer, O., Jordaan, M. A., Carena, G., Bono, T., Shapi, M., & Tuszynski, J. A. (2022). An overview of the biological evaluation of selected nitrogen-containing heterocycle medicinal chemistry compounds. *International Journal of Molecular Sciences*, *23*(15), 8117.

Fayez, R., & Gupta, V. (2023, May 22). Imipramine. In *Stat Pearls* [Internet]. Stat Pearls Publishing. https://www.ncbi.nlm.nih.gov/books/NBK557656/

Freibüchler, A., & Seifert, R. (2024). Analysis of clinical studies on clozapine from 2012–2022. *Naunyn-Schmiedeberg's Archives of Pharmacology*, *397*(12), 9745–9765.

Gao, B., Fang, Q., &Wang, Y. (2022). Recent advances in biosynthesis strategies of nitrogen-containing natural products. *Natural Product Reports*, *39*(7), 1251–1287.

Ghose, A. K., Ott G. R., & Hudkins R. L. (2017). Technically extended multiparameter optimization (TEMPO) and explainable MPOs for CNS lead selection. *ACS Chemical Neuroscience*, *8*(1), 147–154.

Giri, P., et al. (2025). Rational design of heterocyclic kinase inhibitors: Structural themes and therapeutic directions. *Medicinal Research Reviews*, *45*(2), e20651.

Goel, K. K., Bedi, P. K., Singh, R., et al. (2023). Imidazoles as serotonin receptor modulators: SAR and structural insights. *Journal of Chemical Neuroanatomy*, *130*, 102475.

Goldenson, N. I., Fearon, I. M., Buchhalter, A. R., & Henningfield, J. E. (2021). An open-label, randomized, controlled, crossover study to assess nicotine pharmacokinetics and subjective effects of the JUUL system with three nicotine concentrations relative to combustible cigarettes in adult smokers. *Nicotine & Tobacco Research: Official Journal of the Society for Research on Nicotine and Tobacco*, *23*(6), 947–955.

Gupta, M., Babu, A., & Ramesh, A. (2024). Experimental and computational methods to assess blood–brain barrier permeability: Integration into lead optimisation. *Molecules*, *29*(6), 1264.

Hâncu, I. M., Popescu, C., & Matei, A. (2024). Monoamine oxidase as drug target. *Molecular and Cellular Biochemistry*, *499*(6), 45–62.

Hossain, M., Habib, I., Singha, K., & Kumar, A. (2024). FDA-approved heterocyclic molecules for cancer treatment: Synthesis, dosage, mechanism of action and their adverse effect. *Heliyon*, *10*(1), e23172

Huang, L., et al. (2022). P-glycoprotein aggravates blood–brain barrier dysfunction: Transporter roles and clinical implications. *Frontiers in Neuroscience*, *16*, 946967.

Hughes, R., et al. (2022). Hydrogen bonding in receptor–ligand recognition: Relevance to heterocyclic drug design. *Journal of Medicinal Chemistry*, *65*(18), 12021–12045.

Ibeyaima, A., Yang, S., &Patel, N. (2024). ADMET prediction and Lipinski heuristics. *Computational Toxicology*, *20*, 100214.

Jadala, C., Mishra, S., Potuganti, G. R., Cardoza, S., & Velma, G. R. (2026). Nitrogen-containing FDA-approved drugs in 2025: Synthesis, significance and therapeutic applications. *RSC Advances*, *16*, 21967–22005.

Jampílek, J., & Král'ová, K. (2024). Heterocycles in medicinal chemistry II: An overview. *Molecules*, *29*(5), 1132.

Jordan, C. J., Xi, Z., Lowe, J. R., et al. (2018). Discovery and development of varenicline for smoking cessation. *Journal of Neuroscience*, *38*(33), 7110–7119.

Kerru, N., Gummidi, L., Maddila, S., Gangu, K. K., & Jonnalagadda, S. B. (2020). A review on recent advances in nitrogen-containing molecules and their biological applications. *Molecules (Basel, Switzerland)*, *25*(8), 1909.

Khalid, S., et al. (2021). A physiologically based pharmacokinetic (PBPK) model for diazepam: Route-dependent absorption and disposition. *Pharmaceutics*, *13*(9), 1480.

Kowalska, T., et al. (2023). Structural determinants of indole-based ligands at serotonergic GPCRs: Docking and molecular dynamics insights. *Molecules*, *28*(1), 383.

Kumar, A., Maurya, A., & Singh, M. (2023). Nitrogen-containing heterocycles as anticancer agents: Recent progress. *Pharmaceuticals*, *16*(7), 1012.

Kuns, B., Rosani, A., Patel, P., et al. (2024, Jan 31). Memantine. In *Stat Pearls* [Internet]. Stat Pearls Publishing. https://www.ncbi.nlm.nih.gov/books/NBK500025/

Li, J. C., Miao, C. F., Lei, Y., & Liu, A. L. (2025). Physiologically based pharmacokinetic modeling to predict lamotrigine exposure in special populations to facilitate therapeutic drug monitoring and guide dosing regimens. *Pharmaceuticals (Basel, Switzerland), 18*(5), 637.

Lillethorup, I. A., Hemmingsen, A. V., & Qvortrup, K. (2025). Prodrugs and their activation mechanisms for brain drug delivery. *RSC Medicinal Chemistry, 16*(3), 1037–1048.

Lin, F., Liu, X., & Chen, J. (2017). 3D-QSAR and molecular docking of arylpiperazines for serotonergic modulation. *Molecules, 22*, 1064.

Lin, S. X., Kundu, S., Gopalakrishnan, A., et al. (2020). Pyridine alkaloids with activity in the central nervous system: Therapeutic potential and pharmacology. *Phytochemistry Letters, 34*, 75–89.

Luo, W., Chen, Q., &Zhou, Y. (2024). Nitrogen-containing heterocyclic drug products approved in 2023: A focused analysis of chemical features and regulatory outcomes. *European Journal of Medicinal Chemistry, 260*, 115139.

Machu T. K. (2011). Therapeutics of 5-HT3 receptor antagonists: Current uses and future directions. *Pharmacology & Therapeutics, 130*(3), 338–347.

Mo, X., Li, Y., &Wang, M. (2024). Indole derivatives: A versatile scaffold in modern drug discovery. *Molecules, 29*(3), 621.

Moore, J. J., & Saadabadi A. (2023, Aug 17). Selegiline. In *Stat Pearls* [Internet]. StatPearls Publishing. https://www.ncbi.nlm.nih.gov/books/NBK526094/

Murphy, P. B., Patel, P., & Barrett, M. J. (2025, Sep 29). Morphine. In *Stat Pearls* [Internet]. Stat Pearls Publishing. https://www.ncbi.nlm.nih.gov/books/NBK526115/

Muschong, P., Doveston, R. B., & Feeney, O. (2024). Conquering the beyond rule of five space: An integrated experimental and computational workflow. *Drug Discovery Today, 29*(9), 2043–2055.

Nehra, B., Kumar, M., Singh, S., Chawla, V., Chawla, P., & Grover, P. (2025). Recent developments in the antimicrobial potential of some nitrogenous heterocycles and their SAR studies: A review. *Current Medicinal Chemistry, 32*(41), 9273–9323.

Niu, Z. X., Hu, J., Sun, J, & Wang, Y. (2024). Fluorine in the pharmaceutical industry: Synthetic approaches and application of clinically approved fluorine-enriched anti-infectious medications. *European Journal of Medicinal Chemistry, 271*, 116446.

Nowak, G., et al. (2023). Dopamine and serotonin receptor pharmacology of new-generation heterocyclic antipsychotics. *CNS Drugs, 37*(4), 321–340.

Ohshima, T., & Nakahara, Y. (2024). Advances in the synthesis and application of nitrogen-containing heterocycles. *Frontiers in Chemistry,12*, 1349210.

Oliveira, M., et al. (2022). Advances in monoamine oxidase inhibitors based on nitrogen heterocycles. *Bioorganic & Medicinal Chemistry, 63*, 116092.

Ong, H. W., Tan, J., & Lim, S. (2024). Strategic fluorination to improve metabolic stability and CNS exposure: Case studies and guidelines. *Molecules, 29*(17), 4158.

Ozbek, O., et al. (2024). Advances in physiologically based pharmacokinetic (PBPK) modelling: Integrating machine learning and quantitative systems approaches. *ACS Pharmacology & Translational Science*, 5(4), 567–584.

Pal, R., Kumar, B., Akhtar, M., & Chawla, P. (2021). Voltage gated sodium channel inhibitors as anticonvulsant drugs: A systematic review on recent developments and structure activity relationship studies. *Bioorganic Chemistry, 115*, 105230.

Pal, S., & Singh, P. (2023). Nitrogen heterocycles: Synthetic strategies and biological importance. *Current Medicinal Chemistry, 30*(14), 1621–1645.

Patel, M., et al. (2023). Transporter interactions of CNS-active heterocycles: Implications for efficacy and toxicity. *Pharmaceutics, 15*(5), 1279.

Petrovic, D., et al. (2023). Electronic and steric modulation of nitrogen heterocycles in therapeutic agents. *Chemical Reviews,123*(2), 1451–1492.

Pibiri, I. (2024). Recent advances: Heterocycles in drugs and drug discovery. *Molecules, 29*(2), 350.

Podgorski, M. N., Smith, J., & Lane, D. (2023). Oxidation of heteroatom-rich scaffolds by oxygenases: Lessons for heterocyclic drug metabolism. *Chemistry – A European Journal, 29*(42), e202301371.

Rajkumar, R., & Parkes, L. (2021). Mechanistic insights into antidepressant action on monoamine transporters. *Neuropharmacology, 194*, 108627.

Rizzo, C., Amata, S., Pibiri, I., Pace, A., Buscemi, S., & Palumbo Piccionello, A. (2023). FDA-Approved fluorinated heterocyclic drugs from 2016 to 2022. *International Journal of Molecular Sciences, 24*(9), 7728.

Seo, H. J., Park, E. J., Kim, M. J.et al. (2011). Design and synthesis of arylpiperazine–imidazole derivatives targeting 5-HT2A receptors and SERT. *Journal of Medicinal Chemistry*, *54*, 6305–6318.

Shahwan, M., Al-Khatib, H., &Sayed, M. (2024). High-affinity MAO-B inhibitors. *Frontiers in Pharmacology*, *15*, 1422080.

Shearer, J., Castro, J. L., Lawson, A. D. G., MacCoss, M., & Taylor, R. D. (2022). Rings in clinical trials and drugs: Present and future. *Journal of Medicinal Chemistry*, *65*(13), 8699–8712.

Singh, D., & Saadabadi, A. (2024). Varenicline. In *Stat Pearls*. Stat Pearls Publishing.

Sohel, A. J, Shutter, M. C, Patel, P., et al. (2024, Feb 28). Fluoxetine. In *Stat Pearls* [Internet]. Stat Pearls Publishing. https://www.ncbi.nlm.nih.gov/books/NBK459223/

Spielvogel, C. P., Schindler, N., Schröder, C., Stellnberger, S. L., Wadsak, W., Mitterhauser, M., Papp, L., Hacker, M., Pichler, V., & Vraka, C. (2025). Enhancing blood-brain barrier penetration prediction by machine learning-based integration of novel and existing, in silico and experimental molecular parameters from a standardized database. *Journal of Chemical Information and Modeling*, *65*(6), 2773–2784.

Stellnberger, S. L., Kopsidas, G., & Gedeck, P. (2024). Investigating experimental vs predicted pKa values for CNS activity prediction. *Journal of Computer-Aided Molecular Design*, *38*(4), 251–265.

Świderska-Mosiężko, J., Pietraszek-Gremplewicz, K., & Tatarczyk, M. (2025). Indole amide derivatives in Alzheimer's disease: Emerging therapeutic approaches. *Discover Chemistry*, *6*, 14.

Tsakalozou, E., et al. (2022). Physiologically based pharmacokinetic modeling to support regulatory submissions and dose selection. *Clinical Pharmacology & Therapeutics*, *111*(3), 657–669.

Urbina, F., Wager, T. T., &Abboud, K. A. (2021). Comparing the Pfizer central nervous system multiparameter optimization (CNS MPO) and related workflows: Performance and practical considerations. *Journal of Chemical Information and Modeling*, *61*(12), 6222–6238.

Wang, Y., Liu, S., & Chen, L. (2023). Structural determinants of nitrogen heterocycles in drug–target interactions: An updated analysis. *European Journal of Medicinal Chemistry*, *257*, 115684.

Ward, M., & O'Boyle, N. (2025). Analysis of the structural diversity of heterocycles amongst European medicines agency approved pharmaceuticals (2014–2023). *RSC Medicinal Chemistry*, *16*, 4540–4570.

Wilson, T. K., & Tripp, J. (2023, Jan 17). Buspirone. In *Stat Pearls* [Internet]. Stat Pearls Publishing. https://www.ncbi.nlm.nih.gov/books/NBK531477/

Wu, D., Pan, J., & Wang, H. (2023). The blood–brain barrier: Structure, regulation and drug delivery strategies. *Signal Transduction and Targeted Therapy*, *8*, 245.

Xu, J., & Fang, Y. (2020). Aromatic interactions in drug–protein complexes: Implications for heterocyclic pharmacology. *Molecular Pharmaceutics*, *17*(9), 3452–3465.

Yamada, K., et al. (2025). Nitrogen heterocycles in modern therapeutics: Pharmacodynamic and structural considerations. *Drug Discovery Today*, *30*(1), 102–118.

Yuan, T., Smith, A., & O'Donnell, M. (2024). Recent advances in ring-editing and ring-expansion strategies for drug discovery: Implications for scaffold optimization. *Organic & Biomolecular Chemistry*, *22*.

Zhao, M., Wang, S., & Zhang, H (2021). Cytochrome P450 enzymes and drug metabolism in humans: An updated review. *International Journal of Molecular Sciences*, *22*(23), 12808.

Zheng, Q., et al. (2021). Small-molecule modulation of ion channels: Mechanistic roles of heterocyclic scaffolds. *Pharmacology & Therapeutics*, *226*, 107868.

Index

For Product Safety Concerns and Information please contact our EU representative GPSR@taylorandfrancis.com Taylor & Francis Verlag GmbH, Kaufingerstraße 24, 80331 München, Germany

Batch number: 10731074

Printed by Printforce, the Netherlands